Manifold Mirrors

The Crossing Paths of the Arts and Mathematics

FELIPE CUCKER

City University of Hong Kong

CAMBRIDGE
UNIVERSITY PRESS

CAMBRIDGE
UNIVERSITY PRESS

University Printing House, Cambridge CB2 8BS, United Kingdom

One Liberty Plaza, 20th Floor, New York, NY 10006, USA

477 Williamstown Road, Port Melbourne, VIC 3207, Australia

314-321, 3rd Floor, Plot 3, Splendor Forum, Jasola District Centre, New Delhi - 110025, India

79 Anson Road, #06-04/06, Singapore 079906

Cambridge University Press is part of the University of Cambridge.

It furthers the University's mission by disseminating knowledge in the pursuit of
education, learning and research at the highest international levels of excellence.

www.cambridge.org
Information on this title: www.cambridge.org/9780521728768

First published 2013

A catalogue record for this publication is available from the British Library

Library of Congress Cataloging in Publication data
Cucker, Felipe, 1958–
   Manifold mirrors : the crossing paths of the arts and mathematics /
Felipe Cucker, City University of Hong Kong.
      pages cm
   Includes bibliographical references.
   ISBN 978-0-521-42963-4 (Hardback) – ISBN 978-0-521-72876-8 (pbk.)
1. Arts–Mathematics.   I. Title.
   NX180.M33C83 2013
   700.1'05–dc23

                                                                    2012046405

ISBN 978-0-521-42963-4 Hardback
ISBN 978-0-521-72876-8 Paperback

CONTENTS

MATHEMATICS: USER'S MANUAL

Launch not beyond your depth, but be discreet,
And mark that point where sense and dullness meet.

A. Pope (1966: An Essay on Criticism)

The writer of any book dealing with mathematics who wishes to reach a broad audience invariably faces a dilemma: How to describe the mathematics involved. No matter how well motivated the intervening notions, nor how lengthily described, the question that eventually will pose itself is what to do regarding proofs.

Working mathematicians are generally reluctant to dispense with them, and I am no exception. For, on the one hand, a proof of a statement shows its necessity, its truth with respect to an underlying collection of assumptions. And, on the other hand, in doing so, it usually conveys an intuition on the nature of the objects occurring in the statement. This intuition is of the essence. It decreases the confusion that the alternation of definitions and statements in the mathematical discourse naturally creates.

Occasionally, however, the understanding afforded by a proof does not compensate for the effort of its reading. This may be so because one already has a form of the intuition mentioned above (and would, therefore, feel annoyed by having to "prove the evident") or because the proof is too involved and fails to convey any intuition. In these cases the task of following the proof's details becomes boring.

In this trade-off between boredom and confusion[1] different readers find different solutions by choosing subsets of proofs to be read that best suit their circumstances. To make these choices possible, this book gives proofs for many of its (mathematical) statements. To further make it easier, some observations are now given.

The mathematical development of this book is, essentially, self-contained and relies on knowledge widely taught in secondary school courses. In this sense, any person having benefited from these courses will be able to read what follows. Mathematical content, while present throughout the book, is concentrated in Chapters 2, 3, 7, 9, 12 and 13. The first three of these chapters are, essentially, self-contained, in the sense that almost all results therein are proved. For the last three, in contrast, we could not proceed in a like manner without unduly increasing the proportion of mathematics in our exposition.

---

[1] Another form of this trade-off will be central to the arguments in this book.

Mathematical statements do not share the same conceptual importance. Common practice in mathematical writing describes as "theorems" those results whose statements are goals in themselves. Stepping stones toward the proof of a theorem are called either "Propositions" (when the statement is nevertheless of independent interest) or "Lemmas" (when the statement is understood as subordinate and of a technical nature only [2]). It is left to each individual reader to decide what degree of attention to pay to the mathematical details in the book. An extreme choice would be to read every mathematical result with its proof, provided such a proof is given. The other extreme would be – of course paying attention to the formal definitions and general description of the notions at hand – to skip or skim everything except for the theorems' statements. (In this case, for instance, Sections 2.5, 2.6, 3.3 and 7.4, as well as § 3.5.2 and § 9.2.3, would reduce to a single, simple statement.) An intermediate strategy would be to proceed with an initial reading following the latter choice and then return to the skipped details if the need arises in later chapters. In the choice of this degree of attention the reader is encouraged to keep in mind Pope's advice and fix it at that personal point "where sense and dullness meet".

---

[2] Sometimes, however, an author writes as a lemma a result which time proves to be of crucial importance. There are numerous examples of lemmas that eventually become worthy of the status of Theorem.

Appetizers

appetizer, *n.* ('æpɪtaɪzə(r)) [...] ► *fig.* Something intended to arouse interest in what follows; a sample of what may be expected in the future.

OXFORD ENGLISH DICTIONARY edited by
Simpson & Weiner (1986). Definition of "appetizer".
By permission of Oxford University Press.

A.1 Martini

> Without question, the undisputed king of cocktails is the Martini.
>
> Anistatia Miller and Jared Brown
> Mixellany Limited. www.mixellany.com

At the beginning of the fourteenth century, Giotto di Bondone was sowing in Florence the seed of the Italian Renaissance in painting. His works attempted to abandon the stiffness of the Byzantine style and replace it with a more naturalistic approach with a marked emphasis on the representation of space and volume. In the nearby city of Siena, the intent was different. "Their greatest master of Giotto's generation, Duccio, had tried – and tried successfully – to breathe new life into the old Byzantine forms instead of discarding them altogether" (Gombrich, 1989: 160). So did Duccio's most renowned pupil, Simone Martini.

Little is known with certainty about Martini's early years. He was born in the first half of the 1280s, probably in the town of San Gimignano,[1] where he had an early contact with the craft of painting as his father specialized in the preparation of the first coat (*arriccio* in Italian) applied to wall surfaces on which a fresco was going to be painted. He subsequently became a pupil of Duccio, and by the early 1310s he was producing his first works.

The favourable view of these works is evidenced by the fact that, in 1315, he was commissioned to paint a *Maestà* (Figure A.1) in the Sala del Mappamondo of the Palazzo Pubblico (the town hall) in Siena. Only a few years before, the same city of Siena had commissioned another Maestà, this one to be installed at the Cathedral of Siena, to Duccio (the commission dating from 1308 and finally installed in 1311) and the fact

---

[1] A small town close to Siena which could have been, in a sense, the Manhattan of the Middle Ages. The town traded with textiles, which were dyed and dried in tall towers built for that purpose. The view of such a cluster of towers must have produced at the time a feeling not unlike the twentieth-century view of Manhattan.

Figure A.1 *Maestà*, 1315 (fresco) by Simone Martini (1284–1344). (Palazzo Pubblico, Siena, Italy/The Bridgeman Art Library.)

that the chosen one was now Duccio's pupil points to the recognition that Martini must have enjoyed by 1315.

The word *Maestà*, meaning "majestic" in Italian, denotes a depiction of the enthroned Virgin Mary with Jesus. It may or may not be accompanied by saints and angels. The former is the case in both Duccio's and Martini's frescoes. The common representation of the Virgin and the Child provides examples for the latter. The word "majestic" is entirely appropriate for Martini's *Maestà*. More interestingly, some attributes of medieval painting, such as a certain hieratic manner or a marked solemnity, are only partly satisfied, and one may say that the fresco "breathes new life" to these attributes.

Martini's fresco is large; it occupies the whole end wall of the Sala del Mappamondo. It depicts the Virgin and Child surrounded by a multitude of angels and saints. We see at the Virgin's right (our left) Saints Catherine of Alexandria, John the Evangelist and Mary Magdalene, the Archangel Gabriel and Saint Paul (whom we recognize by his sword). Also, in almost identical positions but at the Virgin's left, Barbara, John the Baptist, Saint Agnes, the Archangel Michael and Saint Peter (identified by the key he is holding). Martini included in the scene the four patron saints of Siena: Ansano, Bishop Savino, Crescenzio and Vittore, who are kneeling in front accompanied by two angels offering flowers to Mary.

The composition of the whole reveals an interest in space and volume that had only recently reached medieval painting, as we mentioned, through the works of Giotto and Duccio. This interest is evinced in the poles supporting the canopy, which are drawn in correct perspective, as well as in the placement of the host of saints and angels. The predominant composition for such a host in the Byzantine style would have positioned these figures in both the left and right groups in parallel rows, with equal spaces between figures, standing in the same stance. Martini, in contrast, does not impose such a strict collocation of the figures and allows different ones in the scene to have a bearing (and an appearance, for Martini is also attributed to have introduced individual portraiture in medieval painting) appropriate to his or her role in the Sacred Story. Yet, unordinate as each of the two groups may be, there is a law in their joint display which is given by the fact that the positions of the figures at the right are replicated, as if mirrored, by the positions of the figures at the left. This arrangement, which is going to appear in the chapters to come, increases the feeling of solemnity and drives the attention towards the central pair of Virgin and Child.

In 1336 Martini moved to Avignon, where the Papal See had recently been established. A meeting point of artists coming from different places in Europe (but mostly from northern Italy) in search of the Pope's favour, the exchange of ideas and techniques at the Papal Court would evolve into a style known today as International Gothic, of which Martini, who incorporated a number of features of northern European art into his own work, is presently considered a precursor.

A.2 On their blindness

John Milton is nowadays considered one of the highest exponents in English literature, mostly due to his epic poem *Paradise Lost*. Born in 1608, Milton pursued studies at Cambridge between 1625 and 1632. With a broad set of interests, encompassing theology, philosophy, history, politics, literature and science, he is also considered one of the most learned English poets. Milton had an unusual knowledge of both classic and modern languages, with a command of Hebrew, Greek, Latin, Spanish, French and Italian. The last two he used extensively in his tour to France and (mostly) Italy between May 1638 and summer 1639.

In a moment of religious and political turmoil, Milton, who had very definite religious opinions, became a strong critic of some facets of Christianity (notably, of episcopacy). This aspect of Milton's life was prominent among his fellow citizens and, remarkably, most of the references to Milton reaching us from the seventeenth century "testify to his public life as a polemicist rather than his relatively unnoticed career as a poet" (Campbell, 2001: 483).

Milton's existence was plagued by a number of familiar misfortunes, as well as by the ups and downs that the changes in the English government during his lifetime had to bring forward to him as a consequence of his marked political and religious views. Neither was he immune to physical calamity.

During the early 1650s Milton had been losing his eyesight and by March 1652 he was totally blind (Flannagan, 2002: 75). More than a decade later, the narrator of *Paradise Lost* would be sightless, as Milton was, and compared "to the blind epic poet Homer or blind prophets such as Tiresias, given the gift of second sight to compensate for the loss of eyesight" (Flannagan, 2002: 75). But soon after the loss, Milton has a less assertive stance. Around 1655 he wrote a poem – formally *Sonnet 19*, but later called *On His Blindness* by the compilers of his work – in which he wonders whether his impairment would affect his capacity to serve God and, in particular, to use his talents for poetry as he had done until then. To pose the last question he refers to the "Parable of the talents" (Matthew, 25:14–30)[2] and points to "that one talent which is death to hide lodged with me useless". It is his patience (which he capitalizes to personify it) that reminds him that God is in no need of having men working for Him and that he may serve Him in ways different than those he had carried out until then. In the last verse, Patience even suggests that he might not need to do anything, since "they also serve [Him] who only stand and wait". This is the poem in its entirety:

> When I consider how my light is spent
> Ere half my days, in this dark world and wide,
> And that one talent which is death to hide
> Lodged with me useless, though my soul more bent
>
> To serve therewith my Maker, and present
> My true account, lest He returning chide;
> "Doth God exact day-labour, light denied?"
> I fondly ask. But Patience, to prevent
>
> That murmur, soon replies: "God doth not need
> Either man's work, or His own gifts, who best
> Bear His mild yoke, they serve Him best. His state
> Is kingly; thousands at His bidding speed
> And post o'er land and ocean without rest;
> They also serve who only stand and wait."

---

[2] A master gives three servants some talents (a unit of weight for money in Biblical times) to hold for him while he is away on a trip. After returning, he rewards the two servants who increased the money given to them and cast into darkness the third one, who had buried the talent given to him.

Milton would return shortly after to the subject of his blindness in *Sonnet 22*.

Ironically, Milton's best poetry would be written after the onset of his blindness. *Paradise Lost* was dictated by Milton to transcribers. Helped by a remarkable memory, "critics still marvel at the fact that the blind Milton could remember what he had dictated in Book 2 of *Paradise Lost* even as he was dictating lines in Book 11" (Flannagan, 2002: 76).

Three centuries later, and much to the south, Jorge Luis Borges, one of the greatest writers in Spanish of the twentieth century, was going to meet a fate similar to Milton's.

Borges grew up in Palermo, at that time a suburb of Buenos Aires. According to his recollections (in *Evaristo Carriego*) he "grew up in a garden, behind a speared railing, and in a library of unlimited English books". Just as Milton, his knowledge of languages was extraordinary (his translations into Spanish of works originally in Old English or Norse give evidence of this fact). And just as Milton as well, his life was not exempt of ups and downs, in his case motivated by the moving wheels of Argentinian political life. With the ascension to power of Juan Domingo Perón in 1946, Borges was dismissed from his position at a Buenos Aires municipal library and given a new appointment as poultry inspector for the municipal market (an appointment he immediately resigned).

Probably the most important coincidence in the lives of the two writers, though, is the fact that Borges also became blind during adulthood. Alberto Manguel (2006: 15–16), one of the people who read to Borges after this event, provides the following description: "His was a particular kind of blindness, grown on him gradually since the age of thirty and settled in for good after his fifty-eighth birthday. It was a blindness expected since his birth, because he always knew he had inherited feeble eyesight from his English great-grandfather and his grandmother [...]". Graham Greene also mentions Borges' blindness in a reminiscence talk. He had been spending a day with Borges at Buenos Aires. "After an agreeable lunch, he sat on a sofa and quoted large chunks of Anglo-Saxon. That, I'm afraid, I was not able to follow. But I looked at his eyes as he recited and I was amazed at the expression in those blind eyes. They did not look blind at all. They looked as if they were looking into themselves in some curious way, and they had great nobility" (Greene, 1990: 306–307).

Not unexpectedly, Borges would turn blindness into a poetic subject. Probably the first work where he does so is the *Poema de los Dones* (Poem of the gifts), written around 1960. Other poems followed. In which is probably the last one on the subject, he made an unambiguous reference to Milton in giving it the title *On His Blindness* (in English).[3] It appears in

---

[3] This is actually the second poem by Borges with this title, the first having been published in 1972, in the book *El Oro de los Tigres*.

the book *Los Conjurados* published in 1985, just one year before Borges' demise. This is how it reads.[4]

Al cabo de los años me rodea	I am surrounded, after all these years,
una terca neblina luminosa	by a stubborn shining mist
que reduce las cosas a una cosa	which reduces all things to one thing
sin forma ni color. Casi a una idea.	without shape or form. Merely an idea.
La vasta noche elemental y el día	The vast elemental night and the day
lleno de gente son esa neblina	full of people are this mist
de luz dudosa y fiel que no declina	of dubious, loyal, light, without decay
y que acecha en el alba. Yo querría	and watching at dawn. I wish
ver una cara alguna vez. Ignoro	to see sometimes a face. I know not
la inexplorada enciclopedia, el goce	the unexplored encyclopedia, the transport
de los libros que mi mano reconoce,	of the books my hand recalls,
las altas aves y las lunas de oro.	the golden moon and the birds aloft.
A los otros les queda el universo:	For the others remains the universe:
a mi penumbra, el hábito del verso.	to my penumbra the habit of the verse.

Borges coincided with Milton in considering blindness as a fate imposed on him by God. But, unlike Milton, who began wondering whether he would be able to continue serving God and eventually fully accepted his fate, the passage of time shows an increasing bitterness in Borges' poems. Thus, he begins the *Poema de los Dones* describing his blindness as "God's irony" and making clear that there is "neither tear nor reproach" in this description, a description that otherwise compares blindness with the night that severed him from his all-encompassing library:

Nadie rebaje a lágrima o reproche	Let neither tear nor reproach besmirch
esta declaración de la maestría	this declaration of the expertise
de Dios, que con magnífica ironía	of God who, with magnificent irony,
me dio a la vez los libros y la noche.	granted me at once the books and the night.

In contrast with this apparent lack of tear or reproach, in the poems that will come later, adjectives such as "vano" (vane), "mero" (mere) or "inútil" (useless), which point towards a lack of reality, have an increasing presence. In the same line, objects lose their individuality and are referred to with nouns such as "cosa" (thing) or "forma" (shape). In his last poem on the subject (see above) even this diminished individuality

[4] *On His Blindness* by Jorge Luis Borges, currently collected in *El Oro de los Tigres* and *Los Conjurados*. Copyright © 1995 by Maria Kodama. Used by permission of The Wylie Agency LLC and courtesy of Random House Mondadori, SA. The translation to English at the right is mine. It has no pretensions besides conveying some sense from the original Spanish.

is further enhanced, as different things are now "reduced to one thing". The act of writing, that in Milton was to serve God, in Borges becomes a shelter. The two last lines of *On His Blindness* leave no doubts.

Putting aside the differences in the way Milton and Borges faced the personal tragedy of blindness, it is worth noting the coincidences in the 1655 and 1985 namesake poems. Both are sonnets. That is, both are made of 14 lines, each of which having a fixed number of syllables (10 in Milton's, 11 in Borges'). In both cases the final syllables of these lines repeat their sound forming a well-defined pattern. Using the same letter to denote the same sound, Milton's poem has the pattern ABBA ABBA CDE CDE and Borges' ABBA CDDC EFF EGG. There is a strange evenness in the small differences in these patterns. The last lines in Milton's correspond to a pattern common in Italian and Spanish poetry. Instead, the last lines in Borges' are usually referred to either as "Shakespearean" or as "English".

Actually, Milton's poem is said to be a *Petrarchan sonnet*, since it possesses a rhyme structure introduced in the fourteenth century by Francesco Petrarca, the great Italian poet. Petrarca repeatedly used this structure in the love poems he wrote for Laura de Noves. Incidentally, these poems mention a portrait of Laura in the possession of Petrarca which, art historians argue, may have been one of the first portraits with an intention of likeness (Gombrich, 1989: 161). According to Vasari (1991: 42) this portrait was executed by Simone Martini.

A.3 The *Musical Offering*

The picture in Figure A.2 shows Johan Sebastian Bach in 1748 in a portrait painted by Elias Gottlieb Haussmann. Widely regarded today as the greatest composer of all times, it was not as such but as a virtuoso that Bach was known during his life (just as Milton was known as a polemicist rather than as a poet during his own). And so much so that even Frederick the Great, King of Prussia, had long desired to have Bach pay a visit to his court in Potsdam. Frederick had actually made suggestions to that effect to Bach's son, Carl Philipp Emanuel, who at that time was employed at the court as choirmaster. In addition to Bach's reputation as an organ player, he was also celebrated because of his ability to improvise. This explained Frederick's desire, as the King was the happy owner of several Silbermann pianos, instruments that, in the middle of the eighteenth century, were just making their debut, and Frederick wanted Bach to improvise on them.

Be it because of the mediation of his son, or because of some other reason, the fact is that on the evening of 7 May 1747, Bach paid his long sought-after visit to Frederick's court. The following description of that

Figure A.2 **Bach painted by Elias Gottlieb Haussmann. (Erich Lessing/Art Resource, NY.)**

evening is by Johann Nikolaus Forkel (1920: 25), one of Bach's earliest biographers:

Accompanied from room to room by the King and the musicians, Bach tried the instruments and improvised upon them before his illustrious companion. After some time he asked the King to give him a subject for a Fugue, that he might treat it extempore. The King did so, and expressed his astonishment at Bach's profound skill in developing it. Anxious to see to what lengths the art could be carried, the King desired Bach to improvise a six-part Fugue. But as every subject is not suitable for polyphonic treatment, Bach himself chose a theme and, to the astonishment of all who were present, developed it with the skill and distinction he had shown in treating the King's subject. [...] On his return to Leipzig he developed the King's theme in three and six parts, added *Canones diversi* upon it, engraved the whole under the title "Musikalisches Opfer" and dedicated it to the royal author of the theme.

The dedication deserves attention. Translated into English, it would read "At the King's Command, the Song and the Remainder Resolved with Canonic Art". Here, Bach is playing with words, since "Canonic" alludes to the meaning "with canons" as much as it does to the meaning "in the best possible way". But the play on words goes beyond that. The actual dedication, in Latin, reads as follows:

Regis Iusſu Cantio Et Reliqua Canonica Arte Reſoluta.

The initials of it form the word "RICERCAR", an Italian word meaning "to seek" which was also used as a synonym for "fugue". Such a word structure is called an *acrostic*. Actually "ricercar" was the original term for a fugue (the Italian word "fuga" being of more recent use) but, by Bach's time, it had earned a more erudite usage. Hofstadter (1979) notes that "a similar usage survives in English today: the word 'recherché' means, literally, 'sought out', but carries the same kind of implication, namely of esoteric or high-brow cleverness". In the *Musical Offering* Bach used the word "ricercar" to name its fugues. Following Hofstadter, we may say that this choice of words was itself recherché.

To understand that evening's exploit one needs to grasp the difficulty of composing (not to say improvising!) a canon or a fugue. Without entering into details, those are musical forms in which a melody is played against itself. Whereas this may seem easy, there is a basic snag: when the different copies of the primary melody are played together one is likely to obtain unpleasant sounds. And such sounds, called *dissonant chords*, are to be avoided for a canon to be considered well formed.

Accompanying the *Musical Offering* was also the following dedicatory letter in which Bach appears to dismiss all the complexity of the *Musical Offering* and put the emphasis on the theme given to him by Frederick the Great. Musical pundits ever since have, notwithstanding, disagreed with him.

𝔐𝔒𝔖𝔗 𝔊𝔯𝔞𝔠𝔦𝔬𝔲𝔰 𝔎𝔦𝔫𝔤!

In deepest humility I dedicate herewith to Your Majesty a musical offering, the noblest part of which derives from Your Majesty's own August Hand. With awesome pleasure I still remember the very special Royal Grace when, some time ago, during my visit in Potsdam, Your Majesty's Self deigned to play to me a theme for a fugue upon the clavier, and at the same time charged me most graciously to carry it out in Your Majesty's Most August Presence. To obey Your Majesty's command was my most humble duty. I noticed very soon, however, that, for lack of necessary preparation, the execution of the task did not fare as well as such an excellent theme demanded. I resolved therefore and promptly pledged myself to work out this right Royal theme more fully and then make it known to the world. This resolve has now been carried out as well as possible, and it has none other that this irreproachable intent, to glorify, if only in a small point, the fame of a Monarch whose greatness and power, as in all the sciences of war and peace, so especially in music, everyone must admire and revere. I make bold to add this most humble request: may Your Majesty deign to dignify the present modest labor with a gracious acceptance, and continue to grant Your Majesty's Most August Royal Grace to

> Your Majesty's most humble
> and obedient servant
>
> The Author

Leipzig, July 7, 1747

Forkel (1920: 26) also writes that by late 1749 or early 1750 Bach completely lost his eyesight. A British surgeon named John Taylor, who had been the oculist of George II, operated twice on Bach, both times without success. In contrast with Milton or Borges, Bach was seriously affected by this condition, "sank gradually for full half a year, and expired on the evening of July 30, 1750".

A.4 The garden of the crossing paths

> Mighty is geometry; joined with art, resistless.
>
> Attributed to Euripides[5]

The preceding sections are somehow unrelated, at least at a first glance. They narrate episodes in the lives of artists who lived in different places and different times and even the artworks involved are of a different nature.

A closer look at these artworks reveals, however, a common feature which we may describe as a sense of order, a feeling that the constituents of the work, be it a painting, a poem or a musical piece, fit together following a pattern, an underlying law. A goal of the next few chapters is to unveil some of these laws, notably those having a geometric nature. These laws are doubtless meeting points in the paths of art and mathematics. But they are not the only points where these paths cross, because occasionally art has returned in kind the influence it received from mathematics and taken a role in shaping its development. Another goal in this book is to give an account of some of these occasions. To accomplish these goals we must first discuss geometry.

---

[5] The attribution is faulty. In verse 884 of *Hecuba*, the character with that name replies to the question posed by Agamemnon on how will women defeat men with the words δεινὸν τὸ πλῆθος σύν δόλῳ τε δύσμαχον which Morwood (Euripides, 2000) translates as "Women in numbers are formidable, and when allied with trickery, hard to fight against". I am amazed at the semantic derives that turned "numbers" into "geometry" and – doubtless through "artfulness" – "trickery" into "art". But I have read the quotation in a (large) number of places and, empowered by the might of this number, I have found the temptation to use it, as it were, resistless.

1 Space and geometry

O God! I could be bounded in a nut-shell, and count myself a king of infinite space [...]

W. Shakespeare, Hamlet, II, 2

1.1 The nature of space

Figure 1.1 **Over the edge of the world. (Kean Collection/Archive Photos/Getty Images.)**

The engraving in Figure 1.1 shows the edge of the world as it was conceived in ancient times. The known land was surrounded by a vast expanse of sea close to whose end waters were inhabited by sea serpents, dragons and other fearsome creatures.[1]

Monsters aside, a fact at odds with our contemporary perception of the world is that the surface of the world was – waves, mountains and other accidents of this surface ignored – flat. This feature is inconsistent with the images of the Earth that satellites bring daily to us but was perfectly consistent with the information available to, say, the ancient Greeks. Indeed, moving north one was led to an increasingly cold landscape and moving south to an increasingly hot one. And at both east and west boundaries of the known land mass, oceans extended which were navigable only close to the land (and hence whose end could neither be seen nor refuted). In addition, it was not possible to distance oneself enough from the surface of the Earth to appreciate its curvature. All these enclosures created a confinement within which the flatness of the Earth was as likely as its spherical curvature.

At the centre of this phenomenon there are issues of locality and scale. When I look at some objects – for example, an orange or a tomato can – I can perceive them in their entirety. For some others – the Earth being an example – I cannot do so. Admittedly, I cannot perceive the whole of an orange at a single instant, since part of it will be behind its contour. But turning the fruit in my hands I can explore the entirety of its surface in a matter of seconds. The difference between an orange and the Earth is, we now know, a matter of scale. But the difference between a sphere and a plane is not. No matter how much I distance myself from the plane, nor how much I change my viewpoint, I cannot comprehend the whole plane. The collection of local perceptions I gather of a plane makes possible the emergence of a global picture of the plane but allows neither for the verification of this picture nor for its refutation.

---

[1] In the first movie of the *Pirates of the Caribbean* trilogy, Captain Barbossa warns Jack Sparrow "You're off the edge of the map now, Jack! Here there be Monsters!"

Another issue in the assumption of a flat Earth of a philosophical nature is the call to an infinite regress. The waters at the edge of the world are falling down in Figure 1.1 but the Earth is not. What keeps it in place? Asimov (1983: 15) explains that

The Hindus placed it on four pillars [...But ...] On what were the four pillars standing? On elephants! And on what were the elephants standing? On a gigantic turtle! And the turtle? It swam in a gigantic ocean! And this ocean —

Locality, scale and infinite regress will play an important role in many of the discussions in this book. They also played a critical role in the development of geometry.

1.2 The shape of things

> The wrong of unshapely things is a wrong too great to be told.
>
> W.B. Yeats (2001: *The Lover Tells of the Rose in His Heart*)

Art and geometry are linked in a primordial manner: the way one visually perceives things presupposes a geometry which decodes this perception.

First, such a geometry gives an account of the properties of the space where the perceived things, as well as the perceiver, live. As we noted in Section 1.1, however, a verification of these properties may not be at hand owing to the local character of our perceptions coupled with an issue of scale. A fictional account of this hindrance is offered in the novella *Flatland* by Edwin Abbott (1987). Originally written to satirize Victorian society, it has attained an enduring character in its description of our limitations to intuit higher-dimensional space. In the novella, characters live in a two-dimensional world. The narrator, called A. Square, is visited by a sphere who introduces him to Spaceland and, thus guided, he conceives the existence of three-dimensional space. He then attempts to convey his discovery to Flatland's inhabitants but he fails. This inability to grasp a higher-dimensional space is a recurrent theme in the book. Not only Flatlanders are hindered by it but also Linelanders – whose monarch cannot conceive any extent beyond the line where he lives – the solipsistic king of Pointland, and even the Sphere – who refuses to accept four-dimensional space.

Second – and in addition to the nature of the space where perception takes place – there is a context which, so to speak, chooses a geometry in this space. To illustrate this statement we may elaborate on the notion of shape. In the (online version of the) *Oxford English Dictionary* one can find the following definition:

shape, *n.* (ʃeɪp)   External form or contour; that quality of a material object (or geometrical figure) which depends on constant relations of position and proportionate distance among all the points composing its outline or its external surface; a particular variety of this quality.[2]

Useful as this definition might be, it is of little help, at this juncture, in describing the shape of a given object, or even the simpler task of deciding whether two objects have the same shape. In fact, the second task is simpler but not simple, since even this question, whose answer is either "yes" or "no", appears to depend on the context in which it is posed. The fact that the second letter in

is slanted with respect to the first may be disregarded to conclude that both are the same letter. The same slant in the following tiles

would be harder to disregard when tiling the walls of my kitchen. The following two portraits

show the same shape with a difference in scale which is not essential. Such a difference in scale would, in contrast, be most detrimental for my tiling work.

This dependence on the context is reflected in the way the notion of shape is dealt with in geometry. Roughly speaking, a number of measurable features – distances between points, angles between lines, alignment, proportions, etc. – are considered and objects sharing some of these features (in a way we will soon describe) are given odd-sounding names – congruent, similar, homeomorphic, etc. – depending on which of these features are shared. The set of features which are shared captures the properties considered as essential in a given context and we may say that defines a geometry. But this is a recent idea. For centuries, the word

[2] OXFORD ENGLISH DICTIONARY edited by Simpson & Weiner (1986). Definition of "shape". By permission of Oxford University Press.

"geometry" meant a specific nature of space endowed with an equally specific set of relevant features. It is with this notion of geometry that we will be involved for some time to come.

1.3 Euclid

Tradition has it that at the gate of Plato's academy an engraving read "Let no one ignorant of geometry enter" (ΑΓΕΩΜΕΤΡΗΤΟΣ ΜΗΔΕΙΣ ΕΙΣΙΤΩ). The geometry Plato was referring to was a corpus of statements dispersed in the works of Pythagoras, Theaetetus and Eudoxus, among others. The building of the edifice of geometry that would integrate these statements took place less than a century after the building of Plato's Academy in the form of Euclid of Alexandria's *Elements* (Euclid, 1956).

A collection of 13 books, the *Elements* summarize the knowledge of geometry, number theory and elementary algebra in Euclid's time. The contents on geometry are remarkable, in the sense that they still largely overlap the contents of many secondary schools' courses. The style of the *Elements* is, if possible, even more remarkable. Geometric statements in previous works had often been given a justification that we can call a proof. But these justifications relied on assumptions which, even though agreed upon, had an ad hoc character. In contrast with this practice, Euclid introduced what today is called the *axiomatic* (or *deductive*) *method*.

A mathematical proof, to put it in the simplest manner, is a deduction of some statement from some known true statements. Such a description immediately calls for a careful scrutiny.

First, where do the "known true statements" come from? They may come from previously proved (and hence true) statements. But then, where do the ones used in these proofs come from? One feels an infinite regress not unlike the search of the ultimate fulcrum in a flat Earth conception of the world.

Second, statements assert properties of objects, and in order to be able to agree on a deduction we must, more often than not, agree on the meaning of these properties and objects. The obvious solution is that these meanings are previously defined. But definitions can describe a notion only in terms of other, previously agreed upon, notions. Infinite regress strikes again.

Third, how exactly is a statement deduced from known ones? Or, more precisely, when do we agree that a given deduction is valid?

The solution given by Euclid to these three points was a single one. For the first one, we set a family of statements to be considered as true ones (we call them *axioms* or *postulates*). For the second one, we agree on a family of notions whose meaning is considered to be known and shared by everybody (we call them *primitive notions*). And for the third one, we agree on a family of logical rules, allowing one to infer a statement from a collection of other statements.

For instance, in the *Elements*, the notions of *point, straight line, distance angle,* and *equality* are primitive.[3] From these, one defines further notions such as *right angle, circle, triangle* and *parallel*.

In addition, Euclid takes as axioms or postulates the following five statements (Cederberg, 2001: 391):

P1 To draw a straight line from any point to any point.
P2 To produce a finite straight line continuously in a straight line.
P3 To describe a circle with any centre and distance.
P4 That all right angles are equal to one another.
P5 That, if a straight line falling on two straight lines makes the interior angles on the same side less than two right angles, the two straight lines, if produced indefinitely, meet on that side on which the angles are less than the two right angles.

Postulates and primitive notions play roles that complement each other. It is through the common acceptance of the postulates' truth that a shared meaning for the primitive notions can be asserted. For instance, we may say that by accepting as true postulates P1, P2, P4 and P5 we agree on the nature of straight lines in Euclidean geometry. This nature is implicit in these four postulates and any property of a straight line (e.g. being green) not derived from them is merely accidental. This is the situation in all fields of contemporary mathematics. As it happens, however, it is precisely in the case of the nature of straight lines that problems arose as geometry developed. And this was so for a number of reasons.

The first reason – and quite an understandable one for the first exposition ever using the axiomatic method – was a certain vagueness in the statement of the postulates. For instance, P1 asserts the existence of a straight line passing through any two points. If these two points coincide then there are many such lines. But what if they do not? At Euclid's time it was assumed that in that case the joining line was unique. But this is never stated in P1. Similarly, P2 was generally understood as stating that the straight line containing any finite straight line (that is, any segment such as ——) could be infinitely extended in both directions. But whether this entailed that straight lines are infinite or unbounded was

---

[3] Actually, in the text of the *Elements* Euclid attempts to define these notions, but the resulting definitions – seen with modern eyes – are less than satisfying. Compare, for instance, the one for point ("a *point* is that which has no parts") with that for right angle ("When a straight line set up on a straight line makes the adjacent angles equal to one another, each of the equal angles is *right*, and the straight line standing on the other is called *perpendicular* to that on which it stands"). The latter has a precision which the former is completely lacking.

left in the dark. To understand the difference involved in these seemingly purposeless subtleties, imagine that the two-dimensional space where these straight lines live, what we call the *plane*, is the surface of a large sphere. Say the Earth, to fix ideas. What would a straight line be in this case? Well, given two different points in this surface, say Paris and Rio de Janeiro, there is an obvious candidate for the segment between these points, namely the shortest trajectory between them. This is an arc of a circle which, when extended in both directions, yields a great circle on the sphere (a circle like the Equator line but tilted so that it passes through both cities). So a straight line on this surface is just a great circle. But this notion of line fails to satisfy both the uniqueness in P1 and the infinity in P2. Indeed, there are many great circles passing through two antipodal points (for instance, many lines of longitude passing through the poles). Also, each great circle is unbounded (it has no end points) but not infinite. The difference between the common understanding of P1 and P2 and the rather unusual one arising from the example above lies in the possible curvature of the plane where the lines live. Although never explicitly stated, P1 and P2 assume these lines live in a flat plane and the example above makes them live on a spherical plane.

Modern mathematics, with its reliance in a formal language as opposed to natural languages, is built in such a way that ambiguities such as those described above do not occur. There is a price to pay for this precision, though: abstraction. And even though Euclid's *Elements* represent a gigantic step towards our contemporary manners in mathematics, the detachment of the latter from any consideration beyond formal correctness is far from being there. Hartshorne (1997: 10) notes that

Euclid's geometry is still tied to the real world [...] For example, Euclid does not hesitate to use arguments [...] that seem perfectly acceptable in view of our experience of the real world, yet are not logical consequences of his initial assumptions.

These ties between the real world and Euclid's geometry were not questioned for many centuries. And its first four postulates were commonly accepted with their intended, if not explicit, meaning. The fifth postulate, however, was a cause of concern much sooner. The difference between P5 and the other postulates seems to stand from a number of reasons. Clearly, its statement is longer and more complicated than those of the previous postulates. But also, its content touches the issues of locality and scale we mentioned before in a way that the previous postulates do not. To see why, let us compare P5 with, say, P4. While we cannot empirically prove P4, since we cannot confirm its truth for all pairs of right angles, we certainly can confirm this truth for each pair we start with. In contrast, to verify P5 on a given instance requires us to extend two lines until they meet, and there is no *a priori* estimate of when, if ever, this is going to occur. This feature is also seen in the following equivalent form of P5 (which explains why P5 is usually known as the *parallels postulate*):

P5′ For each point P and each line ℓ there exists at most one line through P parallel to ℓ.

Indeed, to verify P5′ one needs to show that, for any two lines passing through P, one of them, when sufficiently extended, will intersect ℓ.

The truth of the fifth postulate, it should be noted, was not questioned. At least, it was not until modern times. The reason P5 became a cause of concern was because it was suspected, and many efforts were devoted to prove this suspicion, that it could be derived from the other four postulates. But the nature of the geometry described in the *Elements* was commonly agreed upon and its faithfulness to the real world widely accepted. Euclidean geometry, even today, is the embodiment of geometry for most people. Some of its proofs are, in addition, of a striking beauty. The statements below correspond to two basic results in Euclidean geometry, present in all high school curricula for geometry. We give a proof of the first one to convey the beauty we just mentioned.

Theorem 1.1 (Pythagoras) *In all triangles with a right angle the square of the side opposite to this right angle equals the sum of the squares of the other two sides.*

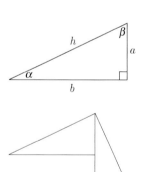

Proof Let us call α and β the other two angles, h the side facing the right angle and a and b the sides facing α and β respectively, so that the triangle looks like the top drawing on the left.

Now take a (rotated) copy of the triangle and place it as in the middle drawing.

Since the angles of a triangle add up to $180°$ the sum of α and β must be $90°$. Therefore, the angle at the top in this drawing is a right angle. Adding two more copies of the triangle we obtain the bottom drawing.

We know that the side of the large square is h, so its area equals h^2. But this area can be written as the sum of the areas of the four triangles (each of them being $\frac{ab}{2}$) plus the area of the small square inside whose side is $b - a$. Therefore:

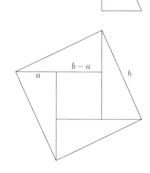

$$h^2 = 4\left(\frac{ab}{2}\right) + (b - a)^2 = 2ab + (b^2 + a^2 - 2ab) = b^2 + a^2. \quad \square$$

The second classical result we state is the following.

Theorem 1.2 (Thales) *When three parallel lines (ℓ, ℓ', ℓ'' in the picture) are cut by two transversals (s and t) the ratio between the two segments determined in one of the transversals ($\overline{AB}$ and $\overline{BC}$) is the same as the ratio of the segments determined in the other ($\overline{PQ}$ and $\overline{QR}$).*

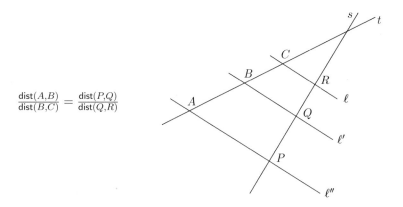

$$\frac{\text{dist}(A,B)}{\text{dist}(B,C)} = \frac{\text{dist}(P,Q)}{\text{dist}(Q,R)}$$

We have used in the statement of Thales' theorem the notation $\text{dist}(A, B)$ to denote the distance between two points A and B on the plane. We will keep doing so in the rest of the book.

1.4 Descartes

> I am sorry that I had to torture you with these elements of analytic geometry. The purpose of this invention of Descartes' is nothing but to give *names* to the points X in a plane by which we can distinguish and recognize them.
>
> H. Weyl (1952: 95)

Any reasoning involving geometric objects (points, lines, etc.) is greatly simplified if we give names to these objects (as we have done in the proof of Pythagoras' theorem). For instance, if our pondering involves three points, one may name them Alice, Bob and Carol. Tradition (and an appreciation for simplicity) induced mathematicians to name them simply A, B and C. Such a naming is at the core of many advances in geometry and underlies Euclid's exposition. Yet, the complete freedom in its choice of names obscured the possible relations (distance between points, angles between lines, etc.) these objects could have. A different naming system bringing to the light these relations was introduced by René Descartes (1954) in the 1630s. To understand its functioning, let us begin by considering points in a straight line.

We may fix two points on this line and call them 0 and 1:

0   1

By doing so, a number of other points on the line naturally get names. For instance, it is clear which points are named -1, 2 or 3:

It is also clear that we do not have enough names if we restrict them to be integer numbers. For instance, some points on the line will have to be named with fractions:

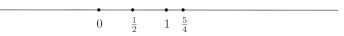

And for some others it turns out that we will need names that cannot be expressed as fractions, and realizing this was a major step in the history of mathematics:

An example of a relation between points on the line that now becomes visible is the notion of distance. Once we fix as the unit for measurement the distance between the points 0 and 1, the distance between any other pair of points follows. Thus, the distance between 4 and -1 is 5 and the distance between 2 and $\frac{7}{2}$ is $\frac{3}{2}$. More generally, the distance between points p and q is the absolute value of their difference, i.e. $\mathrm{dist}(p, q) = |p - q|$.

To each point in the line there corresponds a name. We will denote by $\mathbb{R}$ the set of all these names and refer to it as the *set of real numbers*. There are too many points in the line (and hence too many numbers in $\mathbb{R}$) for us to be able to actually use all these names. But, as it happens, we will never need to use most of them. We do not deal with the real numbers one by one but rather with the set $\mathbb{R}$ itself.

To name points on the plane, Descartes extended the previous ideas as follows. Take any two non-parallel lines, each with their distinguished points 0 and 1, so that they intersect at their 0s and they make a right angle.

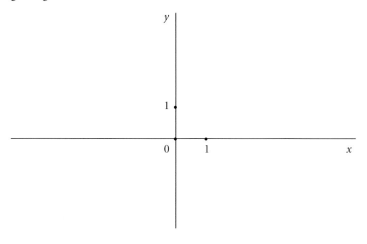

We call these two lines the *axes of coordinates* and distinguish them by calling one the *x*-axis and the other the *y*-axis. Usually, the horizontal one is called the *x*-axis and the vertical the *y*-axis.

We can give a name to any point on the plane as follows. Take two lines on the plane, passing throught that point and parallel to the axes. The vertical line will intersect the *x*-axis at one point and so will the horizontal one and the *y*-axis. Since each of these points has a name (in the corresponding axis) we can take the pair of these names as the name of the point.[4] For instance, the point $(3, 2)$.

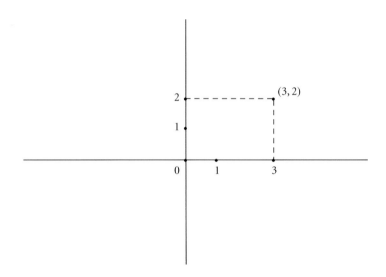

Conversely, given a name we can identify the only point on the plane having this name. If we are given the name $(3, 2)$ we can locate the point 3 on the *x*-axis and draw a vertical line passing through it. Similarly, we locate the point 2 on the *y*-axis and draw a horizontal line parallel to the *x*-axis through it. The point $(3, 2)$ is the intersection of these two lines.

Again, the notion of distance – now between points on the plane – becomes visible for this system of names. Indeed, if P and Q are points in the plane with coordinates (P_x, P_y) and (Q_x, Q_y) respectively, then it follows from Pythagoras' theorem that

$$\mathrm{dist}(P, Q) = \sqrt{(P_x - Q_x)^2 + (P_y - Q_y)^2}. \qquad (1.1)$$

---

[4] By convention, we write first the name for the *x*-axis (called the *x-coordinate* of the point) and then that for the *y*-axis.

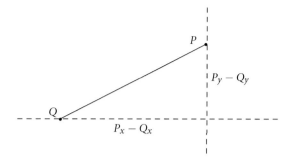

Implicit in the previous reasoning is the fact that Descartes' ideas did not intend to cast a doubt on Euclidean geometry; they rather relied on it. This is witnessed in the occurrence of statements – such as the uniqueness of the lines through $(3, 2)$ parallel to the coordinate axes – whose truth is the very essence of Euclidean geometry. This reliance is more explicit when we give names to more complicated objects such as straight lines.

How does one give names to straight lines on the plane? As a first step toward an answer, assume that the line is vertical and cuts the x-axis at the point 7. Then, every point on this line will have coordinates $(7, y)$ for some value of y. Conversely, for every value y the point $(7, y)$ is on the line. The "essence" of the line is that $x = 7$. We can therefore name the line by this expression and refer to it as its *equation*. One may define this line with other equations (for instance, $3x = 21$); in this sense, we consider all equations defining the same line as equivalent. We next consider a non-vertical line passing through the origin of coordinates (i.e. the point $(0, 0)$). Take any point (x, y) on this line (different from the origin) and let $m = y/x$. The value of m is independent of which point we took on the line. Indeed, for any other point (x', y') we have $y'/x' = y/x$ because of Thales' theorem (an instance of it where ℓ passes through the intersection of s and t and, therefore, $C = R$; see Theorem 1.2).

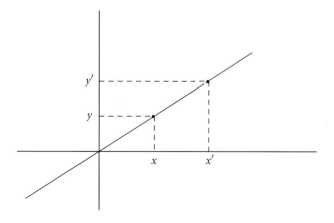

The equation $y = mx$ therefore describes this line in the same sense as above: any solution of this equation is (the name of) a point in this line and, conversely, any point on the line is a solution of the equation. This includes the point $(0, 0)$.

We finally consider arbitrary lines (non-vertical, since we already have equations for these lines). Let ℓ be one such line. Since it is not parallel to the y-axis it intersects this axis at a point $(0, q)$. Consider the line ℓ' parallel to ℓ and passing through the origin. As we have just seen, the equation of ℓ' is $y = mx$ for some real number m. For any point P in ℓ, consider the line through P parallel to the y-axis and let P' be the intersection of this line with ℓ'. The four-sided figure with vertices $(0, 0)$, P', P and $(0, q)$ is a parallelogram (i.e. opposite sides are parallel). Therefore, opposite sides are equal and it follows that if the coordinates of P are (a, b) then those of P' are $(a, b - q)$.

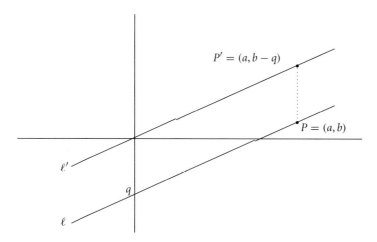

But then we must have $b - q = ma$ or, equivalently, $b = ma + q$. That is, the coordinates (a, b) of the point P satisfy the equation $y = mx + q$. A similar reasoning shows that every point satisfying this equation lies on ℓ. Note that we have implicitly proved the following result.

Proposition 1.3 *Two lines with equations $y = mx + q$ and $y = m'x + q'$ are parallel if and only if $m = m'$.*

So, points are named with a pair of numbers (their coordinates) and lines, which are sets of points, with an equation that is satisfied by precisely the points in this set and by no others. But once we have used equations to name lines, there is no obstacle to do the same for other sets of points. Take for instance a circle with radius R centred at the origin of coordinates. By Pythagoras' theorem, for every point (a, b) on the circle we must have $a^2 + b^2 = R^2$ and, conversely, a point satisfying this equality is at a distance R from the origin, and hence on the circle.

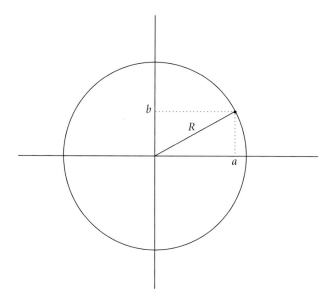

We conclude that this circle has equation $x^2 + y^2 = R^2$. A reasoning with parallelograms would allow us to derive equations for all circles (not necessarily centred at the origin). But we will stop here.

A major benefit of *analytic geometry* (as Descartes' invention came to be known in contrast with Euclid's exposition, which in this context is called *synthetic geometry*) is an increase in our capacity to derive proofs by translating geometry into algebra. Take, for example, the fact that a straight line cuts a circle in at most two points. This appears to be true in every single drawing we do,[5] but it demands a substantial amount of work to be derived from Euclid's postulates P1–P5. Compare with the following proof.

Proposition 1.4 *Every line cuts a circle in at most two points.*

Proof Choose the y-axis to be the only line parallel to the given line and passing through the centre of the circle. Choose the x-axis to also pass through this centre and (needless to say) perpendicular to the y axis. Then the equation of the line is $x = d$ for some real number d and that of the circle is $x^2 + y^2 = R^2$, where R is the radius of the circle.

Replacing $x = d$ in the circle's equation we find

$$d^2 + y^2 = R^2, \quad \text{i.e.} \quad y^2 = R^2 - d^2.$$

As one could expect, the number of solutions is zero when $|d| > R$, one when $d = \pm R$ (and in this case $y = 0$), and two when $|d| < R$. $\qquad \square$

---

[5] It is, in addition, so "evident" that we feel a certain resistance to the need to prove it. Not being an axiom, however, we must do so.

The intersection of two circles can be dealt with in a similarly straightforward way.

Proposition 1.5 *Two different circles cut in at most two points. In addition, if they cut in two points, the segment joining these points cuts the line joining the centres of the circles in a right angle and the distances from these points to this intersection are the same.*

Proof Let C_1 and C_2 be the two circles and R and r be their respective radii. Choose the x-axis to be the line passing through their centres and the y-axis to be the line perpendicular to it and passing through the centre of C_1. Then the centre of C_2 has coordinates $(a, 0)$ for some a in $\mathbb{R}$. Furthermore, C_1 has equation $X^2 + Y^2 = R^2$ and C_2 has equation $(X - a)^2 + Y^2 = r^2$. If a point (x, y) is in the intersection of C_1 and C_2, then we have both

$$x^2 + y^2 = R^2 \tag{1.2}$$
$$(x - a)^2 + y^2 = r^2$$

and, therefore, subtracting the first equation from the second:

$$(x - a)^2 - x^2 = r^2 - R^2$$

or yet

$$a^2 - 2ax = r^2 - R^2.$$

It follows from here that the x-coordinate must satisfy

$$x = \frac{R^2 - r^2 + a^2}{2a}. \tag{1.3}$$

In addition, from the first equation in (1.2), we know that $y^2 = R^2 - x^2$. Depending on the values of R, r and a, the difference $R^2 - x^2$ can be positive, negative or zero. We conclude that

$$\text{number of values of } y = \begin{cases} 2 & \text{if } R^2 - x^2 > 0 \\ 1 & \text{if } R^2 - x^2 = 0 \\ 0 & \text{if } R^2 - x^2 < 0. \end{cases}$$

This shows the first statement. For the second, note that if there are two solutions (x, y) these have different values of y but the same value of x, namely that given in (1.3). Therefore, the line joining them has equation $X = (R^2 - r^2 + a^2)/2a$. It follows that this line is vertical and, hence, perpendicular to the line joining the centres of C_1 and C_2. $\quad\square$

We briefly return to the issue of names. The vast majority of mathematical statements make assertions on generic objects. Pythagoras' theorem is valid for any right-angled triangle, Thales' theorem is valid for any triple of parallel lines and any pair of transversals to them.

Reasoning about these objects is simplified by giving them names (which do not carry any particular meaning). Analytic and synthetic geometry are similar in this regard. A difference occurs, however, in the names given to objects derived along the proof. In synthetic geometry we will still give arbitrary names and separately book-keep the relations of these derived objects with the original ones. In analytic geometry the name of the derived object will explicitly (as much as it can) convey these relations. A simple example is the name for the midpoint of a segment $\overline{PQ}$. In synthetic reasonings we will give it a name, say m, and keep in mind that $\text{dist}(P, m) = \text{dist}(m, Q)$. In analytic geometry, if $P = (x_P, y_P)$ and $Q = (x_Q, y_Q)$ the "name" of m is determined, namely

$$m = \left(\frac{x_P + x_Q}{2}, \frac{y_P + y_Q}{2} \right).$$

This is so because, by Thales, theorem,

$$\frac{\text{dist}(P, m_x)}{\text{dist}(m_x, O)} = \frac{\text{dist}(P, m)}{\text{dist}(m, Q)} = 1.$$

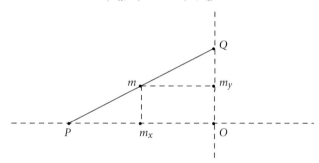

This shows that m_x is the midpoint of the segment $\overline{PO}$ and, therefore, that $m_x = (x_P + x_Q)/2$. And similarly for m_y. Note that the coordinates x_P, y_P, x_Q, y_Q for P and Q are arbitrary, but those for m are not: they are a specific expression of these four numbers, which carries over the fact that m is the midpoint of $\overline{PQ}$.

A slightly more complicated situation involves two non-parallel lines and their intersection point. In the synthetic approach we call the lines ℓ and ℓ' and the intersection point P. In addition, we keep in mind that P is the intersection of ℓ and ℓ'. In the analytic approach we refer to the lines by their equations $y = mx + q$ and $y = m'x + q'$ (with $m \neq m'$ as the lines are not parallel) and we deduce that the x-coordinate of the intersection point must satisfy

$$mx + q = m'x + q', \quad \text{i.e.} \quad x = \frac{q' - q}{m - m'}$$

and therefore that this point is $\left(\frac{q'-q}{m-m'}, m\frac{q'-q}{m-m'} + q \right)$. Again, the numbers m, m', q, q' for the original lines are arbitrary but their intersection point is not: its coordinates are a specific expression of these four numbers, which carries over the fact that the point is at the intersection of the lines.

The contents of the present section, along with that of the preceding one, have laid down the defining characteristics of a geometry which, for centuries, has been considered the only one possible. They have also equipped us with some notions and ideas that will allow us to further develop this geometry in Chapter 2 and explore some of the points where its path crosses that of artistic creation – a tour starting in Chapter 3 whose meanders extend through subsequent chapters.

2 Motions on the plane

How could we certify that two figures on the plane have the same shape? An approach that seems to be simple is to move one figure to place it on top of the other and, once this is done, check that they match, i.e. that their overlapping is perfect. In a bricolage implementation of this idea, one of the figures would be drawn on a fixed, say white, board and the other on a transparency. In this implementation, "to move one figure" is equivalent to moving the transparency.

If the transparency is initially placed on top of the white board, to each point of the former there corresponds a point on the board, namely the point "below". After moving it to attempt a matching, each point on the transparency corresponds to *another* point on the board. This correspondence induces, in turn, a correspondence between points on the board. Indeed, for every point P_b on the board we consider the point P_t above it on the transparency. Then we move the transparency and consider the new point P'_b which is now below P_t. A way (admittedly convoluted looking at first) to describe the movement itself would consist of describing the transformation that associates to each point P_b on the board the corresponding P'_b. This chapter pursues this idea. In doing so, we will eventually prove that there are, essentially, just a few ways to move figures on the plane.

2.1 Translations

Arguably the simplest of the transformations on the plane, a translation maps every point on the plane to the only point at a given distance in a given direction. A reader not engaged in the everyday practice of mathematics may find himself at a loss with a number of words in this description. The next few paragraphs, which introduce some basic notions and notations, should assist. A *vector* on the plane is a line segment oriented in a way that one of its extremities is its origin and the other its end. They are usually represented by an arrow like this ⟍⟋. If these points are O and E respectively, we write $\overrightarrow{OE}$ to denote the corresponding vector. Any vector $\overrightarrow{v}$ has a *length* (the distance between its endpoints, denoted by $\text{length}(\overrightarrow{v})$) and a *direction*. Vectors sharing both length and direction are considered to be the same. That is, particular endpoints are not seen as essential.

The other technical words in the first paragraph above ("transformation", "map") relate to a very basic notion. A *function* or *map f* between two sets A and B is an association which to any element x in A (we write

$x \in A$) assigns a unique element in B. We write $f : A \rightarrow B$ to denote the map and $f(x) = y$ to denote that y is the particular element in B assigned to $x \in A$. For instance, the "square" map assigns to any real number its square. If we denote this map by Sq, we will have Sq : $\mathbb{R} \rightarrow \mathbb{R}$, Sq$(-3) = 9$, and Sq$(5) = 25$. We say that 9 is the *image* of -3 (and 25 that of 5) under the function Sq. In the case of the square function the description of the assignment is simple enough to be carried out with words. But this is generally not the case. So, more often than not, the assignment is described on a variable (representing an arbitrary element of A) as in, for instance, Sq$(x) = x^2$. We can now define translations on the plane which, in all that follows, will be denoted by $\mathbb{E}$ (to emphasize we are referring to the Euclidean plane).

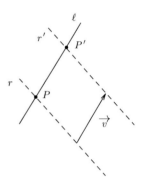

Consider a vector $\vec{v}$. A *translation* with vector $\vec{v}$ is a function trans$_{\vec{v}}$: $\mathbb{E} \rightarrow \mathbb{E}$ which associates, with every point P, the point P' obtained as follows. Let ℓ be the line passing through P and parallel to $\vec{v}$, r be that passing through P and the origin of $\vec{v}$, and r' be that passing through the end of $\vec{v}$ and parallel to r. We take P' to be the intersection of ℓ and r'.

Since r is parallel to r' and ℓ parallel to $\vec{v}$, the quadrangle with vertices P, P' and the endpoints of $\vec{v}$ is a parallelogram. It follows that dist$(P, P') = $ length$(\vec{v})$. Therefore, P' is the only point in ℓ at a distance length$(\vec{v})$ from P in the direction of $\vec{v}$. This shows that P' only depends on the length and direction of $\vec{v}$ and not on particular endpoints.

Let us briefly return to a general function $f : A \rightarrow B$. Since every element in A is assigned an element in B, then to a subset S of A (we write $S \subseteq A$ meaning that every element in S is an element of A as well) there will correspond a subset of B which we denote by $f(S)$. This is obtained by "putting together" the images of all the elements in S. Using standard notation for sets we would write

$$f(S) = \{f(x) \mid x \in S\}$$

and read "$f(S)$ is the set of all the points $f(x)$ with x in S" or, in brief, $f(S)$ is the *image* of S under f. The left-hand part of the figure shows a possible set S and the right-hand part is its image $f(S)$ under a translation with the vector shown:

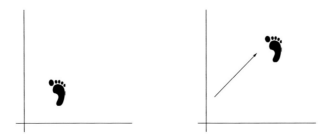

2.2 Rotations

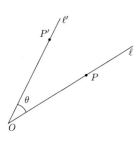

Consider a point O on the plane and an angle θ. A *rotation* of centre O and angle θ associates to each point P on the plane the point P' obtained as follows. If $P = O$ then $P' = O$ as well. If not, consider the half-line ℓ passing through P with origin at O and let ℓ' be the only half-line with origin at O such that the angle[1] $\widehat{\ell\ell'}$ is θ. We take P' to be the only point in ℓ' whose distance to O is $\text{dist}(O, P)$.

As with translations, this association defines a function $\text{rot}_{O,\theta} : \mathbb{E} \to \mathbb{E}$ and we can consider the image under this function of objects on the plane. The left-hand part of the following figure shows the same object we considered in Section 2.1 and the right-hand part is its image under a rotation of $270°$ whose centre O is marked in the figure:

2.3 Reflections

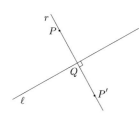

Consider now a line ℓ on the plane. A *reflection* with axis ℓ associates with every point P the point P' defined as follows. If $P \in \ell$ then $P' = P$. If not, let r be the only line perpendicular to ℓ and passing through P and let Q be the intersection of r and ℓ. Then P' is the only point in r satisfying $P' \neq P$ and $\text{dist}(P, Q) = \text{dist}(P', Q)$.

We denote this function by $\text{refl}_\ell : \mathbb{E} \to \mathbb{E}$. The image under this function of our (by now familiar) object is shown in the next figure (the axis of reflection being the dotted line):

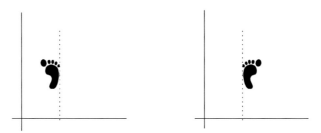

---

[1] By convention, the notation $\widehat{\ell\ell'}$ denotes the angle between ℓ and ℓ' starting at ℓ and moving counterclockwise. One could equally convene on the other possibility, namely moving clockwise, but the former is the one commonly adopted.

This image is equally familiar to us, but it appears to have suffered a change. Indeed, the original footprint corresponded to a right foot whereas the reflected one corresponds to a left foot. The reflection has changed its *orientation*. We have implicitly dealt with orientation in footnote 1. We can be more precise now.

The way we measure angles relies on a convention, namely on whether we measure $\widehat{\ell\ell'}$ moving counterclockwise (as we agreed in Footnote 1 in this chapter) or clockwise. There is nothing essential in this choice, and it is sometimes convenient to change it. Indeed, consider the angle between the thumb and the index finger in the following hand on the left:

The image of this hand under a reflection around the vertical line is the image to the right. We would like to understand this transformation as *preserving* the angle between the thumb and index finger but *changing* the orientation so that the left hand becomes a right hand instead of the other way round. We will do so by agreeing that reflections change the convention on the sense used to measure angles.

2.4 Glides

Fix a line ℓ and a vector $\overrightarrow{v}$ parallel to ℓ. The *glide* of axis ℓ and vector $\overrightarrow{v}$ is the function $\text{glide}_{\ell,\,\overrightarrow{v}} : \mathbb{E} \to \mathbb{E}$ mapping any point P to the point P' obtained by first reflecting P with respect to ℓ and then translating the point thus obtained with vector $\overrightarrow{v}$.

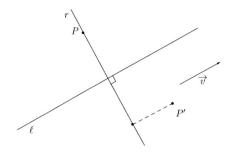

The following figure shows the image under a glide (whose axis ℓ is the dotted line and vector $\vec{v}$ is shown) of the footprint:

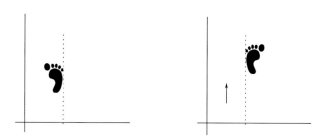

Just as reflections do, glides reverse orientation.

2.5 Isometries of the plane

Translations, rotations, reflections and glides are collectively known as isometries (a word that aptly describes these transformations, as it translates from the Greek ἴσος, meaning equal, and μέτρον, meaning measure). In this section we will see a number of basic properties of isometries. These properties will be helpful in Section 2.6 to see that any possible isometry is one of the four described in Sections 2.1–2.4.

Definition 2.1 An *isometry* is a function $\varphi : \mathbb{E} \to \mathbb{E}$ satisfying that, for all points $P, Q \in \mathbb{E}$, $\text{dist}(P, Q) = \text{dist}(\varphi(P), \varphi(Q))$. Two subsets A and B of the plane are said to be *congruent* (or *isometric*) when there exists an isometry φ such that $\varphi(A) = B$.

The simplest example of isometry is the identity function Id which maps every point into itself. That is, $\text{Id}(P) = P$ for all points P on the plane. We will see soon enough that translations, rotations, reflections and glides are isometries. Before doing so, we prove some general properties of these transformations.

Proposition 2.2 *Any isometry transforms lines into lines and preserves midpoints of segments.*

Proof We need to show that for any three points P, Q, R in a line the images P', Q', R' under the isometry are aligned as well. Assume this is not the case. Then P', Q', R' are the vertices of a triangle. Without loss of generality, we can assume that P and R are such that Q is in the segment $\overline{PR}$. Then $\text{dist}(P, R) = \text{dist}(P, Q) + \text{dist}(Q, R)$ and, therefore, $\text{dist}(P', R') = \text{dist}(P', Q') + \text{dist}(Q', R')$, since isometries preserve distances.

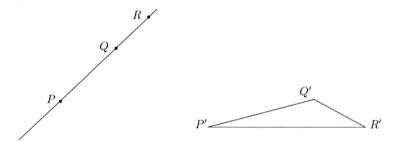

But this is not possible, since in a triangle the sum of the lengths of two sides is always greater than the third side. This shows the first statement.

To show that midpoints are preserved, let m be the midpoint of segment $\overline{PQ}$ and let P', Q', m' be the images of these three points. Then $\text{dist}(P, m) = \text{dist}(P', m')$ and $\text{dist}(Q, m) = \text{dist}(Q', m')$. Since $\text{dist}(P, m) = \text{dist}(Q, m)$ we deduce that $\text{dist}(P', m') = \text{dist}(Q', m')$ and, therefore, that m' is the midpoint of $\overline{PQ}$. □

Proposition 2.3 *The image of a line under an isometry is determined by the images of two points in this line.*

Proof Let P, Q be two points on the line ℓ and let P', Q' be their images. Choose a system of coordinates in ℓ setting P as 0 and let q be the coordinate of Q. Without loss of generality, we may assume $q > 0$. For any other point R on ℓ there is a unique real number r so that the coordinate of R is r. This means that $\text{dist}(P, R) = |r|$ and $\text{dist}(Q, R) = |q - r|$.

But then we must have $\text{dist}(P', R') = |r|$ and $\text{dist}(Q', R') = |q - r|$. There are only two possibilities for R' to satisfy the first equality (by Proposition 1.4), and from the two only one will satisfy the second equality (and it will be the one preserving the relative ordering of P, Q and R). □

Proposition 2.4 *Isometries preserve parallelism. That is, if ℓ and r are parallel lines and ℓ' and r' are their images under an isometry then ℓ' is parallel to r'.*

Proof Assume that ℓ' and r' are not parallel. Then they interesect at a point S. Let Q be any point in ℓ and Q' be its image in ℓ'. Let $d = \text{dist}(Q', S)$ and let P_1 and P_2 be the two points in ℓ at distance d from Q (we are using Proposition 1.4). Then S is either the image of P_1 or it is the image of P_2 (because there are two points in ℓ' at a distance d from Q' and one of them is S). Say it is the image of P_1.

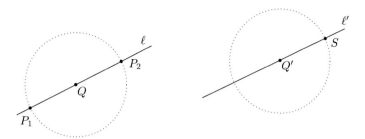

A similar argument shows that there is a point R in r such that S is the image of R. But since ℓ and r are parallel, $\text{dist}(R, P_1) \neq 0$, which contradicts the fact that $\text{dist}(S, S) = 0$. $\qquad\square$

Proposition 2.5 *An isometry is determined by the image of three non-collinear points. That is, two isometries coinciding on three such points must be the same.*

Proof Let O, P, Q be three non-collinear points and O', P', Q' be their images under the isometry. We will show that, for any other point $R \in \mathbb{E}$, the image R' of R is determined by O', P' and Q'. To do so, let ℓ_Q and ℓ_P be the lines extending $\overline{OQ}$ and $\overline{OP}$ respectively and r_Q and r_P be the lines through R parallel to ℓ_Q and ℓ_P respectively. Finally, let R_Q be the intersection of r_P and ℓ_Q and R_P the intersection of r_Q and ℓ_P.

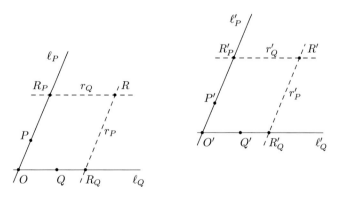

The image R'_Q of R_Q is determined by those of O and Q by Proposition 2.3 and so is the image R'_P of R_P by those of O and P. Using Proposition 2.4 we deduce that the image r'_P of r_P is also determined, since it must be the only line through R'_Q parallel to ℓ'_P and similarly for the image r'_Q of r_Q. Since R' must be the intersection of r'_P and r'_Q the proof is finished. $\qquad\square$

Lemma 2.6 *Let OPQ be a triangle and let P' and Q' be the midpoints of the segments $\overline{OP}$ and $\overline{OQ}$ respectively. Then the line through P and Q is parallel to that through P' and Q'.*

Proof We use analytic methods. Take the x-axis to be the PQ side of the triangle and the y-axis to pass through P so that $P = (0,0)$ and $Q = (q,0)$ for some $q \in \mathbb{R}$. Let (r,s) be the coordinates of O in this system of coordinates. Then the coordinates of P' and Q' are

$$P' = \left(\frac{r+0}{2}, \frac{s+0}{2}\right) = \left(\frac{r}{2}, \frac{s}{2}\right) \quad \text{and} \quad Q' = \left(\frac{r+q}{2}, \frac{s}{2}\right).$$

The equation of the line passing through P' and Q' is therefore $y = s/2$ and this line is parallel to the x-axis (whose equation is $y = 0$, a requirement incompatible with $y = s/2$). $\qquad \square$

Definition 2.7 We say that a point P is *fixed* under an isometry φ when $\varphi(P) = P$. We say that a subset $S \subset \mathbb{E}$ is *fixed* under φ when $\varphi(P) = P$ for all the points $P \in S$. Finally, we say that S is *invariant* under φ when $\varphi(S) = S$.

Note that if S is fixed under φ then it is also invariant. But S can be invariant under φ without being fixed (for instance, a circle rotated $90°$ around its centre remains invariant but is not fixed).

Proposition 2.8 *If an isometry has no fixed points then it has an invariant line.*

Proof Let $\varphi : \mathbb{E} \to \mathbb{E}$ be an isometry without fixed points. Let $A \in \mathbb{E}$, $A' = \varphi(A)$, $A'' = \varphi(A')$ and $A''' = \varphi(A'')$. If $A'' = A$ then, by Proposition 2.2, the midpoint of the segment $\overline{AA'}$ is fixed. It follows that $A'' \neq A$. Similarly, $A''' \neq A'$. Finally, $A''' \neq A$. Indeed, if $A''' = A$ then the triangle $AA'A''$ is equilateral. A rotation of $60°$ centred at the centre of this triangle would take A to A', A' to A'', and A'' to A. Therefore, by Proposition 2.5, it would coincide with φ. But this cannot be, since a rotation has a fixed point (its centre).

Since φ is an isometry, $\mathrm{dist}(A, A') = \mathrm{dist}(A', A'') = \mathrm{dist}(A'', A''')$. Also, $\mathrm{dist}(A, A'') = \mathrm{dist}(A', A''')$. Therefore, the quadrangle with vertices A, A', A'' and A''' is a parallelogram. Now take the midpoints m of $\overline{AA'}$, m' of $\overline{A'A''}$ and m'' of $\overline{A''A'''}$. By Proposition 2.2, $\varphi(m) = m'$ and $\varphi(m') = m''$.

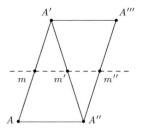

Let ℓ, ℓ', r and r' be the lines extending the segments $\overline{mm'}$, $\overline{m'm''}$, $\overline{AA''}$ and $\overline{A'A'''}$ respectively. Using Lemma 2.6 one can see that ℓ is parallel to r and ℓ' is parallel to r'. But r is parallel to r' and there is only one line through m' parallel to them (because of P5). We deduce that $\ell = \ell'$ and, since $\varphi(\overline{mm'}) = \overline{m'm''}$ and using the fact that φ preserves lines (by Proposition 2.2), we finally conclude that $\varphi(\ell) = \ell' = \ell$. $\qquad \square$

Theorem 2.9 *Translations, rotations, reflections and glides are isometries on the plane.*

Proof

Translations Let $\text{trans}_{\vec{v}}$ be a translation of vector $\vec{v}$ and P, Q be two points on the plane. Let ℓ be the line through them and let r_P, r_Q be the lines parallel to $\vec{v}$ through P and Q respectively. Let $P' = \text{trans}_{\vec{v}}(P)$ and ℓ' be the line through P' parallel to ℓ. Finally, let Q' be the intersection of ℓ' and r_Q. By construction, $Q' = \text{trans}_{\vec{v}}(Q)$.

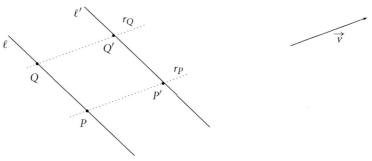

The quadrangle with vertices P, P', Q and Q' is a parallelogram, since opposite sides are parallel. This implies that $\text{dist}(P, Q) = \text{dist}(P', Q')$.

Rotations Let $\text{rot}_{O,\alpha}$ be a rotation of centre O and angle α. For two arbitrary points P and Q we denote $P' = \text{rot}_{O,\alpha}(P)$ and $Q' = \text{rot}_{O,\alpha}(Q)$. Also, let p, p', q and q' be the lines passing through O and, respectively, P, P', Q and Q'.

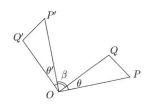

Let θ be the angle between p and q and similarly define θ'. Finally, let β be the angle between q and p'. Then $\alpha = \theta + \beta$ (the outer arc in the figure) and $\alpha = \beta + \theta'$ (the inner arc in the figure). It follows that $\theta = \theta'$. Also, $\text{dist}(O, P) = \text{dist}(O, P')$ and $\text{dist}(O, Q) = \text{dist}(O, Q')$. We conclude that the triangles POQ and $P'OQ'$ are equal and, therefore, $\text{dist}(P, Q) = \text{dist}(P', Q')$.

Reflections We now consider a reflection refl_{ℓ} of axis ℓ and, as before, two arbitrary points P and Q together with their images P' and Q'. In this case it is convenient to use analytic methods. Choose the y-axis to be ℓ and the x-axis to pass through Q. Then the coordinates of Q are $(q, 0)$ for some $q \in \mathbb{R}$ and those of P are (p, s) for some $p, s \in \mathbb{R}$. Furthermore, the coordinates of Q' will be $(-q, 0)$ and those of P' will be $(-p, s)$.

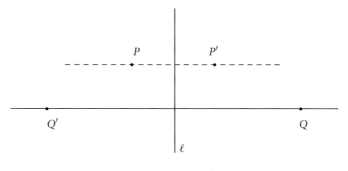

Using (1.1), it follows that

$$\text{dist}(P, Q) = \sqrt{(p-q)^2 + s^2}$$

and

$$\text{dist}(P', Q') = \sqrt{[(-p) - (-q)]^2 + s^2}$$

and, therefore, $\text{dist}(P, Q) = \text{dist}(P', Q')$, since $[(-p) - (-q)]^2 = [-(p-q)]^2 = (p-q)^2$.

Glides A glide is obtained by first applying a reflection and then a translation. Since both reflections and translations preserve distances, so must glides. □

2.6 On the possible isometries on the plane

In this section we prove a single result showing the limits of what isometries can be. We have already seen that translations, rotations, reflections and glides are isometries. They are actually the only possible ones.

Theorem 2.10 *Any isometry is either a translation, a rotation, a reflection, or a glide.*

Proof Let $\varphi : \mathbb{E} \to \mathbb{E}$ be an isometry. We divide into cases according to the number of fixed points of φ.

(i) Three non-collinear fixed points. In this case φ must be the identity Id, since both isometries coincide on three non-collinear points and we may use Proposition 2.5.

(ii) Two fixed points. Let $A, B \in \mathbb{E}$ be fixed by φ and let ℓ be the line determined by these two points. Let P be any point not in ℓ and let $P' = \varphi(P)$. Since $\text{dist}(A, P) = \text{dist}(A, P')$ it follows that P' belongs to the circle C_A with centre A and radius $\text{dist}(A, P)$. By the same token, it belongs to the circle C_B with centre B and radius $\text{dist}(B, P)$. So, both P and P' are in the intersection of C_A and C_B.

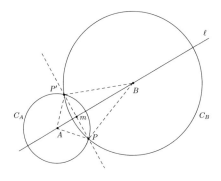

If $P' = P$ we have three fixed points and we are in case (i).[2] Therefore, we can assume $P \neq P'$. By Proposition 1.5, C_A and C_B intersect in P and P' and the point m where ℓ crosses the segment $\overline{PP'}$ joining these two points bisects the segment. Furthermore, $\overline{PP'}$ is perpendicular to ℓ. It follows that the reflection with axis ℓ takes P to P'. Moreover, it leaves A and B fixed. Another application of Proposition 2.5 shows that φ is this reflection.

(iii) One fixed point. Let A be fixed by φ, P be any other point and $P' = \varphi(P)$. If $P' = P$ we are in case (ii) so we can assume $P' \neq P$. Let $P'' = \varphi(P')$. We divide into cases. If $P'' = P$ we consider the middle point m of the segment joining P and P' and the line ℓ passing through A and m.

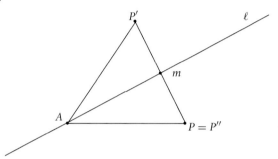

Since $\text{dist}(A, P) = \text{dist}(A, P')$ and $\text{dist}(m, P) = \text{dist}(m, P')$, it follows that the triangles APm and $AP'm$ are equal. Therefore, the two angles at m need to be the same, and since their sum is $180°$ each of them is $90°$. Using Proposition 2.5 one shows that φ is a reflection with axis ℓ.

We now assume that $P'' \neq P$. Then, since φ is an isometry, $\text{dist}(A, P) = \text{dist}(A, P') = \text{dist}(A, P'')$ and the points P, P', P'' are in the circle of centre A and radius $\text{dist}(A, P)$.

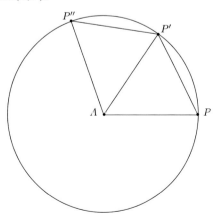

---

[2] If you feel that P' cannot coincide with P since $P \notin \ell$, you are right. But, as it happens, we do not need to prove this fact.

Again, since $\text{dist}(A, P) = \text{dist}(A, P'')$ and $\text{dist}(P, P') = \text{dist}(P', P'')$, it follows that the triangles APP' and $AP'P''$ are equal. Therefore, the two angles at A need to be the same and one more application of Proposition 2.5 shows that φ is a rotation with this angle centred at A.

(iv) No fixed points. Since φ has no fixed points, by Proposition 2.8 there exists a line ℓ invariant under φ. Let A be any point in ℓ, and take $A' = \varphi(A)$ and $A'' = \varphi(A')$. One has that $\text{dist}(A, A') = \text{dist}(A', A'')$ and that $A'' \neq A$, since otherwise the middle point of $\overline{AA'}$ would be fixed under φ. Let q be the line through A perpendicular to ℓ and let B be any point in q different from A. Also, let q' be the line parallel to q passing through A'. Then, q' is also perpendicular to ℓ.

Since isometries preserve right angles and $\varphi(A) = A'$ we have $\varphi(q) = q'$. Therefore, $\varphi(B)$ is either the point B' in q' at distance $d(A, B)$ from A' and in the same half-plane as B or the point B'' in q' at distance $d(A, B)$ from A' and in the opposite half-plane.

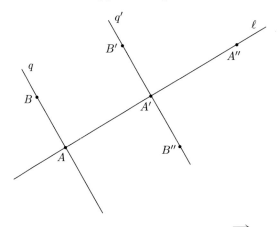

If $\varphi(B) = B'$ then φ must be the translation with vector $\overrightarrow{AA'}$ (use Proposition 2.5 plus the fact that $\varphi(A) = A'$ and $\varphi(A') = A''$). If $\varphi(B) = B''$ then φ must be the glide with axis ℓ and vector $\overrightarrow{AA'}$ (again, use Proposition 2.5).

□

The many symmetries of planar objects

The composition of a temple is based on symmetry, whose principles architects should take the greatest care to master. Symmetry derives from proportion [. . .]. Proportion is the mutual calibration of each element of the work and of the whole, from which the proportional system is achieved. No temple can have any compositional system without symmetry and proportion, unless, as it were, it has an exact system of correspondence to the likeness of a well-formed human being.

Vitruvius (1999: III.1.1)

What immortal hand or eye
Could frame thy fearful symmetry?

W. Blake (1970: Tiger 4)

The first epigraph opening this chapter is taken from Vitruvius' *De Archi-tectura*, a compendium of architectural practice from the first century BCE with enduring influence in Western architecture. Vitruvius states that architects "should take the greatest care to master" symmetry while designing a temple. A precise definition of symmetry is, however, missing from this text. We are instead told that the respect of symmetry passes through observing some proportions between the constituent parts of what is designed. The instance chosen by Vitruvius to exemplify this observance of proportions was the human body, and the proportions respected by nature, in the case of "a well-formed human being", have the form of simple ratios between lengths in the body, as in "the face, from the chin to the top of the forehead [. . .] should be one-tenth [of the total height of the body]". This set of canonical proportions was the source of one of the most celebrated drawings of Leonardo da Vinci. In one of his journals, around 1487, Leonardo drew a male figure in two super-imposed positions illustrating them (Figure 3.1). Curiously, Leonardo wrote the notes accompanying the drawing, like many of his personal notes, in what is called "mirror script". This means that the writing is reversed in such a way that it becomes plain script when placed in front of a mirror. That is, mirror script is a reflection (with respect to a vertical axis) of ordinary writing.

Returning to the issue of symmetry, the problem remains that no explanation is given of what symmetry is other than the observance

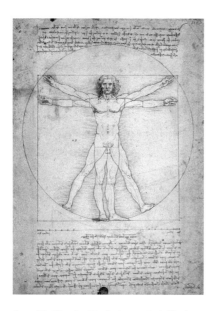

Figure 3.1 *Vitruvian Man* by Leonardo da Vinci, *c.* 1492 (pen & ink on paper). (Galleria dell' Accademia, Venice, Italy/The Bridgeman Art Library.)

of certain proportions. A look at Leonardo's drawing can, nevertheless, offer a hint.

Leaving aside minor details – such as the lateral view of the left feet as opposed to the front view for the right ones and the derived difference in the legs' torsions – the figure exhibits what we will call bilateral symmetry. If we trace an imaginary vertical line through the centre of the image, the resulting halves are mirror images of one another. A consequence of this feature is the one-to-one ratio between left and right parts of the body, a proportion which, maybe because it is too obvious, was not explicitly stated by Vitruvius. And yet, this is the very feature responsible for the feeling of balance in the Vitruvian man.

A possibly stronger exhibition of symmetry is present in the plans of both the temples whose forms made the substance of *De Architectura* and those which were built to follow its dicta.

The groundplan of the Temple of Diana at Ephesos (Figure 3.2) also displays bilateral symmetry (the imaginary line now being horizontal). But it exhibits other regularities as well. The columns of the temple, which in the figure are represented by ◎, are placed in parallel lines and equally spaced within these lines, thus inducing a sense of order in their repetition.

Both the uniformity of the columns' placements and the bilateral symmetry can be precisely described in terms of invariance under certain

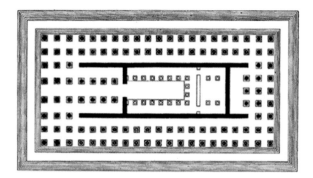

Figure 3.2 **Groundplan of the Temple of Diana at Ephesos.**

isometries, a fact that naturally leads us to a general definition of symmetry. Furthermore, the classification of isometries we obtained in Chapter 2 yields a basic classification of symmetries. It is the extension of this basic classification that will unfold the many symmetries of planar objects.

3.1 The basic symmetries

To describe the effect of the different isometries we consider subsets S of the plane $\mathbb{E}$ and their images under these isometries. In what follows we will call any such subset of the plane a *planar figure* (or simply a *figure*).

3.1.1 Bilateral symmetry: the straight-lined mirror

The most frequent use of the word symmetry in common speech refers to the presence of bilateral symmetry. Briefly described, this is the invariance of a figure under a reflection. A precise definition is the following.

Definition 3.1 We say that a figure S has *bilateral symmetry* when there exists a line ℓ such that $\text{refl}_\ell(S) = S$.

It is important to notice that S is not fixed by refl_ℓ. The only possible fixed points (i.e. points p in S for which $\text{refl}_\ell(p) = p$) are those lying in ℓ. But it is invariant under refl_ℓ; that is, the figure S is indistinguishable from its reflection $\text{refl}_\ell(S)$ (even though most points in S changed their position). Thus, if we reflect the word bod with respect to a vertical line passing through the centre of the "o"

bod

we observe that the word has bilateral symmetry. In the reflection defining the symmetry the letter "o" is invariant but the letters "b" and "d" are not. In fact, the reflection turns one into the other while making them swap places.

Whereas bilateral symmetry is pervasive both in nature and in different art forms, its ubiquitousness is maybe best seen in architecture. We have already mentioned it as a Vitruvian canon and shown the ground-plan of a classical temple as an example. A collection of groundplans, elevations and transverse sections for a number of buildings is given in Figure 3.3. Bilateral symmetry is apparent in all of them.

3.1.2 Rotational symmetry

Invariance under an isometry naturally induces forms of symmetry other than bilateral by considering isometries other than reflections. In the case of rotations we obtain rotational symmetry.

Definition 3.2 We say that a figure S has *rotational symmetry* when there exist a point O and an angle θ such that $\text{rot}_{O,\theta}(S) = S$.

Rotational symmetry is also common in architecture. One can see it in domes, floors, fountains, and windows. The latter, because of their resemblance to roses, came to be known in the middle ages as *rosettes*. Examples of rosettes from the cathedrals of Strasbourg and Lyon are shown in Figure 3.4.

Rotational symmetry is found in various contexts beyond architecture. Occasionally, inspiration from architectural design is evident, as in the Tabriz carpet in Figure 3.5.

Other objects, like the photograph in Figure 3.6 by Jerry Matchett, have no direct links with architectural design.

3.1.3 Central symmetry: the one-point mirror

A particular case of rotation is that with an angle of 180°. Such a rotation is usually called a *half-turn*. In this instance the image P' of a point P lies on the line determined by P and the centre O of the rotation, and is "opposed" to P.

We can think of such a rotation as the effect of a mirror consisting of a single point. Unlike the line-shaped mirror, though, orientation is preserved. Yet another difference is that the one-point mirror does not partition the plane as the line-shaped mirror does. Consequently, there is no division between "this side and the other side" of the mirror. The symmetry associated with 180° rotations is known as *central symmetry*.

A well-known centrally symmetric figure is the symbol representing the yin–yang principle, of Taoism. With its absence of boundaries,

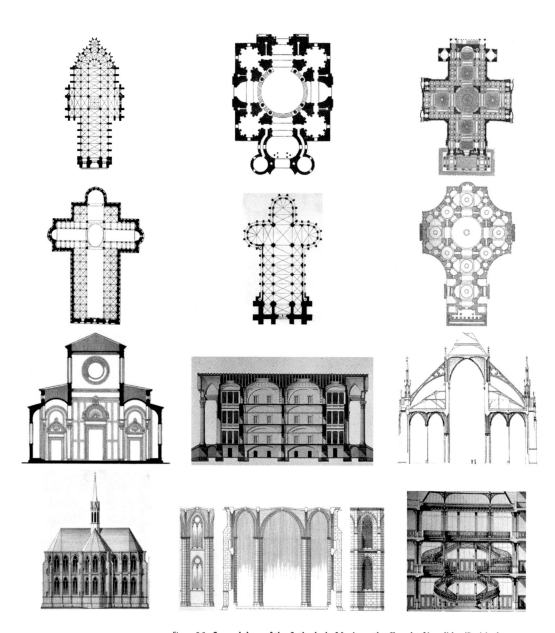

Figure 3.3 **Groundplans of the Cathedral of Amiens, the Church of Invalides (Paris), the Pantheon (Paris), Pisa Cathedral, the church of St Elizabeth (Marburg) and St Paul's Cathedral (London). Elevations or sections of the Church of Santo Spirito (Florence), Girard College (Philadelphia), the Cathedral of Ulm, the church of St Elizabeth (Marburg) (two plates) and the Magasins Bon Marché (Paris).**

Figure 3.4 **Design of rosettes in the Cathedrals of Strasbourg (Benutzer: Sansculotte CC-BY-SA-2.0) (left) and Lyon (right).**

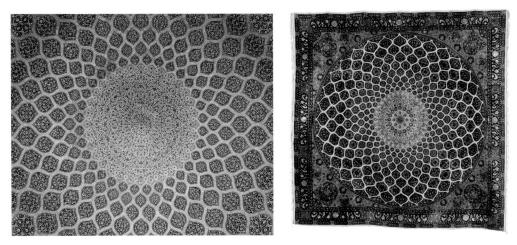

Figure 3.5 **The Dome of the Sheikh Lotf Allah Mosque in Isfahan, Iran (©Phillip Maiwald/Wikimedia Commons/CC-BY-SA-3.0/GFDL) (left), and a Tabriz carpet with a dome design (right).**

central symmetry provides a fitting representation of the yin-yang principle, which postulates that forces that appear to be opposite are bound together, intertwined and interdependent in the universe, giving rise to each other in turn.

3.1.4 Translational symmetry: repeated mirrors

Definition 3.3 We say that a figure S has *translational symmetry* when there exists a vector $\vec{v}$ such that $\text{trans}_{\vec{v}}(S) = S$.

Figure 3.6 *Rosette 22* by Jerry Matchett.
© 2005 J.W. Matchett.

How would a translational invariant figure look? Assume (just for graphical simplicity) that the translation vector is horizontal and points to the right, like $\longrightarrow$. Assume, for simplicity as well, that the figure contains a letter "*A*". Since the whole figure is invariant by translation, the letter *A* will "move" to a position already containing an *A*. That is, the figure contains the pair *A   A*. But repeating this reasoning produces a third *A*, and a fourth, a fifth, and so on. Furthermore, the space "left empty" by our first *A* needs to be filled by another *A* which had to be originally located at its left. And again this reasoning shows the necessity of an infinite sequence of *A*s to the left of the one we started with. We conclude that the figure must have (if we start with no more than the *A*) the following aspect:

A A A A A A A A A A A A A A A A A A A A A A A A A A A A A A A A A

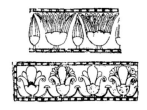

Figure 3.7 **Friezes at the Temple of Hathor, Denderah, Egypt.**

It endlessly repeats a motif by producing translational copies of it. Strictly speaking, therefore, we would be at pains to find translationally symmetric figures in any form of art. Human creations have the limitation of finiteness. But infinity can be suggested in finite fragments of translationally symmetric figures. Examples of such fragments from ancient Egypt are shown in Figure 3.7.

As with rotational invariance, architecture provided a name to these translationally symmetric patterns. Although having an originally narrower meaning, in modern architecture terminology a *frieze* is a horizontal broad band which is occupied by sculpture. It is not required that the sculpted design will be translationally invariant, but it often is. Hence the use of the expression "frieze patterns" to denote translationally symmetric designs, such as the pavement on the Greek island of Rhodes in Figure 3.8.

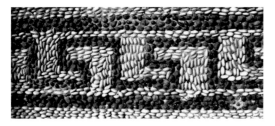

Figure 3.8 **Pavement at Rhodes, Greece (Hannes Grobe/ CC-BY-SA-2.5).**

3.1.5 Glidal symmetry

The next definition is the natural close to our family of basic symmetries.

Definition 3.4 We say that a figure S has *glidal symmetry* when there exists a line ℓ and a vector $\overrightarrow{v}$ parallel to ℓ such that $\text{glide}_{\ell, \overrightarrow{v}}(S) = S$.

The footprint we encountered in Chapter 2, conveniently repeated here, may give us a first idea of the visual appearance of a glidally symmetrical figure:

Glidal symmetry can also be found in different forms of artwork. For example, it is observable in the Maori design shown in Figure 3.9. It is also present (with a vertical direction) in the Japanese pattern shown in Figure 3.10.

Figure 3.9 **Maori Kowhaiwhai pattern in the Ngaru style, New Zealand.**

A number of differences between these three patterns deserve emphasis. The pattern in Figure 3.9 has translational symmetry for the translation with vector $\overrightarrow{v}$ (the vector associated with the glide), whereas the set of footprints does not. It does have translational symmetry, but the vectors associated with translations leaving the footprints invariant always have a length which is an even multiple of the length of $\overrightarrow{v}$ (intuitively, I need to do an even number of steps to translate both my left

Figure 3.10 **Seigaiha (waves from the blue ocean) pattern, Japan. (www. mikworks.com. Used with permission of Mikio Inose.)**

and right feet). This is not by chance. Every figure having glidal symmetry is a frieze. Also, the Maori pattern has bilateral symmetry (around a horizontal line at its centre), whereas the footpath does not.

Finally, the Seigaiha in Figure 3.10 has two possible translations with different directions. This fact allows one (and in theory forces one if the pattern is going to have both symmetries) to extend the design all over the plane. Figures having two translational symmetries – with non-parallel translation vectors – are called *wallpapers*. The architectural glossary appears again, this time through the humble activity of wall decoration.

3.2 The arithmetic of isometries

Although both windows in Figure 3.4 exhibit rotational symmetry, the design on the left produces a feeling of balance that the right one fails to achieve. The reason is all too obvious. The Strasbourg rosette has, in addition to the rotational invariance, bilateral symmetry. Plane figures may posses invariance with respect to various isometries. Thus, Figures 3.9 and 3.10 have glidal, translational and bilateral invariance, and the latter possesses translational invariance with respect to two non-parallel vectors.

The family of isometries leaving a figure invariant has, in addition, some structure that makes it more than a loose collection of transformations. To understand this structure a few definitions are necessary.

Definition 3.5 Let $f, g : \mathbb{E} \to \mathbb{E}$ be two functions on the plane. The *composition* $f \circ g : \mathbb{E} \to \mathbb{E}$ is the function associating with any $x \in \mathbb{E}$ the point $f(g(x))$. That is, we first compute the image $g(x)$ of x under g and then the image $f(g(x))$ of $g(x)$ under f:

$$x \mapsto g(x) \mapsto f(g(x)).$$

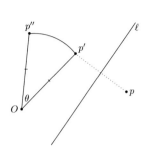

For instance, given O, θ and ℓ, the composition $\mathrm{rot}_{O,\theta} \circ \mathrm{refl}_\ell$ associates with any point p on the plane the point p'' obtained by first reflecting p with respect to ℓ and then rotating the resulting point, say p', an angle θ around the centre O.

Restricted to isometries, we may regard composition as an "operation" between them much in the same manner as we regard addition as an operation between numbers. The addition of two numbers produces a third one and, similarly, the composition of two isometries produces a third one. Indeed, if φ and ψ are two isometries and $p, q \in \mathbb{E}$, then

$$\mathrm{dist}((\varphi \circ \psi)(p), (\varphi \circ \psi)(q))$$

$$= \mathrm{dist}(\varphi(\psi(p)), \varphi(\psi(q))) = \mathrm{dist}(\psi(p), \psi(q)) = \mathrm{dist}(p, q).$$

Here, the second equality holds because φ, preserves distances and the last one because ψ does. Hence, $\varphi \circ \psi$ preserves distances and is, therefore, an isometry.

The number 0 has no effect when added to any other number. That is, one has $0 + x = x + 0 = x$ for all numbers x. An isometry with a similar behaviour (with respect to composition) is the identity Id. Indeed, one can immediately check that, for any isometry φ and any point $p \in \mathbb{E}$,

$$(\mathrm{Id} \circ \varphi)(p) = \mathrm{Id}(\varphi(p)) = \varphi(p)$$

and, similarly, $(\varphi \circ \mathrm{Id})(p) = \varphi(p)$. That is, $\mathrm{Id} \circ \varphi = \varphi \circ \mathrm{Id} = \varphi$.

The analogy between $+$ and $\circ$ does not end here. For any number x there exists another number y such that $x + y = y + x = 0$. Simply take y to be $-x$. To prove that the corresponding statement for composition of isometries holds true we can rely on the classification of isometries we obtained in Theorem 2.10. Indeed, it is easy to check that:

(i) for any line ℓ, $\mathrm{refl}_\ell \circ \mathrm{refl}_\ell = \mathrm{Id}$;
(ii) for any point O and any angle θ, $\mathrm{rot}_{O,\theta} \circ \mathrm{rot}_{O,360-\theta} = \mathrm{rot}_{O,360-\theta} \circ \mathrm{rot}_{O,\theta} = \mathrm{Id}$;
(iii) for any vector $\overrightarrow{v}$, $\mathrm{trans}_{\overrightarrow{v}} \circ \mathrm{trans}_{-\overrightarrow{v}} = \mathrm{trans}_{-\overrightarrow{v}} \circ \mathrm{trans}_{\overrightarrow{v}} = \mathrm{Id}$; and
(iv) for any line ℓ and vector $\overrightarrow{v}$ parallel to ℓ, $\mathrm{glide}_{\ell,\overrightarrow{v}} \circ \mathrm{glide}_{\ell,-\overrightarrow{v}} = \mathrm{glide}_{\ell,-\overrightarrow{v}} \circ \mathrm{glide}_{\ell,\overrightarrow{v}} = \mathrm{Id}$.

Here, $-\overrightarrow{v}$ denotes the vector whose endpoints are those of $\overrightarrow{v}$ but with the opposite orientation.

We therefore see that for any isometry φ there exists an isometry ψ such that $\varphi \circ \psi = \psi \circ \varphi = \text{Id}$. We call ψ the *inverse* of φ and denote this inverse by φ^{-1}.

Yet another structural coincidence between addition of numbers and composition of isometries – whose verification is easy and left to the reader – is the fact that for all isometries φ, ψ, ξ one has

$$\varphi \circ (\psi \circ \xi) = (\varphi \circ \psi) \circ \xi.$$

That is, it is irrelevant whether we first compute $\psi \circ \xi$ and then compose φ with the result or, instead, we first compute $\varphi \circ \psi$ and then compose the result with ξ.

A feature breaking down the analogy between addition of numbers and composition of isometries is the fact that in the former the order of the operands is irrelevant, whereas in the latter it is not. For any numbers x and y we have $x + y = y + x$. In contrast, for the composition described right after Definition 3.5, a brief observation shows that $(\text{rot}_{O,\theta} \circ \text{refl}_\ell)(p) \neq (\text{refl}_\ell \circ \text{rot}_{O,\theta})(p)$.

It is convenient to summarize the discussion above, and mathematics provides the right notion to do so. Before proceeding, though, we introduce some notation. For sets A, B we will denote by $A \times B$ the set of all ordered pairs (a, b) with $a \in A$ and $b \in B$.

Definition 3.6 A *group* is a set G endowed with a function $* : G \times G \to G$ and a designated element e satisfying the following properties:

associative – for all $x, y, z \in G$, $x * (y * z) = (x * y) * z$;
neutral element – for all $x \in G$, $x * e = e * x = x$;
inverse – for all $x \in G$ there exists $x^{-1} \in G$, such that $x * x^{-1} = x^{-1} * x = e$.

If, in addition, one has that $x * y = y * x$ for all $x, y \in G$, then we say that G is *commutative*.

The set $\text{Isom}(\mathbb{E})$ of plane isometries endowed with the composition is a group. It is not commutative. The set $\mathbb{Z}$ of integers (i.e. $\{\ldots, -3, -2, -1, 0, 1, 2, 3, \ldots\}$) endowed with the addition is a commutative group. So are the sets $\mathbb{Q}$ and $\mathbb{R}$ of rational (i.e. fractions of integers) and real numbers respectively. The last three examples are not unrelated. Indeed, every integer number is a rational number as well (simply write integer p as the fraction $\frac{p}{1}$) and the addition of two integers yields the same result whether we consider these numbers as integers or as fractions.

Definition 3.7 We say that H is a *subgroup* of G when $H \subseteq G$ and H is a group when endowed with $*$. This means that $h * g \in H$ for all $h, g \in H$ and that the three properties of Definition 3.6 are satisfied when replacing G by H.

We can restate our remark above by saying that $\mathbb{Z}$ is a subgroup of $\mathbb{Q}$ (with the addition). We also remark that $\mathbb{Q}$ is a subgroup of $\mathbb{R}$. Other examples of subgroups, now of $\mathsf{Isom}(\mathbb{E})$, are the group $\mathsf{Transl}(\mathbb{E})$ of translations on the plane and, for each point O, the group $\mathsf{Rot}(\mathbb{E}, O)$ of rotations with centre O.

Given a figure S on the plane we denote by $\mathsf{Sym}(S)$ the set of isometries leaving S invariant. A fact whose importance is not apparent in the ease of its proof is the closure of $\mathsf{Sym}(S)$ under composition (the statement of the next proposition).

Proposition 3.8 If φ, ψ are in $\mathsf{Sym}(S)$ then so is $\varphi \circ \psi$. In other words, if $\varphi(S) = \psi(S) = S$ then $(\varphi \circ \psi)(S) = S$.

Proof $(\varphi \circ \psi)(S) = \varphi(\psi(S)) = \varphi(S) = S.$ □

It follows from this proposition (together with the fact that if $\varphi(S) = S$ then $\varphi^{-1}(S) = S$) that $\mathsf{Sym}(S)$ is a subgroup of $\mathsf{Isom}(\mathbb{E})$. We will call it the *group of symmetries* of S. In the barest case, it is the *trivial group* consisting on the identity Id only; no other isometry leaves S invariant. If this is not so, we say that S *has (non-trivial) symmetries*. For instance, both patterns in Figure 3.11 have a group with four elements.

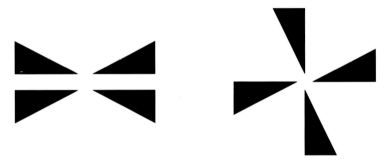

Figure 3.11 **Two patterns with three non-trivial symmetries each.**

For the design on the left, these are the reflections refl_v and refl_h with vertical and horizontal axes respectively, the rotation rot_{180} and the identity Id. For the one on the right, they are the rotations $\mathsf{rot}_{90}, \mathsf{rot}_{180}, \mathsf{rot}_{270}$ and the identity. The group structure in each case – that is, the way these isometries relate to each other under composition – can be succinctly expressed by the following tables:

$\circ$	Id	refl_v	refl_h	rot_{180}
Id	Id	refl_v	refl_h	rot_{180}
refl_v	refl_v	Id	rot_{180}	refl_h
refl_h	refl_h	rot_{180}	Id	refl_v
rot_{180}	rot_{180}	refl_h	refl_v	Id

$\circ$	Id	rot_{90}	rot_{180}	rot_{270}
Id	Id	rot_{90}	rot_{180}	rot_{270}
rot_{90}	rot_{90}	rot_{180}	rot_{270}	Id
rot_{180}	rot_{180}	rot_{270}	Id	rot_{90}
rot_{270}	rot_{270}	Id	rot_{90}	rot_{180}

These two tables show different "structures" for the isometries of the two patterns above. This is apparent on the fact that every isometry in the group for the pattern on the left gives the identity when composed with itself (that is, $\text{refl}_v^2 = \text{refl}_h^2 = \text{rot}_{180}^2 = \text{Id}$), but such a property fails to occur for the other group with rot_{90} and rot_{270}.

We note, in addition, that two figures may have the same group structure and yet have different kinds of symmetry. The simplest example is given by the groundplan in Figure 3.2 and the yin-yang symbol. Both have a symmetry group with two elements and a table

$\circ$	Id	φ
Id	Id	φ
φ	φ	Id

but in the first case φ is a reflection (yielding bilateral symmetry) and in the second a rotation of $180°$ (yielding central symmetry).

One of the many ways to specify a subgroup of a group $(G, *)$ is by means of generators. In the case of a single generator $x \in G$ the subgroup is the set

$$\langle x \rangle = \{x^k \mid k \in \mathbb{Z}\}$$

of all possible powers of x. Here, $x^k = \underbrace{x * x * \cdots * x}_{k \text{ times}}$ if $k > 0$, $x^k = (x^{-1})^{-k}$ if $k < 0$, and $x^0 = e$ (the neutral element of G).

When $G = \text{Isom}(\mathbb{E})$ the structure of $\langle \varphi \rangle$ depends on what $\varphi \in \text{Isom}(\mathbb{E})$ is. If φ is a reflection then $\langle \varphi \rangle = \{\text{Id}, \varphi\}$ since $\varphi^2 = \text{Id}$. In contrast, if φ is a translation then the elements φ^k are different from one another. Indeed, we have, for all $k \in \mathbb{Z}$,

$$\text{trans}_{\vec{v}}^k = \text{trans}_{k\vec{v}}.$$

Here, for $k \geq 0$, $k\vec{v}$ denotes the vector with the same origin and direction as $\vec{v}$ but whose length is k times that of $\vec{v}$. Also, for $k < 0$, $k\vec{v} = -(|k|\vec{v})$.

Since, in addition, $\varphi^k \circ \varphi^q = \varphi^{k+q}$ we find that in this case $\langle \varphi \rangle$ has the same structure as the set of integers $\mathbb{Z}$ with the addition $+$. If φ is a glide then $\langle \psi \rangle$ also has the structure of $(\mathbb{Z}, +)$, but it alternates translations and glides as

$$\text{glide}_{\vec{v}}^k = \begin{cases} \text{trans}_{k\vec{v}} & \text{if } k \text{ is even} \\ \text{glide}_{k\vec{v}} & \text{if } k \text{ is odd.} \end{cases}$$

The case of rotations is less clear-cut. For a rotation $\text{rot}_{O,\theta}$ the group $\langle \varphi \rangle$ is finite if and only if there exist integers n, m, with $m \neq 0$, such that $\theta = \frac{n}{m} \times 360$. The case of interest to us is that of $\langle \text{rot}_{O,\theta} \rangle$ finite. We will have a closer look at it in Section 3.4.

We have just seen that a number of isometries φ (all reflections and some rotations) have the property that, for some positive integer n, $\varphi^n =$ Id. For an element x in a group G, we call the *order* of x the smallest positive integer n such that $x^n = e$. In the case no such integer exists we say that x has order infinity. Thus, for instance, a reflection has order 2, a 45° rotation has order 8, and a (non-trivial) translation has order infinity.

3.3 A representation theorem

The arithmetic of isometries allows one to show that, ultimately, all isometries reduce to reflections.

Theorem 3.9 *Every isometry can be represented as the composition of at most three reflections.*

The next two propositions are stepping-stones towards the proof of Theorem 3.9. The non-commutative character of the composition of isometries is patent in these propositions.

Proposition 3.10 *The composition* $\text{refl}_{\ell_1} \circ \text{refl}_{\ell_2}$ *of two reflections with parallel edges is a translation whose vector is perpendicular to the edges, has length twice the distance between them, and points from ℓ_2 to ℓ_1.*

Proof We will show that $\text{refl}_{\ell_1} \circ \text{refl}_{\ell_2}$ coincides with the claimed translation on three non-collinear points and then use Proposition 2.5. To this end, choose P and Q in ℓ_2 and R in ℓ_1. Let d be the distance between ℓ_1 and ℓ_2. Then P' and Q' are at a distance $2d$ from P and Q respectively and "on the other side" of ℓ_1.

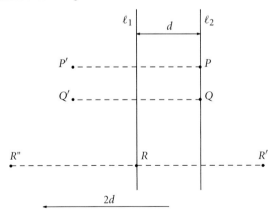

Also, $R' = \text{refl}_{\ell_2}(R)$ is at a distance $2d$ from R but "on the wrong side" of ℓ_1. The reflection around ℓ_1 yields $R'' = (\text{refl}_{\ell_1} \circ \text{refl}_{\ell_2})(R)$ also at a distance $2d$ but now on the correct side. $\qquad\square$

Proposition 3.11 *The composition* $\text{refl}_{\ell_1} \circ \text{refl}_{\ell_2}$ *of two reflections whose edges meet at a point O is a rotation with angle $2\widehat{\ell_2 \ell_1}$.*

Proof The proof is similar to the one in the preceding result, again using Proposition 2.5.

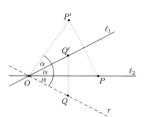

The reader can check that the images of the points O, P and Q chosen as shown above (where $r = \text{refl}_{\ell_2}(\ell_1)$ and $\alpha = \widehat{\ell_2 \ell_1}$) under $\text{refl}_{\ell_1} \circ \text{refl}_{\ell_2}$ are their rotation around O with angle $2\widehat{\ell_2 \ell_1}$. $\qquad \square$

Proof of Theorem 3.9 The claim is obviously true for a reflection. A translation can be written as the composition of two suitably choosen reflections by Proposition 3.10 and a similar fact holds true for a rotation by Proposition 3.11. Finally, since a glide is a composition of a reflection and a translation, the claim also holds for glides. $\qquad \square$

The next two propositions exhibit the results of composing a translation with a rotation and a reflection, respectively.

Proposition 3.12 *Let θ be an angle, O a point and $\overrightarrow{v}$ a vector. Then* $\text{rot}_{O,\theta} \circ \text{trans}_{\overrightarrow{v}} = \text{trans}_{\overrightarrow{v'}} \circ \text{rot}_{O,\theta}$, *where* $\overrightarrow{v'} = \text{rot}_{O,\theta}(\overrightarrow{v})$.

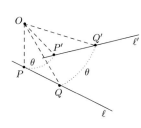

Proof We will show that, for any point P, $(\text{rot}_{O,\theta} \circ \text{trans}_{\overrightarrow{v}})(P) = (\text{trans}_{\overrightarrow{v'}} \circ \text{rot}_{O,\theta})(P)$. Let ℓ be the line through P parallel to $\overrightarrow{v}$ and $Q = \text{trans}_{\overrightarrow{v}}(P)$ (so that $Q \in \ell$). Let ℓ' be the image of ℓ under $\text{rot}_{O,\theta}$, $P' = \text{rot}_{O,\theta}(P)$ and $Q' = \text{rot}_{O,\theta}(Q)$. Since $\overrightarrow{v} = \overrightarrow{PQ}$ we have $\overrightarrow{v'} = \overrightarrow{P'Q'}$. So, $Q' = \text{trans}_{\overrightarrow{v'}}(P') = (\text{trans}_{\overrightarrow{v'}} \circ \text{rot}_{O,\theta})(P)$.

On the other hand, $Q' = \text{rot}_{O,\theta}(Q) = (\text{rot}_{O,\theta} \circ \text{trans}_{\overrightarrow{v}})(P)$. This shows the statement. $\qquad \square$

Important remark. The proof of the next proposition deserves attention. It makes use of the representation in Theorem 3.9 in a clever way. The idea is that not only can we represent a translation by the composition of two reflections with parallel axes but we can choose any pair of such axes, provided the distance between them is the good one (namely, $\frac{\text{length}(\overrightarrow{v})}{2}$) and so is their direction (perpendicular to $\overrightarrow{v}$). A similar remark holds true for the representation of a rotation. By appropriately chosing these representations we can compute the composition of isometries without resource to Proposition 2.5. We will often do so in the next few sections (without necessarily filling in all the details).

Proposition 3.13 *Let $\overrightarrow{v}$ be a vector and ℓ be a line not perpendicular to $\overrightarrow{v}$. Then* $\text{refl}_\ell \circ \text{trans}_{\overrightarrow{v}} = \text{trans}_{\overrightarrow{v'}} \circ \text{refl}_\ell$ *where* $\overrightarrow{v'} = \text{refl}_\ell(\overrightarrow{v})$.

Proof The proof relies on Theorem 3.9. Let r_1 and r_2 be two lines perpendicular to $\overrightarrow{v}$ and such that $\text{trans}_{\overrightarrow{v}} = \text{refl}_{r_2} \circ \text{refl}_{r_1}$. Then, $\text{refl}_\ell \circ \text{trans}_{\overrightarrow{v}} = \text{refl}_\ell \circ \text{refl}_{r_2} \circ \text{refl}_{r_1}$.

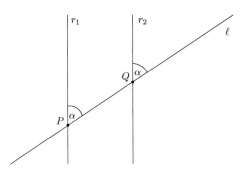

The composition $\mathrm{refl}_\ell \circ \mathrm{refl}_{r_2}$ is a rotation (of angle 2α) with centre at the point Q where ℓ cuts r_2. We may represent this rotation as a composition of any two reflections whose axes meet at Q and such that their angle is α. In particular, we may write

$$\mathrm{refl}_\ell \circ \mathrm{trans}_{\vec{v}} = \mathrm{refl}_{r_3} \circ \mathrm{refl}_\ell \circ \mathrm{refl}_{r_1},$$

where r_3 is obtained by rotating ℓ around Q an angle $-\alpha$. Note that r_3 is the reflection of r_2 with respect to ℓ.

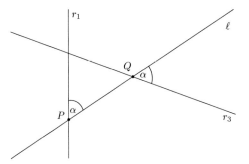

We may now repeat the same idea and replace the rotation $\mathrm{refl}_\ell \circ \mathrm{refl}_{r_1}$ by an equivalent representation $\mathrm{refl}_{r_4} \circ \mathrm{refl}_\ell$ with r_4 obtained in a similar way; that is, as the reflection of r_1 with respect to ℓ.

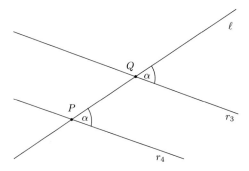

We have thus proved that

$$\mathsf{refl}_\ell \circ \mathsf{trans}_{\overrightarrow{v}} = \mathsf{refl}_{r_3} \circ \mathsf{refl}_{r_4} \circ \mathsf{refl}_\ell.$$

Since r_1 and r_2 are parallel and isometries preserve parallelism (by Proposition 2.4), it follows that r_3 and r_4 are also parallel. Therefore, $\mathsf{refl}_{r_3} \circ \mathsf{refl}_{r_4}$ represents a translation. Moreover, its vector $\overrightarrow{v}$ is twice the vector between r_3 and r_4, which is twice the image under the reflection around ℓ of the vector $\overrightarrow{v}$ between r_1 and r_2. We conclude that $\mathsf{refl}_\ell \circ \mathsf{trans}_{\overrightarrow{v}} = \mathsf{trans}_{\overrightarrow{v'}} \circ \mathsf{refl}_\ell$ with $\overrightarrow{v'}$ as claimed in the statement.

$\square$

3.4 Rosettes and whirls

The class of all possible groups is too large to give an idea of its many possible structures. Even the class of all finite groups (such as, for instance, those given by the three tables in Section 3.2) is so. Finite groups of the form $\mathsf{Sym}(S)$ for some figure S on the plane are, nonetheless, amenable to such a description. As a first step towards this description consider the two three-legged stars on the left.

Both are invariant under two non-trivial rotations, with angles $120°$ and $240°$. The one on the left, in addition, is also invariant under three different reflections (with axes ℓ_1, ℓ_2 and ℓ_3 in the figure). A moment's consideration shows that these are all the isometries leaving these objects invariant. An extra rumination, and the use of Propositions 3.10 and 3.11, yields that these isometries behave under composition as displayed in the following tables:

$\circ$	Id	rot_{120}	rot_{240}	refl_{ℓ_1}	refl_{ℓ_2}	refl_{ℓ_3}
Id	Id	rot_{120}	rot_{240}	refl_{ℓ_1}	refl_{ℓ_2}	refl_{ℓ_3}
rot_{120}	rot_{120}	rot_{240}	Id	refl_{ℓ_3}	refl_{ℓ_1}	refl_{ℓ_2}
rot_{240}	rot_{240}	Id	rot_{120}	refl_{ℓ_2}	refl_{ℓ_3}	refl_{ℓ_1}
refl_{ℓ_1}	refl_{ℓ_1}	refl_{ℓ_2}	refl_{ℓ_3}	Id	rot_{120}	rot_{240}
refl_{ℓ_2}	refl_{ℓ_2}	refl_{ℓ_3}	refl_{ℓ_1}	rot_{240}	Id	rot_{120}
refl_{ℓ_3}	refl_{ℓ_3}	refl_{ℓ_1}	refl_{ℓ_2}	rot_{120}	rot_{240}	Id

$\circ$	Id	rot_{120}	rot_{240}
Id	Id	rot_{120}	rot_{240}
rot_{120}	rot_{120}	rot_{240}	Id
rot_{240}	rot_{240}	Id	rot_{120}

That is, the symmetry groups of the two stars are given by these tables. Groups with either of these tables are usually denoted by D_3 and C_3 respectively.

For each integer $n > 3$ we can consider regular stars (both plain-legged and crooked-legged) with n legs and we obtain as their group of symmetries the group D_n (with n rotations and n reflections) and C_n (with n rotations only). The groups D_2 and C_2 are already known to us, the first being the group of symmetries for the left-hand pattern in Figure 3.11 and the second that for the yin–yang symbol. Finally, we can extend these definitions by taking $C_1 = \{\text{Id}\}$ (the trivial group) and $D_1 = \{\text{Id}, \text{refl}_\ell\}$ (the group of an object having only one bilateral symmetry).

We call C_n the *cyclic group of order* n and D_n the *dihedral group of order* n. Figures S such that $\text{Sym}(S)$ is C_n for some $n \geq 2$ are often called *whirls* (or *vortices*). If, instead, $\text{Sym}(S)$ is a D_n with $n \geq 2$ then we use the term *rosette*.

The collection of all groups D_n and C_n exhausts all possible finite symmetry groups.

Theorem 3.14 *Let S be a plane figure such that* $\text{Sym}(S)$ *is finite. Then there exists $n \geq 1$ such that either* $\text{Sym}(S) = C_n$ *or* $\text{Sym}(S) = D_n$.

A first remark towards the proof of Theorem 3.14 is the fact that $\text{Sym}(S)$ cannot possess translations or glides. A second, less obvious, remark is that rotations with the same centre can be generated by a single rotation.

Lemma 3.15 *Let* $\text{Rot}(O, S)$ *be the subgroup of rotations with centre O of* $\text{Sym}(S)$ *and assume the latter is finite. If* $\text{Rot}(O, S)$ *has n elements then*

$$\text{Rot}(O, S) = \left\{ \text{Id}, \text{rot}_{O, \frac{360}{n}}, \text{rot}_{O, \frac{2 \times 360}{n}}, \ldots, \text{rot}_{O, \frac{(n-1) \times 360}{n}} \right\}.$$

Furthermore, for any line ℓ passing through O and any point P not in ℓ there exists $\varphi \in \text{Rot}(O, S)$ such that ℓ crosses the segment $\overline{P, \varphi(P)}$ (that is, P and $\varphi(P)$ are on different sides of ℓ).

Proof Let P be any point in S, $P \neq O$, and let $P_1, \ldots, P_n$ be the images of P under the elements in $\text{Rot}(O, S)$. Since these points are all rotations of P with centre O they are all in a circle of centre O (and radius $\text{dist}(O, P)$). Reindexing the P_i if necessary we can assume that they are sorted counterclockwise. We claim that the angles between successive pairs of points in this collection are all the same.

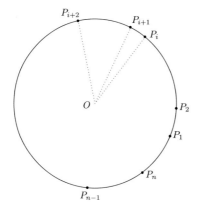

Assume this is not true. Then, there exist i such that $\widehat{P_iOP_{i+1}} < \widehat{P_{i+1}OP_{i+2}}$. Let $\varphi, \psi \in \mathsf{Rot}(O, S)$ be such that $P_i = \varphi(P)$ and $P_{i+1} = \psi(P)$. Then

$$(\psi \circ \varphi^{-1})(P_i) = \psi(\varphi^{-1}(P_i)) = \psi(P) = P_{i+1}.$$

It follows that $\psi \circ \varphi^{-1}$ is a rotation of angle $\theta = \widehat{P_iOP_{i+1}}$. But then $\psi \circ \varphi^{-1}(P_{i+1})$ is a point in between P_{i+1} and P_{i+2}, which is a contradiction. This proves the claim and, with it, the first statement, since it implies that $P_1, \ldots, P_n$ are the rotations of P with angles $\frac{j \times 360}{n}$ for $j = 0, 1, \ldots, n-1$.

For the second statement, note that if there were no images of P on the other side of ℓ then there would be an arc of circle of at least $180°$ without images of P. But we have just proved that this is not possible, since two such consecutive images are separated by an arc with angle $360/n$. □

A last remark is that any two non-trivial rotations in $\mathsf{Sym}(S)$ must have the same centre.

Lemma 3.16 *Let O_1, O_2 be points on the plane and θ_1, θ_2 be non-zero angles such that $\mathsf{rot}_{O_1,\theta_1}$ and $\mathsf{rot}_{O_2,\theta_2}$ belong to $\mathsf{Sym}(S)$. If $\mathsf{Sym}(S)$ is finite then $O_1 = O_2$.*

Proof Let k be the number of elements in $\mathsf{Sym}(S)$. That is, $\mathsf{Sym}(S) = \{\varphi_1, \ldots, \varphi_k\}$. Also, let P be any point in S. Any isometry leaving S invariant must move P to one of the points $\{\varphi_1(P), \ldots, \varphi_k(P)\}$.

Now assume that $O_1 \neq O_2$ and let $\varphi_1, \ldots, \varphi_q$ be the powers of $\mathsf{rot}_{O_1,\theta_1}$ and, similarly, $\psi_1, \ldots, \psi_m$ the powers of $\mathsf{rot}_{O_2,\theta_2}$.

Let ℓ be the line passing through O_1 and O_2 and let r_1 and r_2 be the lines perpendicular to ℓ and passing through O_1 and O_2 respectively. The point P cannot be in both r_1 and r_2 (since these are parallel lines). Assume, without loss of generality, that P is not in r_1 and, to fix ideas, that it is at the right of r_1. Then, there is $j \leq q$ such that $\varphi_j(P)$ is at the left of r_1 by Lemma 3.15. Denote $\varphi_j(P)$ by P'. Then

$$\mathsf{dist}(O_1, P) = \mathsf{dist}(O_1, P') < \mathsf{dist}(O_2, P'). \qquad (3.1)$$

Since P' is at the left of r_1 it is also at the left of r_2. Hence, there exists $j \leq m$ such that $P_1 = \psi_j(P')$ is at the right of r_2.

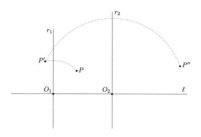

It follows that

$$\mathsf{dist}(O_2, P') = \mathsf{dist}(O_2, P_1) < \mathsf{dist}(O_1, P_1). \tag{3.2}$$

Putting (3.1) and (3.2) together we deduce that $\mathsf{dist}(O_1, P) <$ $\mathsf{dist}(O_1, P_1)$. But now P_1 is at the right of r_2 and, consequently, at the right of r_1 as well, just as at the beginning of our reasoning. Repeating this reasoning, we can find two more rotations φ and ψ in $\mathsf{Sym}(S)$ with centres in O_1 and O_2 respectively such that $P_2 = (\psi \circ \varphi)(P_1)$ is at the right of r_1 as well and $\mathsf{dist}(O_1, P_1) < \mathsf{dist}(O_1, P_2)$. Repeating this reasoning $k - 2$ more times we end up with points $P = P_0, P_1, \ldots, P_k$, which are all obtained as images of P by isometries in $\mathsf{Sym}(S)$ and such that

$$\mathsf{dist}(O_1, P) < \mathsf{dist}(O_1, P_1) < \cdots < \mathsf{dist}(O_1, P_{k-1}) < \mathsf{dist}(O_1, P_k).$$

But this implies that these points are all different (since if two were equal then so would their distance to O_1). This is in contradiction with $\mathsf{Sym}(S)$ having k elements. □

Proof of Theorem 3.14 Assume $\mathsf{Sym}(S)$ is finite. Then, there are no translations or glides on $\mathsf{Sym}(S)$. In addition, because of Lemma 3.16, all the rotations in $\mathsf{Sym}(S)$ have the same centre, say O. They therefore form a subgroup $\mathsf{Rot}(O, S)$ of $\mathsf{Sym}(S)$. Lemma 3.15 now ensures that this subgroup is actually C_n, where n is the number of elements in $\mathsf{Rot}(O, S)$.

If there are no reflections in $\mathsf{Sym}(S)$, then $\mathsf{Sym}(S) = \mathsf{Rot}(O, S)$ and the result is proven. Assume, to conclude, that there are reflections in $\mathsf{Sym}(S)$ and let ℓ be the axis of one such reflection. We claim that ℓ passes through O. Assume that this is not the case and consider the isometry

$$\psi = \mathsf{refl}_\ell \circ \mathsf{rot}_{O,\theta} \circ \mathsf{refl}_\ell,$$

where $\mathsf{rot}_{O,\theta} \in \mathsf{Rot}(O, S)$, $\mathsf{rot}_{O,\theta} \neq \mathsf{Id}$. This composition preserves orientation (is the composition of four reflections) and it is, therefore, either a rotation or a translation. But we haved ruled out translations in $\mathsf{Sym}(S)$. Therefore, ψ is a rotation and, because of Lemma 3.16, it must have its centre at O. We then have $\psi(O) = O$. But this implies that $\mathsf{refl}_\ell(\psi(O)) = \mathsf{refl}_\ell(O)$ or yet,

$$\mathsf{rot}_{O,\theta}(\mathsf{refl}_\ell(O)) = \mathsf{refl}_\ell(O).$$

Since the only fixed point of $\mathsf{rot}_{O,\theta}$ is O it follows that $\mathsf{refl}_\ell(O) = O$ and, therefore, that $O \in \ell$. Our claim is therefore proved; that is, all reflections in $\mathsf{Sym}(S)$ have axes passing through O.

The only thing to prove now is that there are exactly n such reflections and that their axes are equally spaced (i.e. such that two consecutive axes make an angle of $360/n$).

It is easy to check that, for all $j = 0, 1, \ldots, n - 1$, the composition

$$\mathsf{rot}_{O, \frac{j \times 360}{n}} \circ \mathsf{refl}_\ell = \mathsf{refl}_{\ell_j},$$

where $\ell_j = \mathrm{rot}_{O,\frac{j \times 360}{n}}(\ell)$. This immediately produces n reflections as desired in $\mathsf{Sym}(S)$. To show that there cannot be any other, we simply note that if there was one, say refl_r, then there must exist j such that the angle $\widehat{\ell_j r}$ is smaller than $360/2n$. But this implies (use Proposition 3.11) that the composition $\mathsf{refl}_{\ell_j} \circ \mathsf{refl}_r$ is a rotation of angle smaller than $360/n$; a contradiction. $\qquad\square$

Definition 3.17 Let S be a rosette or whirl with group $\mathsf{Sym}(S)$ of symmetries D_n or C_n respectively. We call a *motif* any subset M of S such that any two images in M, $\mathrm{rot}_{\frac{360}{n}}(M)$, $\mathrm{rot}_{\frac{2 \times 360}{n}}(M)$, ..., $\mathrm{rot}_{\frac{(n-1) \times 360}{n}}(M)$ have no common points (except, if it is in S, the centre of rotation), and the union of all of them yields S.

For instance, the two figures on the left are motifs for the crooked-legged star at the beginning of this section and for the whirl at the right of Figure 3.4, respectively.

3.5 Friezes

3.5.1 The seven friezes

In §3.1.4 we mentioned that figures exhibiting translational symmetry are called friezes. We now want to be more precise. For, on the one hand, we want to exclude wallpapers such as the one in Figure 3.10 and, on the other hand, we also want to exclude objects such as the following barcode:

In both cases we have too many translations leaving the figure invariant. In a wallpaper, this is because the object is invariant under translations with non-parallel vectors. And in the barcode, that is because invariance occurs for all translations with a horizontal vector, no matter the length of the vector. In the former the motif spreads in all directions; in the latter, being thin in the extreme, it fails to be perceived.

Definition 3.18 We will call *frieze* (or *frieze pattern*) a subset S of the plane possessing translational invariance for a vector $\vec{v}$ and such that for any vector $\vec{w}$ satisfying $\mathsf{trans}_{\vec{w}}(S) = S$ we have $\vec{w} = k\vec{v}$ for some $k \in \mathbb{Z}$. We call $\vec{v}$ the *basic vector* of the frieze.

We remark that $-\vec{v}$ could also be taken as the basic vector. We remark as well that the definition above amounts to saying that the group of translations leaving S invariant is $\langle \mathsf{trans}_{\vec{v}} \rangle$.

Take any line ℓ perpendicular to the basic vector $\vec{v}$ and let $\ell' = \text{trans}_{\vec{v}}(\ell)$. The part of S lying between ℓ and ℓ' is called the *unit cell* of the frieze. For instance, for the Maori pattern in Figure 3.9, we can isolate a unit cell as follows:

ℓ $\qquad$ ℓ'

Remark 3.19 It is apparent that different choices of position for the original line ℓ will yield different unit cells but these are all equivalent in that they produce the same frieze. For instance, the two squares on the right are possible unit cells for the Maori pattern.

Although not required in Definition 3.18, we usually deal with friezes whose unit cells are bounded. That is, they can be included in a finite rectangle. Unbounded unit cells, while logically possible, are not feasible in practice.

Having conveniently limited our notion of a frieze we can ponder the possible symmetries of friezes besides the translational. These arise from possible symmetries of the unit cell and, we will see, give rise to the following seven possible frieze groups. To symplify the exposition of these groups we will call "horizontal" the direction of the basic vector $\vec{v}$ and, consequently, we will call "vertical" the direction perpendicular to that of $\vec{v}$.

$\boxed{p111}$ We create an instance of this simplest case by repeating a "p". We obtain the following:

p p p p p p p p p p p p p p p p p p p p p

Friezes of this kind are said to be of type *p111*. We have already remarked that the subgroup of translations of Sym(S) is of the form $\langle \text{trans}_{\vec{v}} \rangle$. There is no other apparent symmetry in the frieze above (and we will prove that there is actually none). To obtain additional symmetries one needs to impose them on the unit cell.

$\boxed{p1m1}$ Replacing p by $\frac{p}{b}$ one obtains

p p p p p p p p p p p p p p p p p p p p
b b b b b b b b b b b b b b b b b b b b

a pattern with a visible bilateral symmetry with respect to a horizontal axis (sometimes called the *backbone*) at the middle of the unit cell. Friezes like the one above are said to be of type *p1m1*. A bonus symmetry in these friezes is the glidal one. Indeed, if refl_m denotes the reflection around the backbone, then

$$\text{refl}_m \circ \text{trans}_{\vec{v}} = \text{trans}_{\vec{v}} \circ \text{refl}_m = \text{glide}_{\vec{v}}$$

and Proposition 3.8 ensures that S is invariant under $\text{glide}_{\vec{v}}$. The elements of $\text{Sym}(S)$ generated by the translations and refl_m are, therefore, of the form $\text{trans}_{k\vec{v}}$ or $\text{glide}_{k\vec{v}}$, for some $k \in \mathbb{Z}$, together with refl_m. Group structure is given by the following equalities, for $k, q \in \mathbb{Z}$:

$$\text{trans}_{k\vec{v}} \circ \text{trans}_{q\vec{v}} = \text{trans}_{(k+q)\vec{v}}$$

$$\text{refl}_m \circ \text{trans}_{k\vec{v}} = \text{glide}_{k\vec{v}}$$

$$\text{trans}_{k\vec{v}} \circ \text{refl}_m = \text{glide}_{k\vec{v}}$$

$$\text{refl}_m \circ \text{glide}_{k\vec{v}} = \text{trans}_{k\vec{v}}$$

$$\text{glide}_{k\vec{v}} \circ \text{refl}_m = \text{trans}_{k\vec{v}}$$

$$\text{glide}_{k\vec{v}} \circ \text{trans}_{q\vec{v}} = \text{glide}_{(k+q)\vec{v}}$$

$$\text{trans}_{k\vec{v}} \circ \text{glide}_{q\vec{v}} = \text{glide}_{(k+q)\vec{v}}$$

$$\text{glide}_{k\vec{v}} \circ \text{glide}_{q\vec{v}} = \text{trans}_{(k+q)\vec{v}}.$$

$\boxed{pm11}$ Friezes of type *pm11* are those obtained with a bilaterally symmetrical unit cell with an axis perpendicular to the basic vector. For instance, the cell p q yields

p q p q p q p q p q p q p q p q p q p q

As in the previous case, we get an extra symmetry, but this time of the same type, namely bilateral invariance with respect to a vertical axis. This can be checked in the frieze but can also be easily proved. Consider a line r_1 (such that $\text{refl}_{r_1}(p) = q$) and r_2 (parallel to r_1 and such that the length of $\vec{v}$ is twice the distance between r_1 and r_2):

$$r_1 \quad r_2$$
$$\text{p} \mid \text{q} \mid \text{p}$$

Then, by Proposition 3.10,

$$\text{refl}_{r_2} = \text{refl}_{r_2} \circ \text{Id} = \text{refl}_{r_2} \circ \text{refl}_{r_1} \circ \text{refl}_{r_1} = \text{trans}_{\vec{v}} \circ \text{refl}_{r_1}$$

and it follows that $\text{refl}_{r_2}(S) = S$ since both $\text{trans}_{\vec{v}}$ and refl_{r_1} leave S invariant. In a similar manner we obtain bilateral invariance with respect to all lines bisecting pairs p q or pairs q p. To see why, and to describe all these vertical axes of reflection, fix a reflection axis ℓ_0 (e.g. as r_1 above) and, for all $k \in \mathbb{Z}$, let $\ell_k = \mathrm{trans}_{\frac{k}{2}\vec{v}}(\ell_0)$. We then have

$$\mathrm{trans}_{k\vec{v}} \circ \mathrm{trans}_{q\vec{v}} = \mathrm{trans}_{(k+q)\vec{v}}$$

$$\mathrm{refl}_{\ell_k} \circ \mathrm{trans}_{q\vec{v}} = \mathrm{refl}_{\ell_{(k-q)}}$$

$$\mathrm{trans}_{k\vec{v}} \circ \mathrm{refl}_{\ell_q} = \mathrm{refl}_{\ell_{(k+q)}}$$

$$\mathrm{refl}_{\ell_k} \circ \mathrm{refl}_{\ell_q} = \mathrm{trans}_{(k-q)\vec{v}}.$$

$\boxed{p1a1}$ The fourth kind of frieze (in our sequence) has glidal symmetry without bilateral invariance:

$$\text{p b p b p b p b p b p b p b p b p b p b}$$

Visible elements in its group are of the form $\mathrm{trans}_{k\vec{v}}$ or $\mathrm{glide}_{(k+\frac{1}{2})\vec{v}}$, for $k \in \mathbb{Z}$, and compose as follows:

$$\mathrm{trans}_{k\vec{v}} \circ \mathrm{trans}_{q\vec{v}} = \mathrm{trans}_{(k+q)\vec{v}}$$

$$\mathrm{glide}_{(k+\frac{1}{2})\vec{v}} \circ \mathrm{trans}_{q\vec{v}} = \mathrm{glide}_{(k+q+\frac{1}{2})\vec{v}}$$

$$\mathrm{trans}_{k\vec{v}} \circ \mathrm{glide}_{(q+\frac{1}{2})\vec{v}} = \mathrm{glide}_{(k+q+\frac{1}{2})\vec{v}}$$

$$\mathrm{glide}_{(k+\frac{1}{2})\vec{v}} \circ \mathrm{glide}_{(q+\frac{1}{2})\vec{v}} = \mathrm{trans}_{(k+q+1)\vec{v}}.$$

$\boxed{p112}$ This type of frieze features central symmetries:

$$\text{p d p d p d p d p d p d p d p d p d p d}$$

Again, we add central symmetries with centres in between ps and ds and we find that additional central symmetries with centres in between ds and ps are generated in $\mathrm{Sym}(S)$. Indicating these centres of symmetry with a small filled circle, we have the following situation:

$$\text{p} \cdot \text{d} \cdot \text{p}$$

As we did for *pm11*, fix a centre O_0 of central symmetry and let $O_k = \mathrm{trans}_{\frac{k}{2}\vec{v}}(O_0)$, for $k \in \mathbb{Z}$. The elements of the form $\mathrm{trans}_{k\vec{v}}$ or $\mathrm{rot}_{O_k,180}$, for $k \in \mathbb{Z}$, are all in $\mathrm{Sym}(S)$ and compose as follows:

$$\mathrm{trans}_{k\vec{v}} \circ \mathrm{trans}_{q\vec{v}} = \mathrm{trans}_{(k+q)\vec{v}}$$

$$\mathrm{rot}_{O_k,180} \circ \mathrm{trans}_{q\vec{v}} = \mathrm{rot}_{O_{k+q},180}$$

$$\mathrm{trans}_{k\vec{v}} \circ \mathrm{rot}_{O_q,180} = \mathrm{rot}_{O_{q-k},180}$$

$$\mathrm{rot}_{O_k,180} \circ \mathrm{rot}_{O_q,180} = \mathrm{trans}_{O_{k-q},180}.$$

pma2 | It is also possible to have both central and bilateral symmetries, as in the following pattern:

$$p \; q \; b \; d \; p \; q \; b \; d \; p \; q \; b \; d \; p \; q \; b \; d \; p \; q \; b \; d$$

Note that in this case we also have glidal symmetry (with vector $\frac{1}{2}\vec{v}$) but not bilateral symmetry around the backbone. Drawing dashed lines for axes of bilateral symmetry, dotted lines for the backbone (as a glidal axis) and filled circles for centres of central symmetry, the situation around the unit cell is as follows:

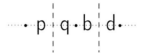

We now use the notations already introduced to describe elements in Sym(S). We fix a centre of half-turn O_0 and a vertical reflection axis ℓ_0. We take the latter at a distance $\frac{\text{length}(\vec{v})}{4}$ of the former and to its right. Then, isometries of the form $\text{trans}_{k\vec{v}}$, refl_{ℓ_k}, $\text{rot}_{O_k,180}$ or $\text{glide}_{(k+\frac{1}{2})\vec{v}}$, for $k \in \mathbb{Z}$, are all in Sym(S). Their compositions obey the equalities we have already written down, namely

$$\text{trans}_{k\vec{v}} \circ \text{trans}_{q\vec{v}} = \text{trans}_{(k+q)\vec{v}}$$

$$\text{refl}_{\ell_k} \circ \text{trans}_{q\vec{v}} = \text{refl}_{\ell_{(k-q)}}$$

$$\text{trans}_{k\vec{v}} \circ \text{refl}_{\ell_q} = \text{refl}_{\ell_{(k+q)}}$$

$$\text{refl}_{\ell_k} \circ \text{refl}_{\ell_q} = \text{trans}_{(k-q)\vec{v}}$$

$$\text{rot}_{O_k,180} \circ \text{trans}_{q\vec{v}} = \text{rot}_{O_{k+q},180}$$

$$\text{trans}_{k\vec{v}} \circ \text{rot}_{O_q,180} = \text{rot}_{O_{q-k},180}$$

$$\text{rot}_{O_k,180} \circ \text{rot}_{O_q,180} = \text{trans}_{O_{k-q},180}$$

$$\text{glide}_{(k+\frac{1}{2})\vec{v}} \circ \text{trans}_{q\vec{v}} = \text{glide}_{(k+q+\frac{1}{2})\vec{v}}$$

$$\text{trans}_{k\vec{v}} \circ \text{glide}_{(q+\frac{1}{2})\vec{v}} = \text{glide}_{(k+q+\frac{1}{2})\vec{v}}$$

$$\text{glide}_{(k+\frac{1}{2})\vec{v}} \circ \text{glide}_{(q+\frac{1}{2})\vec{v}} = \text{trans}_{(k+q+1)\vec{v}},$$

plus the following new ones:

$$\text{rot}_{O_k,180} \circ \text{refl}_{\ell_q} = \text{glide}_{(k-q+\frac{1}{2})\vec{v}}$$

$$\text{refl}_{\ell_k} \circ \text{rot}_{O_q,180} = \text{glide}_{(k-q-\frac{1}{2})\vec{v}}$$

$$\text{glide}_{(k+\frac{1}{2})\vec{v}} \circ \text{refl}_{\ell_q} = \text{rot}_{O_{k+q},180}$$

$$\text{refl}_{\ell_k} \circ \text{glide}_{(q+\frac{1}{2})\vec{v}} = \text{rot}_{O_{k-q},180}$$

$$\text{rot}_{O_k,180} \circ \text{glide}_{(q+\frac{1}{2})\vec{v}} = \text{refl}_{\ell_{k+q}}$$

$$\text{glide}_{(k+\frac{1}{2})\vec{v}} \circ \text{rot}_{O_q,180} = \text{refl}_{\ell_{k+q}}.$$

> $pmm2$ The last type of frieze is the richest in terms of symmetries. It has central, bilateral (both around the backbone and around vertical axes) and glidal symmetries:

pq pq pq pq pq pq pq pq pq pq pq
bd bd bd bd bd bd bd bd bd bd bd

Using the conventions set above, we can represent the symmetries about the unit cell as follows:

p!q ! p!q
b!d ! b!d

Finally, using previous notations we see that isometries of the form $\text{trans}_{k\vec{v}}$, refl_{ℓ_k}, $\text{rot}_{O_k,180}$ or $\text{glide}_{k\vec{v}}$, for $k \in \mathbb{Z}$, as well as refl_m, are all in $\text{Sym}(S)$. We have already described how some of these isometries compose with each other. The compositions we have not described obey (agreeing that ℓ_0 passes through O_0) the following identities:

$$\text{refl}_{\ell_k} \circ \text{glide}_{q\vec{v}} = \text{rot}_{O_{k-q},180}$$

$$\text{glide}_{k\vec{v}} \circ \text{refl}_{\ell_q} = \text{rot}_{O_{k+q},180}$$

$$\text{rot}_{O_k,180} \circ \text{glide}_{q\vec{v}} = \text{refl}_{\ell_{k-q}}$$

$$\text{glide}_{k\vec{v}} \circ \text{rot}_{O_q,180} = \text{refl}_{\ell_{k+q}}.$$

The seven types of friezes described above occur in various forms of art across time and cultures. An extensive collection of these occurrences can be found in *The Grammar of Ornament*, first published in 1856, by Owen Jones (2001).

3.5.2 A classification theorem

The description above of the seven types of friezes leaves two open issues. First, we described, for each type, a subgroup of $\text{Sym}(S)$, but the fact that this subgroup is the whole of Sym(S) (or, in other words, that there are no isometries leaving S invariant but not listed by us) was not proved. Admittedly, these additional elements in Sym(S) are nowhere to be seen, but a proof of their non-existence remains a more solid argument.

Second, there is the issue of whether an eighth type of frieze, not described above, does exist. The remainder of this section is devoted to closing these issues.

Theorem 3.20 *Any frieze is of one of the following seven types: p111, p1m1, pm11, p1a1, p112, pma2 and pmm2. The elements of* Sym(S)*, in each case, are those described in §3.5.1.*

To prove Theorem 3.20 we first show the following three propositions which limit the possible rotations, reflections, and glides on a frieze group.

Proposition 3.21

   (i) *If a rotation of angle θ leaves a frieze invariant, then $\theta = 0$ or $\theta = 180$.*

  (ii) *All half-turns in* Sym(S) *have their centres on a horizontal line.*

 (iii) *Any two centres of these half-turns are separated by a multiple of $\frac{\text{length}(\overrightarrow{v})}{2}$.*

Proof Let S be the frieze and $\overrightarrow{v}$ be its basic vector. Let also $\text{rot}_{O,\theta}$ be a rotation such that $\text{rot}_{O,\theta}(S) = S$. Then,

$$S = (\text{rot}_{O,\theta} \circ \text{trans}_{\overrightarrow{v}})(S) = (\text{trans}_{\overrightarrow{v'}} \circ \text{rot}_{O,\theta})(S) = \text{trans}_{\overrightarrow{v'}}(S).$$

Here, the first equality is because both $\text{rot}_{O,\theta}$ and $\text{trans}_{\overrightarrow{v}}$ leave S invariant and the second is by Proposition 3.12 with $\overrightarrow{v'} = \text{rot}_{O,\theta}(\overrightarrow{v})$. But by definition of a frieze, there are no non-parallel translations leaving S invariant. Therefore, $\overrightarrow{v'}$ is parallel to $\overrightarrow{v}$, which implies that $\theta = 0$ or $\theta = 180$.

To prove part (ii) note that we can write each half-turn as a composition of reflections with perpendicular axes by Proposition 3.11. Choosing these axes so that one of them (for both reflections) is the line ℓ passing through the centres O_1 and O_2, it follows that the composition is a translation (now use Proposition 3.10) with direction parallel to ℓ.

This, in turn, implies that ℓ is parallel to $\overrightarrow{v}$ and hence proves (ii).

It also implies (again by Proposition 3.10) that $\text{rot}_{O_1,180} \circ \text{rot}_{O_2,180} = \text{trans}_{2\overrightarrow{O_2O_1}}$. This shows that $\text{length}(2\overrightarrow{O_2O_1})$ is a multiple of $\text{length}(\overrightarrow{v})$ and hence proves (iii). $\qquad\qquad\square$

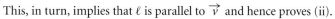

Proposition 3.22

(i) *If a reflection around a line ℓ leaves a frieze invariant, then ℓ is either parallel or perpendicular to the basic vector $\overrightarrow{v}$ of the frieze.*

(ii) *There is at most one such reflection with axis parallel to $\overrightarrow{v}$.*

(iii) *Any two vertical reflection axes are separated by a multiple of $\frac{\mathrm{length}(\overrightarrow{v})}{2}$.*

Proof The proof is similar to that of Proposition 3.21. Let S be the frieze and $\overrightarrow{v}$ be its basic vector. Let also ℓ be a line which we assume is neither parallel nor perpendicular to $\overrightarrow{v}$ and such that $\mathrm{refl}_\ell(S) = S$. Then, we have

$$S = (\mathrm{refl}_\ell \circ \mathrm{trans}_{\overrightarrow{v}})(S) = (\mathrm{trans}_{\overrightarrow{v'}} \circ \mathrm{refl}_\ell)(S) = \mathrm{trans}_{\overrightarrow{v'}}(S).$$

Here, the first equality is because both refl_ℓ and $\mathrm{trans}_{\overrightarrow{v}}$ leave S invariant and the second is by Proposition 3.13 with $\overrightarrow{v'} = \mathrm{refl}_\ell(\overrightarrow{v})$. By the definition of a frieze, there are no non-parallel translations leaving S invariant. Therefore, $\overrightarrow{v'}$ is parallel to $\overrightarrow{v}$, which implies that ℓ must be parallel or perpendicular to $\overrightarrow{v}$; a contradiction.

For part (ii), note that if we have two reflections with axes $\ell \neq \ell'$ which are both parallel to $\overrightarrow{v}$ the composition $\mathrm{refl}_\ell \circ \mathrm{refl}_{\ell'}$ yields, by Proposition 3.10, a translation with vector perpendicular to $\overrightarrow{v}$. Since both refl_ℓ and $\mathrm{refl}_{\ell'}$ leave S invariant, so does their composition; this is in contradiction with the definition of a frieze.

Finally, part (iii) follows from the fact that the vector of the translation $\mathrm{refl}_\ell \circ \mathrm{refl}_{\ell'}$ (for ℓ and ℓ' vertical) has twice the length of the distance between ℓ and ℓ'. $\qquad\square$

Proposition 3.23

(i) *If a glide of vector $\overrightarrow{w}$ leaves a frieze invariant, then $\overrightarrow{w}$ is parallel to $\overrightarrow{v}$.*

(ii) *Any two such glides have the same axis.*

(iii) *Furthermore, $\mathrm{length}(\overrightarrow{w})$ is a multiple of $\frac{\mathrm{length}(\overrightarrow{v})}{2}$.*

Proof We have $\mathrm{glide}_{\overrightarrow{w}} \circ \mathrm{glide}_{\overrightarrow{w}} = \mathrm{trans}_{2\overrightarrow{w}}$. This implies, by Definition 3.18, that $\overrightarrow{w}$ is parallel to $\overrightarrow{v}$ and that its length is a multiple of $\frac{\mathrm{length}(\overrightarrow{v})}{2}$, thus proving (i) and (iii).

To prove (ii) consider two glides, glide_1 and glide_2, leaving S invariant. By part (i) their axes are parallel to $\overrightarrow{v}$. If these axes do not coincide then, it is easy to check, the composition $\mathrm{glide}_1 \circ \mathrm{glide}_2$ yields a translation in a direction not parallel to $\overrightarrow{v}$, which is not possible. $\qquad\square$

Proposition 3.24 *If $\mathrm{Sym}(S)$ possesses half-turns and horizontal reflections (or glides) then the centres of the half-turns are in the axis of reflection (or the axis of the glide).*

Proof Let refl_m be a horizontal reflection and $\mathsf{rot}_{O,180}$ be a half-turn, both in $\mathsf{Sym}(S)$. If O is not in the axis m, then the composition $\mathsf{refl}_m \circ \mathsf{rot}_{O,180}$ yields a glide with vertical axis, which is not possible by Proposition 3.23. Similarly, if $\mathsf{glide}_{\overrightarrow{w}}$ is in $\mathsf{Sym}(S)$, then its axis ℓ is horizontal. If O is not in ℓ, then the composition $\mathsf{glide}_{\overrightarrow{w}} \circ \mathsf{rot}_{O,180}$ yields a translation with vertical axis, which again is not possible. $\qquad\square$

Proof of Theorem 3.20 Because of Propositions 3.21–3.23, translations aside, any isometry leaving the frieze invariant will have to be of one of the following four types: reflection with vertical axis, reflection with horizontal axis, glide (with horizontal axis) and half-turn. There are consequently 16 possibilities for the group according to whether or not an isometry of each of these kinds is present. We display these possibilities in the following table:

	Vertical reflection	Horizontal reflection	Glide	Half-turn	
1	Y	Y	Y	Y	*pmm2*
2	Y	Y	Y	N	
3	Y	Y	N	Y	
4	Y	Y	N	N	
5	Y	N	Y	Y	*pma2*
6	Y	N	Y	N	
7	Y	N	N	Y	
8	Y	N	N	N	*pm11*
9	N	Y	Y	Y	
10	N	Y	Y	N	*p1m1*
11	N	Y	N	Y	
12	N	Y	N	N	
13	N	N	Y	Y	
14	N	N	Y	N	*p1a1*
15	N	N	N	Y	*p112*
16	N	N	N	N	*p111*

Assume a frieze S corresponds to row 1. Then there is a vertical reflection with axis, say, ℓ_0. The composition

$$\mathsf{trans}_{k\overrightarrow{v}} \circ \mathsf{refl}_{\ell_0} = \mathsf{refl}_{\ell_k}$$

shows that $\mathsf{Sym}(S)$ must have the reflections refl_{ℓ_k}, for all $k \in \mathbb{Z}$, where $\ell_k = \mathsf{trans}_{\frac{k}{2}\overrightarrow{v}}(\ell_0)$. Furthermore, Proposition 3.22(iii) shows that no other vertical reflection is possible. A similar reasoning (using Propositions 3.21 and 3.23) shows that $\mathsf{Sym}(S)$ also possesses half-turns $\mathsf{rot}_{O_k,180}$ with centres uniformly spaced with a separation $\frac{\mathrm{length}(\overrightarrow{v})}{2}$ and glides $\mathsf{glide}_{k\overrightarrow{v}}$ for all $k \in \mathbb{Z}$, and that these are all the rotations and glides in Sym(S). Finally, Proposition 3.22(ii) ensures that there is only one horizontal reflection, whose axis we denote by ℓ_m.

In addition to the above, Proposition 3.24 ensures that the centres O_k are in ℓ_m. Furthermore, they need to be at the intersection of ℓ_m with the vertical axes of reflection because the composition of a horizontal reflection with a vertical one gives a half-turn with centre in the intersection of the reflection's axes. The isometries in Sym(S) are exactly those introduced in the description of the type *pmm2* in §3.5.1. We conclude that Sym(S) coincides with the group described for the friezes of this type.

The same reasoning, *mutatis mutandis*, shows that a frieze corresponding to row 5 has vertical reflections, half-turns and glides as above. The difference now is that the centres of the half-turns cannot be on the vertical axes of reflection. This is because the composition of one such half-turn with the corresponding vertical reflection yields a horizontal reflection, and this is banned in row 5. We claim that these centres are in the middle of the lines ℓ_k. Indeed, let O be such a centre, d_+ be the distance from O to the closest axis to the right and d_- the distance to the closest axis to the left. Since $d_+ + d_- = \frac{\text{length}(\overrightarrow{v})}{2}$, one of these two distances must be at most $\frac{\text{length}(\overrightarrow{v})}{4}$. Assume, without loss of generality, that it is d_- and let ℓ be the axis at the left of O:

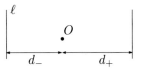

Then, the composition $\text{refl}_\ell \circ \text{rot}_O$ yields a glide whose vector has length $2d_- \leq \frac{\text{length}(\overrightarrow{v})}{2}$. By Proposition 3.23 this length has to be also at least $\frac{\text{length}(\overrightarrow{v})}{2}$. It follows that it is exactly this quantity and, therefore, that $d_- = d_+ = \frac{\text{length}(\overrightarrow{v})}{4}$ as claimed. As before, we conclude that Sym(S) coincides with the group described for friezes of type *pma2*.

In much the same manner, one proves that the group Sym(S) for friezes corresponding to rows 8, 10, 14, 15 and 16 is the one described for friezes of type *pm11*, *p1m1*, *p1a1*, *p112* and *p111* respectively.

To finish the proof we will show that the combinations in the remaining rows are not possible. Rows 2 and 4 are not possible because the composition of a horizontal reflection with a vertical one is a half-turn. Rows 3, 11 and 12 are not possible because the composition of a horizontal reflection with a (horizontal) translation is a glide. Row 6 is not possible because the composition of a (horizontal) glide with a vertical reflection yields a half-turn. Row 7 is not possible because the composition of a half-turn with a vertical reflection yields a glide (or a horizontal reflection). Finally, rows 9 and 13 are not possible because the composition of a half-turn with a horizontal glide yields a vertical reflection. □

Not only does Theorem 3.20 show that there are only seven kinds of frieze, its proof also suggests a way to classify any given one by checking the existence of at most three symmetries (translations aside). The method is succinctly depicted by the following flowchart.

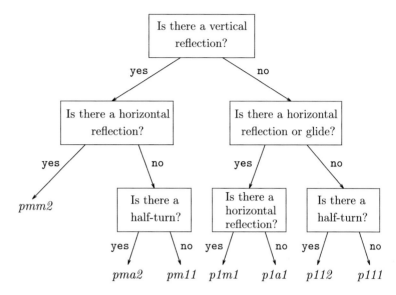

As an example, we see that when applied to the Egyptian friezes in Figure 3.7 we obtain the type *pm11* led by the answers yes, no, no. Similarly, for the pavement in Figure 3.8 we get the answers no, no, yes and, therefore, the type *p112*. Finally, for the Maori pattern in Figure 3.9 the answers are no, yes, yes and hence its type is *p1m1*.

3.6 Wallpapers

3.6.1 The seventeen wallpapers

In the previous sections we have classified all the possible finite groups of symmetries (the dihedral groups D_n and the cyclic groups C_n) as well as all possible groups of friezes. We finally turn our attention to wallpapers. Recall, those are figures which are translationally invariant for at least two non-parallel translation vectors. To set the limits of what exactly we want to consider a wallpaper we become more precise (just as we did with friezes).

Definition 3.25 We call *wallpaper* (or *wallpaper pattern*) a subset S of the plane possessing translational invariance for two non-parallel vectors $\overrightarrow{v_1}$ and $\overrightarrow{v_2}$ and such that for any vector $\overrightarrow{w}$ satisfying $\text{trans}_{\overrightarrow{w}}(S) = S$ there exist two integers n_1, n_2 such that $\overrightarrow{w} = n_1 \overrightarrow{v_1} + n_2 \overrightarrow{v_2}$.

Note that any vector of the form $n_1 \overrightarrow{v_1} + n_2 \overrightarrow{v_2}$ leaves S invariant since $\overrightarrow{v_1}$ and $\overrightarrow{v_2}$ do so. The contents of Definition 3.25 is that these are all vectors leaving S invariant. The figure on the right shows a vector $\overrightarrow{w}$ obtained as $\overrightarrow{w} = -3\overrightarrow{v_1} + 4\overrightarrow{v_2}$.

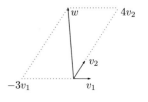

As with friezes, take any point P on the plane and consider the quadrangle with vertices P, $\text{trans}_{\overrightarrow{v_1}}(P)$, $\text{trans}_{\overrightarrow{v_2}}(P)$ and $\text{trans}_{\overrightarrow{v_1}+\overrightarrow{v_2}}(P)$:

We call this quadrangle the *unit cell* of the wallpaper.

Remark 3.26 We observed in Remark 3.19 that the unit cell in a frieze is not uniquely determined. The same occurs with a wallpaper, since neither of the *basic vectors* $\overrightarrow{v_1}$ and $\overrightarrow{v_2}$ are uniquely determined. For instance, one may have $\overrightarrow{v_1}$ and $\overrightarrow{v_2}$ perpendicular and of the same length, which yields a square unit cell. But the pair $\overrightarrow{w_1} = \overrightarrow{v_1}$ and $\overrightarrow{w_2} = \overrightarrow{v_1} + \overrightarrow{v_2}$ is also a pair of basic vectors and yields a unit cell which is not square.

For all possible choices of basic vectors, the unit cell by construction is a parallelogram. But, as we shall see, the presence of other symmetries, besides the translational, may allow for choices of basic vectors that make the unit cell a rhombus or a rectangle, or even a square. In such cases, we will take these choices.

A property of the unit cell is that the whole wallpaper S can be obtained by taking the union (i.e. the putting together) of copies of the unit cell. These copies are the images of the unit cell under all the elements in $\text{Transl}(S)$. In addition, any two such copies do not overlap. It is sometimes convenient to isolate a subset of S satisfying these properties which is not necessarily a parallelogram. For instance, the drawing on the right has this property for the wallpaper *Eight Heads* in Section 4.4. Unlike any possible unit cell, the eight heads that give the name to this engravure are (reasonably) apparent in this drawing. Subsets of wallpapers having the properties above (i.e. they span the wallpaper with translated copies which do not overlap) are said to be *motifs* of the wallpaper. A similar definition can be given for friezes.

Just as the only possible rotations leaving a frieze invariant are the identity and half-turns, rotations leaving a wallpaper invariant must have an angle equal to $0°$, $60°$, $90°$, $120°$ or $180°$ (corresponding to the identity, sixfold, fourfold, threefold and twofold rotations respectively) or, obviously, a multiple of these angles (which yields a rotation belonging to one of these five types). This is known as the *crystallographic restriction*. It is using this property (along with a number of other restrictions on the possible isometries leaving a wallpaper invariant) that a classification theorem, akin to Theorem 3.20 for friezes, is proved showing the existence of exactly 17 different wallpaper groups. The proof of this result, however, is too complicated to be included here (the interested reader may find such a proof in (Baloglou, 2007)). We will instead proceed to simply describe these 17 groups, in the same manner we described the seven frieze groups. Hopefully, our acquired familiarity with friezes might grant us the benefit of this conciseness.

$\boxed{p1}$ This is the simplest wallpaper group. It consists only of translations. There are neither reflections, glides, nor rotations. The two basic vectors may be inclined at any angle to each other.

$\boxed{pg}$ This group contains glides. The axes of these glides are parallel to one basic vector and perpendicular to the other (these axes appear as dashed lines in the picture below). There are neither rotations nor reflections. The red rectangles on the right show (a choice of) unit cells for this pattern.

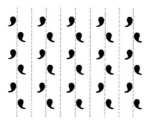
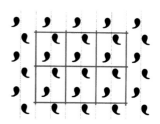

$\boxed{pgg}$ This group contains no reflections, but it has half-turns and two families of glides. The axes for the glides in these families are perpendicular, and the rotation centres – which in the figures that follow we represent with ovals 0, ●, ○, ● – do not lie on these axes. The shaded rectangle with red boundary displays a unit cell.

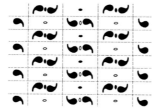
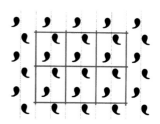

pm This group contains reflections. The axes of these reflections are parallel (and, actually, parallel to one basic vector and perpendicular to the other). There are neither rotations nor non-induced glides.

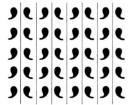   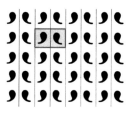

pmg This group contains reflections in only one direction and glides with axes perpendicular to those of the reflections. It has half-turns with centres on the glide axes, halfway between the reflection axes.

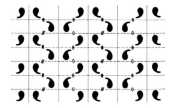   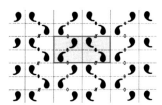

pmm This symmetry group contains reflections in two perpendicular directions and half-turns centred where the axes of these reflections intersect. The unit cell is rectangular.

p2 This group differs from *p1* only in that it contains half-turns. There are neither reflections nor glides. The two basic vectors may be inclined at any angle to each other. Its unit cell is, therefore, a parallelogram.

   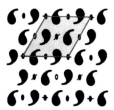

$\boxed{p3}$ This is the simplest group that contains a 120° rotation. It has no reflections or glides. Centres of 120° rotations are indicated with a triangle ▼. The unit cell on the upper part of the figure in the right is a rhombus made up with two equilateral triangles.

 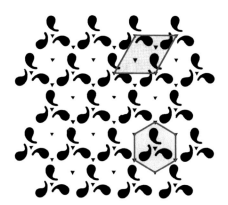

However, for patterns in this group one can choose a hexagonal unit cell, as shown in the lower part of that figure.[1]

$\boxed{p31m}$ This group contains reflections (whose axes are inclined at 60° to one another) and rotations of order 3 (i.e. of 120°). Some of the centres of rotation lie on the reflection axes and some do not. There are some glides. Patterns in this group also admit hexagonal unit cells.

[1] Strictly speaking, a hexagon is not a unit cell in the sense defined above since it is not a quadrangle. Yet, we may equally obtain the whole wallpaper by glueing translational copies of this hexagon and, for most *p3* patterns, it displays the motif better than a quadrangular unit cell.

p3m1 This group is similar to *p31m*, in that it contains reflections and order-3 rotations. The axes of the reflections are again inclined at 60° to one another, but for this group all of the centres of rotation lie on the reflection axes. There are some glides and the unit cell can be taken as hexagonal.

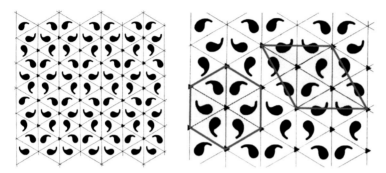

p4 This group has 90° rotations; that is, rotations of order 4 (whose centres we indicate with a lozenge ♦). It also has half-turns. The centres of the latter are midway between the centres of the order-4 rotations. There are no reflections. The unit cell can be chosen to be a square.

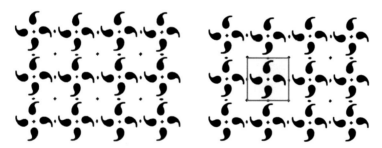

p4g Like *p4*, this group contains rotations of orders 2 and 4. It also contains reflections. There are two perpendicular reflections passing through the centres of the half-turns. In contrast, the centres of the order-4 rotations do not lie on any reflection axis. There are glides in four different directions. The unit cell can be chosen to be a square.

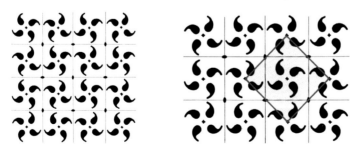

p4m This group also has both order-2 and order-4 rotations. In addition, it has reflections whose axes are inclined to each other by 45° so that four such axes pass through each order-4 rotation centre. Every rotation centre lies on some reflection axis. There are also two glides passing through the centre of each half-turn, with axes at 45° to the reflection axes. It has square unit cells.

p6 This group contains 60° rotations; that is, rotations of order 6. It also contains rotations of orders 2 and 3, but no reflections or glides. Centres of 60° rotations are indicated with a hexagon ⬢. It has hexagonal unit cells.

p6m This group has rotations of order 2, 3 and 6, as well as reflections. The axes of reflection meet at the centres of rotation. At the centres of the order-6 rotations, six reflection axes meet and are inclined at 30° to one another. There are also some glides and the unit cell can be chosen to be hexagonal.

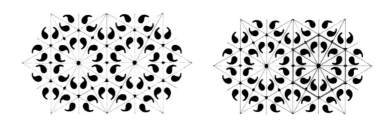

cmm This group has perpendicular reflection axes, as does group *pmm*, but it also has rotations of order 2. The centres of the rotations do not lie on the reflection axes. Unit cells, as for patterns of type *cm*, can be chosen to be rhombic.

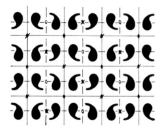 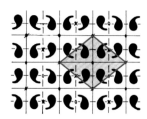

cm This group contains reflections and glides with parallel axes. There are no rotations. The basic vectors may be inclined at any angle to each other, but the axes of the reflections bisect the angle formed by the basic vectors.

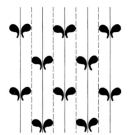 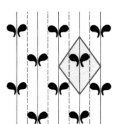

There is at least one glide whose axis is not a reflection axis; it is halfway between two adjacent parallel reflection axes. This group displays symmetrically staggered rows (i.e. there is a shift per row of half the translation distance inside the rows) of identical objects, which have a symmetry axis perpendicular to the rows. Unit cells can be chosen to be rhombic.

3.6.2 A brief sample

As in the case of friezes, the 17 types of wallpaper described above occur in various forms of art across time and cultures. Figures 3.12 and 3.13

Figure 3.12 **Medieval wall diapering (*p1*); ceiling of Egyptian tomb (*p2*), Egyptian tomb at Thebes (*pm*); mat on which Egyptian king stood (*pg*); bronze vessel in Nimroud, Assyria (*cm*); mummy case (presently at the Louvre Museum) (*pmm*); cloth from Sandwich Islands (*pmg*); bronze vessel in Nimroud, Assyria (*pgg*); Egyptian (*cmm*).**

show a (necessarily brief) sample, taken mostly from *The Grammar of Ornament* (Jones, 2001).

3.6.3 Tables and flowcharts

The accumulation of properties in the description of the 17 wallpaper groups may induce confusion: too many details can obscure the

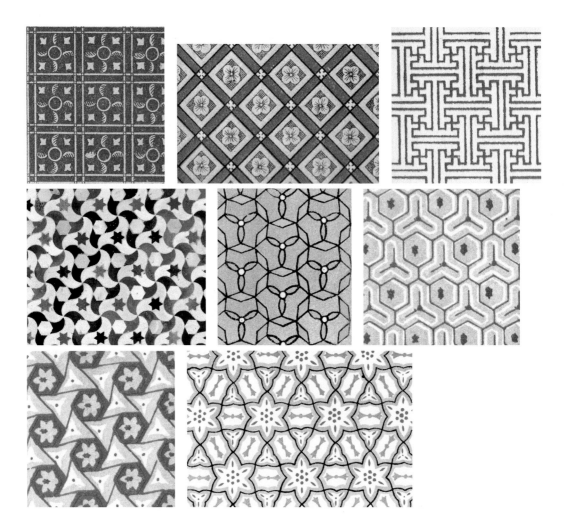

Figure 3.13 **Renaissance earthenware (*p4*); Saint-Étienne Cathedral, Bourges, France (*p4m*); painted porcelain, China (*p4g*); wall tiling in the Alhambra, Spain (*p3*); Persian glazed tile (*p31m*); Persian ornament (*p3m1*); Persian ornament (*p6*); Persian glazed tile (*p6m*).**

general picture. In order to minimize this confusion, two solutions are usually proposed.

Flowcharts allow for the classification of a given wallpaper, in the same spirit as the flowchart for classifying friezes we saw in §3.5.2. The following figure gives one such possible flowchart, classifying a wallpaper with at most four questions:

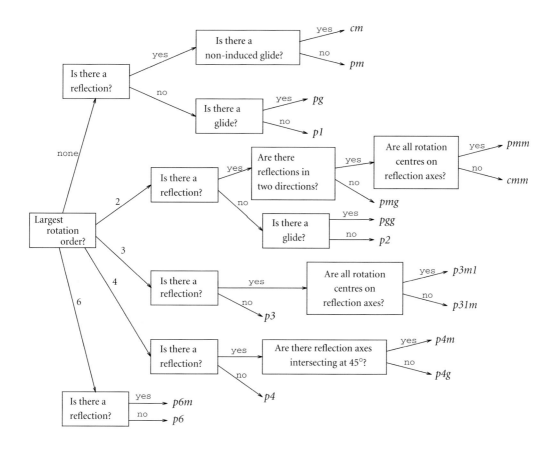

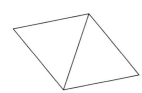

The reader may try to use this flowchart for some of the wallpapers in Figures 3.12 and 3.13.

One may also summarize the different basic features of the 17 groups in tabular form. In the following table, "Rotation order" denotes the highest possible order of a rotation in the group. That is, the highest possible number n for which there is a rotation φ in the group such that $\varphi^n = \mathrm{Id}$ but $\varphi^m \neq \mathrm{Id}$ for all $1 \leq m \leq n-1$. Also, glides are non-induced, in the sense that their component reflection and translation are not themselves elements of the group (we already discussed this in §3.1.5). Finally, we use the word "hexagonal" to denote the possibility of choosing a hexagonal unit cell. Recall, this happens in patterns where another possible choice for unit cell is a rhombus obtained by glueing together two equilateral triangles.

With these notations, the following table shows the salient features of each of the 17 groups at a glance:

Type	Unit cell	Rotation order	Reflections	Glides	Other features
p1	Parallelogram	1	No	No	
p2	Parallelogram	2	No	No	
pm	Rectangle	1	Yes	No	
pg	Rectangle	1	No	Yes	
cm	Rhombus	1	Yes	Yes	
pmm	Rectangle	2	Yes	No	
pmg	Rectangle	2	Yes	Yes	Parallel reflection axes
pgg	Rectangle	2	No	Yes	
cmm	Rhombus	2	Yes	Yes	Perpendicular reflection axes
p4	Square	4	No	No	
p4m	Square	4	Yes	Yes	4-fold centres on reflection axes
p4g	Square	4	Yes	Yes	4-fold centres not on reflection axes
p3	Hexagon	3	No	No	
p3m1	Hexagon	3	Yes	Yes	All 3-fold centres on reflection axes
p31m	Hexagon	3	Yes	Yes	Not all 3-fold centres on reflection axes
p6	Hexagon	6	No	No	
p6m	Hexagon	6	Yes	Yes	

3.7 Symmetry and repetition

> Indeed, if there were no repetition, what then would life be?
>
> S. Kierkegaard (1964: 35)

There is an obvious feeling of repetition in figures having non-trivial symmetries. Translational symmetry amounts to repeating the unit cell once and again, the whirl in Section 3.4 repeats three times the figure on the right and the triangles in both sides of Figure 3.11 are all copies of the same triangle. In the case of translational symmetry, a unit cell captures the smallest subset of the pattern that allows the whole pattern to be generated by putting together translational copies of this subset. This notion extends to the whole group of symmetries of a figure.

Definition 3.27 Let S be a figure with non-trivial symmetries. A subset $R \subset S$ is said to be a *fundamental region* of S when:

(i) the whole of S can be obtained by taking the union of copies of R under different elements in $\mathsf{Sym}(S)$; that is

$$S = \bigcup_{\varphi \in \mathsf{Sym}(S)} \varphi(R).$$

(ii) No proper subset R' of R satisfies condition (i) above.

For example, the subset R on the left is a fundamental region for the Maori pattern in Figure 3.9. Indeed, this is the upper half of the unit cell for this pattern on the right side in Remark 3.19. A reflection around its backbone yields the corresponding lower part and the union of these two parts yields the unit cell. The whole pattern can then be obtained by taking the union of translated copies of this unit cell. Furthermore, no subset R' of R will generate the pattern, since the image of any such R' under an element in the symmetry group of the pattern lies outside R and cannot, therefore, generate R itself.

Figures S having a finite group of symmetries are obtained by glueing exactly n copies of the fundamental region (when $\mathsf{Sym}(S) = C_n$) or $2n$ such copies (when $\mathsf{Sym}(S) = D_n$). In the case of friezes or wallpapers it is the unit cell which is obtained by glueing a finite number of copies of the fundamental region (how many such copies are needed depends on the particular frieze or wallpaper group being considered). The whole figure can then be obtained by putting together copies of the unit cell.

There is repetition in symmetry. There is also order. The following arrangement of rectangles exhibits six copies of a rectangle.

But there is no symmetry in it; no law dictates how to obtain from one of them the remaining five.

3.8 The catalogue-makers

Madamina, il catalogo è questo
Delle belle che amò il padron mio;
Un catalogo egli è che ho fatt'io;
Osservate, leggete con me.[2]

W.A. Mozart, *Don Giovanni*

---

[2] My lady, this is a list of the beauties that my master has loved; a list which I have compiled. Observe, read along with me.

A symmetric pattern is determined by two constituents: its symmetry group and its fundamental region. Once both are given, the entire pattern can be generated. A number of differences between these two ingredients stand out. The first relates to meaning. Essentially, whichever meaning the pattern might carry will reside on its fundamental region; symmetry groups are not meaning bearers. The second relates to number and possibilities. Whereas there are no limits for the possible fundamental regions of a pattern, its possible symmetry groups are restricted to a few choices. Leaving aside the groups of barcodes such as the one in §3.5.1 or of rotational versions of them, such as the one on the right, the possible symmetry groups are not many. We have actually shown the extent of this choice by exhibiting all of them: the groups C_n and D_n, both for $n \geq 1$, the seven groups of friezes and the 17 groups of wallpapers.

The breadth of this catalogue sets a frame within which the creation of patterns will have to take place. In the words of the distinguished art theorist Gombrich (1984: 64), "there is a solid core of fact which restricts the possibilities open to any pattern-maker – I am refering to the laws of geometry".

I am not saying that the pattern-maker needs to be aware of this frame. The opposite is doubtless true and people have produced patterns with all symmetry structures (at least for moderate values of n in the case of C_n and D_n) from sheer ingenuity. The role of mathematics in this context reduces to proving that these symmetry structures are indeed the only possible ones and, maybe, to offer a list with a description of them. Mathematicians thus become catalogue-makers, and in their dialogue with the artist they may say, as Leporello does while unrolling Don Giovanni's deeds to Doña Elvira, "this is a list [...] a list which I have compiled. Observe, read along with me".

4 The many objects with planar symmetries

Greek temples, French department stores, Islamic domes, church windows, Egyptian tombs, Assyrian vessels, Chinese porcelains ... The examples of symmetry in Chapter 3 give evidence of the multitude of its forms and of its ubiquitousness. They were meant only to illustrate different forms of symmetry and neither attempted to focus on specific contexts nor to discuss particular artists.

Such expositions are the goal of this chapter. In its first section we witness early occurrences of symmetry through archaeological findings. We then proceed with a short discussion on Oriental rugs, which are doubtless a colourful source of symmetric patterns. In Section 4.3 the focus is on Chinese lattices. Both Oriental rugs and Chinese lattices spread over a period of more than two millennia (even though most of our examples will be relatively recent). Section 4.4, in contrast, brings the focus to the twentieth century and discusses some of the work of M.C. Escher.

4.1 Origins

A paper by Thomas Wynn (2002) provides an unusual application of the notion of symmetry. A declared goal of the paper is the "reconstruction of aspects of early hominid spatial cognition based on an analysis of artifactual symmetries". Spatial cognition aside, Wynn's paper is of interest to us because of its tracing of early instances of intentional symmetry in human artefacts.

The earliest examples of such artefacts are stone tools approximately 2.5 million years old (from the preceding period of 1.5 million years separating the split between hominids and other African apes and the production of these stone tools there are no archaeological records). But these stone tools do not exhibit any symmetry other than those possibly present in the material or randomly obtained during the knapping process. It is not until 1.4 million years ago that the first specimens with an intentional symmetry appear. Hominids in East Africa, presumably *Homo erectus*, began producing stone tools (Figure 4.1) with an overall two-dimensional shape (therefore called "bifaces"). To do so, they would first detach a flake from a larger stone and then trim it around its margins.

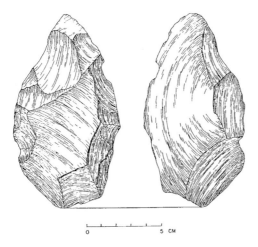

Figure 4.1 **Handaxe from West Natron, Tanzania, approximately 1.4 million years old. (Reproduced with permission from (Wynn, 2002).)**

Wynn notes that "not all bifaces of this age are nicely symmetrical, however, and even the nicest examples look crude compared to the symmetry of later tools". Yet, he believes the symmetry occurring was intended by the knapper. "First, the most symmetrical examples are also the most extensively trimmed, indicating that the knapper devoted more time to production. Second, and more telling, on some bifaces the trimming mirrors a natural shape on the opposite edge (Figure 4.2). Such artefacts do not have the best symmetry, but the economy of means by

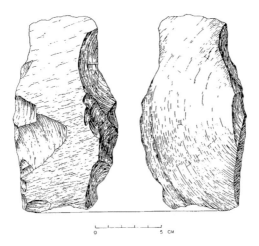

Figure 4.2 **Cleaver from Olduvai Gorge, Tanzania, approximately 1.4 million years old. (Reproduced with permission from (Wynn, 2002).)**

which the symmetry was achieved reveals that some idea of mirroring must have guided the knapper".

By 500,000 years ago, the skills of stone knappers had improved to a degree where symmetry becomes quantitative; in the worked bifaces "the mirrored sides are not just qualitative reversals, but quantitative duplicates, at least to the degree that this is possible given the constraints of stone knapping". Another new development manifests itself in the attention to the third dimension. Early bifaces were, as we mentioned, overall two-dimensional in shape. Symmetry was only present on the section corresponding to these two dimensions. Bifaces of this period attempt symmetry in the profile as well, as shown in the handaxe in Figure 4.3.

The study of more recent periods of time, say the last 100,000 years, allows one to consider perishable materials and, therefore, artefacts made with materials other than stone. This is important, since on these new artefacts the intention of symmetry is better rewarded. For stone, though durable, "is not plastic and shaping can only be done by subtraction". Examples of these new artefacts are the barbed points in Figure 4.4 which exhibit clear translational and glidal symmetries.

As we approach the Lower Palaeolithic the presence of symmetries in artefactual objects becomes less relevant for Wynn's purposes. It also becomes more frequent. A collection of symmetric shapes in various forms of decoration in the Lower Palaeolithic and Neolithic periods is found in the book of Slavik Jablan (2002). The remaining figures

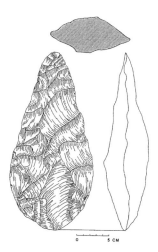

Figure 4.3 **Handaxe from Isimila, Tanzania, approximately 300,000 years old. (Reproduced with permission from (Wynn, 2002).)**

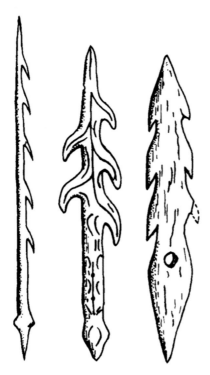

Figure 4.4 **Antler made into barbed points, European Upper Palaeolithic, approximately 17,000 years old. (Reproduced with permission from (Wynn, 2002).)**

in this section are taken from this book.[1] They exhibit different kinds of symmetry. For instance, the drawings in Figure 4.5 possess bilateral symmetry (or, to be more precise, have the intention of bilateral symmetry), in all cases with a vertical edge.

Similarly, the cave paintings in Figure 4.6, found in three different locations in southern Europe, show wallpaper patterns with the symmetry group *pm*. Paintings with patterns featuring other wallpaper groups also exist (and the reader may find some in Jablan (2002)).

Whirls (Figure 4.7, b and f) and rosettes (Figure 4.7, a, c–e and g) also occurred, for instance, in Middle Asia ceramics.

Finally, frieze patterns for all the seven groups are found in a large number of examples. Figures 4.8–4.10 show some of those. In Figure 4.8 all the patterns are of type *p111*. Figure 4.9 shows different patterns of type *p1m1* and Figure 4.10 shows patterns for the type *p112*.

---

[1] They are actually taken from the online version of the book.

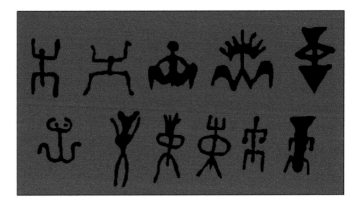

Figure 4.5 **Stylized human figures: Palaeolithic and Neolithic of Italy and Spain.** **(From (Jablan, 2002), ⓒ 2002, World Scientific.)**

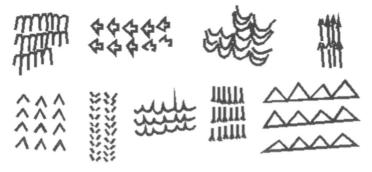

Figure 4.6 **Palaeolithic paintings: Ardales cave (Spain), Gorge d'Enfer (France), Romanelli** **caves (Italy). (From (Jablan, 2002) , ⓒ 2002, World Scientific.)**

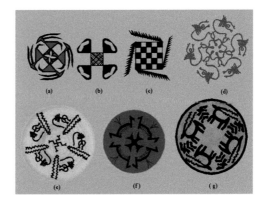

Figure 4.7 **Neolithic ceramics of Middle Asia around** **5500–5000 BCE: (a), (b), (d), (e) and (g), Samara; (c) and (f),** **Susa. (From (Jablan, 2002), ⓒ 2002, World Scientific.)**

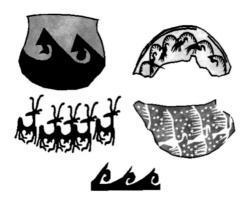

Figure 4.8 **Neolithic ceramics from Hallaf, around 5500–4500 BCE. (From (Jablan, 2002), ©️ 2002, World Scientific.)**

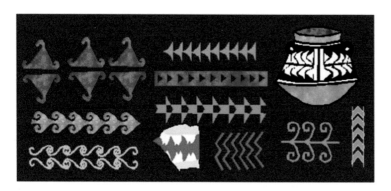

Figure 4.9 **Various decorations in Neolithic art, around 6000–3000 BCE. (From (Jablan, 2002), ©️ 2002, World Scientific.)**

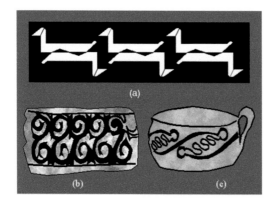

Figure 4.10 **Decorations from: (a) Bakun, Iran, around 5000–4000 BCE; (b) Malta, around 3000 BCE; (c) Crete, around 3000–2500 BCE. (From (Jablan, 2002), ©️ 2002, World Scientific.)**

4.2 Rugs and carpets

The conjunction of temperature fluctuations (both between day and night and between summer and winter) and an early domestication of sheep (resulting in a moderate availability of wool) is likely at the origin of carpet weaving in Central Asia and the Middle East. The inhabitants of these regions would protect themselves from cold by covering floors, walls, doorways and even sofas or cushions with different sorts of rugs. It is certainly beyond our scope to attempt a detailed description of the many varieties of rugs.[2] The following simple taxonomy should suffice for our purposes.

Rugs can be classified according to their weaving technique, the material used, their provenance and their designs. There are three main techniques: knotting, flat weaving and embroidering. In *knotted carpets* the knots are made around a number of parallel strings (the *warps*) tied to a loom. Additional strings perpendicular to the warps (the *weft threads* or *woof*) are used to secure the knots in place. Once the knots are thus secured they are cut to a uniform length and the resulting loose cut ends of the knots are known as the *pile* of the carpet. The *knot density*, usually measured in number of knots per square centimetre (or per square inch), not only provides a measure of the work required to produce a given rug but also sets limits to the possible designs. Knotted carpets correspond to the most extended idea of carpet. In *flat weaved* carpets there is only warp and weft. But their interlacing is more elaborated to allow for the display of coloured designs. The best known kind of such carpets are the Turkish *kilims*. Finally, in an *embroidered* carpet, a base cloth is garnished by the application of stitches.

We next focus on the materials used. The predominant one is wool, of which there are several types and qualities. But other materials are often used, such as cotton, linen or silk, and it is a usual practice to combine materials by taking different ones for pile, warp and weft.

The provenance of a carpet is strongly related to the cultural influences behind its design. We will not go deeper into this matter.

We are thus left with issues of design. A primary remark in this context is the usual division of a rug into *field* and *border*, the former being the central part of the rug and the latter a frame for it (a cursory look at the figures in this section provides clear examples of this division and, in a few cases, of its absence). Although the contents of the field vary, its composition can be classified as belonging to one of the following seven types: medallion, repeated motif, all-over pattern, empty or open field,

---

[2] To have an idea of the level of precision such a description can reach we mention some advice concerning contemporary Chinese carpets which points out that the best of them are the "contract-quality super-washed close-back 90-line $\frac{5}{8}$-pile mill-spun" (Ford, 1989: 32).

compartmentalized or panel, picture or portrait, and prayer.[3] Some of these types have an obvious meaning (e.g. empty or open field, picture or portrait) and some others, even though they may display some form of symmetry, will play no role in our discussion (e.g. all-over pattern, prayer). The compositions that make the best display of symmetry are the medallions and the repeated motifs (as well as the combination of the two). The nature of the motif in both cases will depend on a number of factors, of which the knot density is worth mentioning. In rugs with a high knot density, curvilinear, elaborate, motifs are possible. In those with a low knot density (as well as in kilims), simpler, rectilinear, motifs tend to prevail. A similar observation holds for the motifs adorning the borders. These motifs, almost invariably, draw a frieze around the field.

In the fifth chapter of his book on Anatolian kilims, Peter Davies (2000) offers a description of the visual language of these rugs. He describes what he calls the "lexicon" of this language (consisting of the elementary motifs that recur in Anatolian kilims) as well as its "grammar" (this being the possible overall layouts). In doing so, he emphasizes the marked preference of Anatolian weaving culture for symmetric motifs and, in particular, for those possessing "double mirror symmetry" (i.e. having symmetry group D_2). Asymmetrical motifs are not excluded but "are employed infrequently, and as a result stand out conspicuously in the design when they are used, and are believed to be foreign borrowings" (Davies, 2000: 40). Figure 4.11 shows a minute sample of some common motifs.

Figure 4.11 **Common motifs in Anatolian kilims. (From (Davies, 2002). Copyright © 2000, 1993 by Peter Davies. Used by permission of W. W. Norton & Company, Inc.)**

---

[3] A detailed description with examples of these seven types can be found in (Summers, 1994: Chapter 4).

Symmetry is also present in the overall layouts which, Davies argues, have evolved from a basic striped pattern (and he quotes Barber (1991) to date this pattern back to the fourth millennium BCE). The first four drawings in Figure 4.12 show the possible evolution from a basic striped design (far left) to one with different stripe widths (second) to one with adorned stripes (third) to one where the stripes are no longer present but their order-providing role is (fourth). The remaining three drawings show other possible layouts.

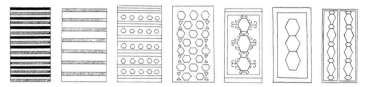

Figure 4.12 **Common layouts in Anatolian kilims. (From (Davies, 2002). Copyright © 2000, 1993 by Peter Davies. Used by permission of W. W. Norton & Company, Inc.)**

The symmetry in Anatolian kilims (and in many antique carpets) is far from perfect and in many cases we should talk about an intention of symmetry, as we did in the preceding section. Yet, Davies (2000: 42) points to a difference which is worth mentioning now and to whose substance we will return in Chapter 6.

A closer examination of a seemingly symmetrical kilim or motif or design usually reveals that within the apparent symmetries there are a host of assymetries created by random insertions, idiosyncratic variations, and color changes. While these variations seem to work against the established symmetries, they actually create an interesting tension between the culturally established lexicon and grammar and what might be described as an improvised subtext created by the weavers themselves.

To which he promptly adds

It is, perhaps, the point at which the individual subtext begins to overwhelm and obscure the symmetries of the main text that we may say that decadence in kilim design begins [. . .] The disintegration of tribal culture [after World War I] has seemingly reduced the potency of the main text allowing the subtexts of the individual weavers to overwhelm it.

The tension between established patterns and individual expression is interesting as long as it is felt as such. If either of the opposites overpowers the other then this will stop being the case.

It is perhaps time now to turn attention to particular rugs. The one in Figure 4.13 is known as the *Pazyryk rug*. Discovered in the Pazyryk valley, in the Altai mountains, during excavations made in the tombs of a Scythian prince in 1949, this is the oldest extant pile carpet, having been

Figure 4.13 **The Pazyryk rug. Hermitage Museum.**

dated around the fifth century BCE. It measures 183 cm by 200 cm and has around 36 knots per Square Centimetre. There is no agreement as to its place of manufacture. The field exhibits a wallpaper of type *pmm* with lotus figures (Figure 4.14).

There are two principal borders alternating with three narrower ones. Two of the narrower borders show a frieze of squares, each of them enclosing a griffin. The third (which is also the middle) displays a highly symmetric pattern with lotus figures like those in the field (its frieze group is *pmm2*). The principal borders show a line of horsemen and another of broad-antlered spotted fallow deer (Figure 4.15).

Figure 4.14 **The Pazyryk rug: detail of the field. Hermitage Museum.**

Figure 4.15 **The Pazyryk rug: detail of the border. Hermitage Museum.**

The rug pictured in Figure 4.16 shows an all-over pattern of concentric rhombic figures composed of alternating latch-hooks. Each of the six central figures exhibits double mirror symmetry, although small differences between these figures are noticeable. An enlargement (rotated) that allows a better appreciation of the detail of the basic figure in terms of both geometry and colour is shown in Figure 4.17. The field is framed by two borders showing patterns of type *pmm2* and *pma2*.

A similar composition (now with four medallions) with very different motifs is shown in the example in Figure 4.18.

The presence of wallpapers in the field is common to rugs from many geographical regions. This is, for instance, the case with the embroidered rug in Figure 4.19 displaying a wallpaper based on a cruciform motif. If the rug had only the disks containing these crosses, the wallpaper group would be *p4m*. But the small black hooked elements between disks are not invariant under 90° rotations, so the wallpaper group is actually of type *pmm*. The field is surrounded by three borders: a broader one with a *pm11* frieze and two narrower with friezes of type *pma2*.

Figure 4.16 **Lori rug, southwest Persia, late nineteenth century. (Reproduced with permission of the Nazmiyal Collection, New York. http://nazmiyalantiquerugs.com.)**

Figure 4.17 **Lori rug, southwest Persia, late nineteenth century: (detail). (Reproduced with permission of the Nazmiyal Collection, New York. http://nazmiyalantiquerugs.com.)**

Figure 4.18 **Kuba rug, East Caucasus, last quarter of nineteenth century. (Reproduced with permission of the Nazmiyal Collection, New York. http://nazmiyalantiquerugs.com.)**

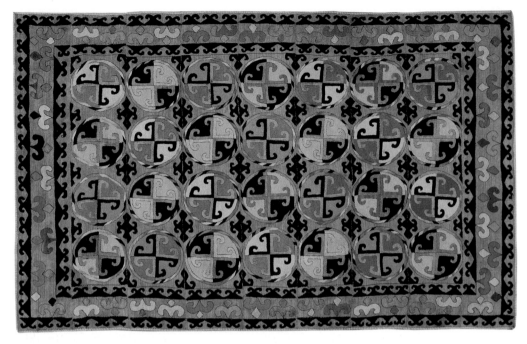

Figure 4.19 **Uzbek embroidered rug, Central Asia, early twentieth century. (Reproduced with permission of the Nazmiyal Collection, New York. http://nazmiyalantiquerugs.com.)**

The statements above rely on a premise which, to avoid confusion, it is of the essence to make explicit. When we say that a figure is invariant under a given isometry, we are disregarding colouring. Were we to require that the image of a point has the same colour of the point itself – otherwise a perfectly valid requirement – the carpet above would not exhibit a proper wallpaper in its field nor a proper frieze in its border (the allocation of the basic colours – cream, brown, black and steel blue – not being sufficiently systematic).[4] The reader might keep this premise in mind when looking at the remaining examples in this section, of which the next four (Figures 4.20–4.23) show different combinations and styles for the field(wallpaper)/border(frieze) structure. The field in the example in Figure 4.23 is so finely patterned that it is close to a texture.

The rug in Figure 4.24 shows a splendid example of the striped pattern we mentioned above.

---

[4] We have already proceeded in this way, without making a point of it, in our choice for wallpaper of type *p3* in Figure 4.13.

Figure 4.20 **Bukhara rug, twentieth century. (Reproduced with permission of the Nazmiyal Collection, New York. http://nazmiyalantiquerugs.com.)**

The last example in this section in Figure 4.25 is a silk embroidery from the eighteenth century. It shows a wallpaper of type *p1* with a naturalistic rendered flower as a motif. At the centre of the field, a medallion displaying a different type of flower with purely central symmetry (i.e. with symmetry group C_2) is imposed on the wallpaper.

Figure 4.21 **Bibkabad rug, nineteenth century. (Reproduced with permission of the Nazmiyal Collection, New York. http://nazmiyalantiquerugs.com.)**

Figure 4.22 **Bibkabad rug, nineteenth century. (Reproduced with permission of the Nazmiyal Collection, New York. http://nazmiyalantiquerugs.com.)**

Figure 4.23 **Tabriz rug, nineteenth century. (Reproduced with permission of the Nazmiyal Collection, New York. http://nazmiyalantiquerugs.com.)**

Figure 4.24 **Caucasian Kilim, nineteenth century. (Reproduced with permission of the Nazmiyal Collection, New York. http://nazmiyalantiquerugs.com.)**

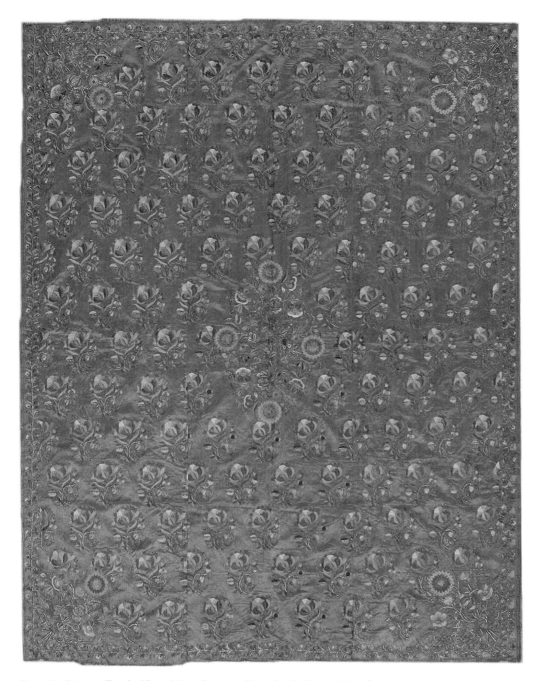

Figure 4.25 **Ottoman silk embroidery, eighteenth century. (Reproduced with permission of the Nazmiyal Collection, New York. http://nazmiyalantiquerugs.com.)**

4.3 Chinese lattices

A distinctive feature in classical Chinese architecture is the use of wood lattices. The lattice openings would be covered with small pieces of paper and the resulting panels used to cover window openings. With time, they would also find their way into various kinds of furniture. Because of the perishable nature of their constituent materials, lattices prior to the eighteenth century are rare. The book of Daniel S. Dye (1974) on Chinese lattices (on which this section relies) points, however, to evidence of their presence around 1000 BCE found in pictographs in sacrificial bronzes of the Zhou Dynasty. More reliable evidence is found in pictures on grave-bricks and in house models in clay placed in graves dating from the Han Dynasty. From the Tang Dynasty onwards, a reasonably detailed picture of the evolution of these lattices both in China and in areas culturally influenced by China (e.g. Korea or Japan) can be inferred. Such a picture is given by Dye (1974: 30–39), together with a large collection of plates illustrating both the wealth of lattice designs in China and the passion Dye developed for them during his more than 40 years in Sìchuān, the very place that Dye unhesitatingly identifies as the apex of this activity.

Chinese lattices invariably possess some form of symmetry. There is often bilateral, central or "double mirror" symmetry. Wallpaper designs are also frequently found and sometimes the lattice possesses a border which, needless to say, displays a frieze. Occasionally, as we have already seen in some carpets in Section 4.2, a "medallion" displaying some non-translational symmetry stands in the centre of a lattice, surrounded by a wallpaper pattern. The examples in the following figures are taken from Dye (1974). We display only a few frieze and wallpaper patterns, starting with the latter (Figures 4.26–4.32). For these, we note that we

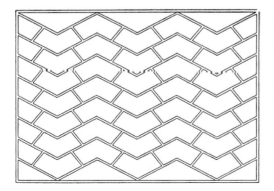
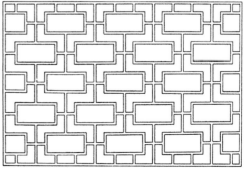

Figure 4.26 **Lattices from Chéngdū, Sìchuān, c. 1875 (*pmg*) and Kienwei, Sìchuān, c. 1750 (*pmm*). (From (Dye, 1974) with permission.)**

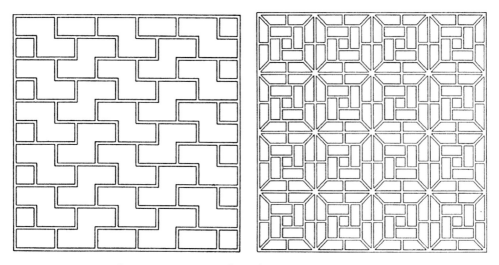

Figure 4.27 **Lattices from Shànghǎi, *c.* 1900 (*pgg*) and Chéngdū, Sìchuān, *c.* 1800 (*p4*).**
(From (Dye, 1974) with permission.)

Figure 4.28 **A lattice with unknown provenance (*p31m*) and one from Chéngdū, Sìchuān,**
***c.* 1800 (*p1*). (From (Dye, 1974) with permission.)**

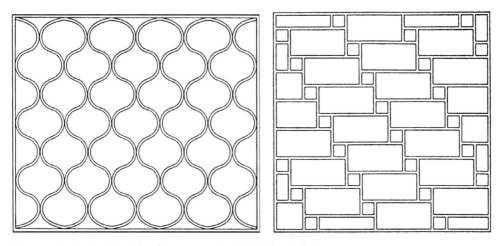

Figure 4.29 **Lattices from Penghsien, Sìchuān, *c.* 1900 (*cmm*) and Chéngdū, Sìchuān, *c.* 1850**
(*p2*). (From (Dye, 1974) with permission.)

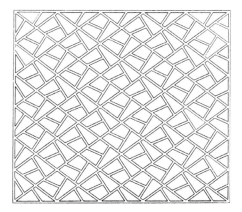
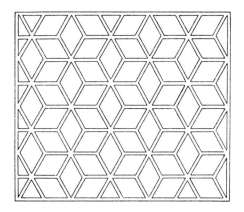

Figure 4.30 **Lattices from Dye, 1930, after a design from Sìchuān (*pg*) and Chéngdū, Sìchuān, *c.* 1875 (*p6m*). (From (Dye, 1974) with permission.)**

Figure 4.31 **Lattices from Mount Omei, Sìchuān, *c.* 1875 (*p4m*) and Shànghǎi, *c.* 1900 (*p6*). (From (Dye, 1974) with permission.)**

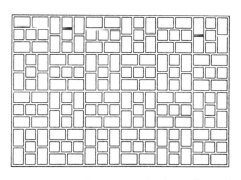
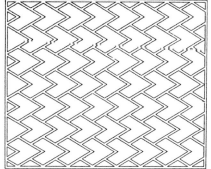

Figure 4.32 **Lattices from Chéngdū, Sìchuān, *c.* 1850 (*p4g*) and *c.* 1900 (*cm*). (From (Dye, 1974) with permission.)**

have not found in Dye's book patterns with groups *pm*, *p3* and *p3m1*.
I believe that lattices with these patterns could be found in China, but
the fact that Dye offers no example suggests that they are rare.

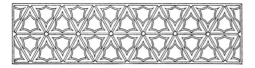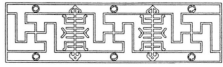

Figure 4.33 **Border lattices from near Shàoxīng, Zhèjiāng, c. 1700 (*pmm2*) and Chéngdū,
Sìchuān, c. 1875 (*p111*). (From (Dye, 1974) with permission.)**

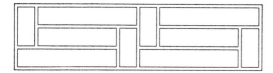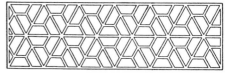

Figure 4.34 **Border lattices from Sìchuān, c. 1800 (*p112*) and Wǔhàn, Húběi, c. 1650 (*p1m1*).
(From (Dye, 1974) with permission.)**

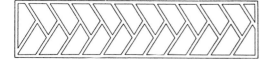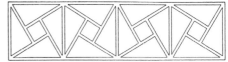

Figure 4.35 **Border lattices from Chéngdū, Sìchuān, c. 1875 (*p1a1*) and Sìchuān, c. 1800 (*pma2*).
(From (Dye, 1974) with permission.)**

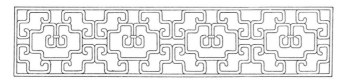

Figure 4.36 **Border lattice from Wǔhàn, Húběi, c. 1650 (*pm11*). (From (Dye, 1974)
with permission.)**

Figures 4.33–4.36 show border lattices. In this context, it is not difficult
to find examples for each of the seven frieze groups.

4.4 Escher

The work of the Dutch graphic artist Maurits Cornelis Escher (1898–
1972) is a long love affair with symmetry. Some of his drawings adhere
strictly to various forms of symmetry. In some others, symmetry is a

Figure 4.37 **M.C. Escher's *Snakes*, woodcut, 1969 © 2012 The M.C. Escher Company-Holland. All rights reserved. www.mcescher.com.**

principle from which he departs in a manner no less rigorous than his treatment to the former. In addition to these works, Escher also produced some pieces where the main role was played by other formal aspects of pictorial representation, notably by perspective. Our focus in this section will be on (a few of the) graphic works whose emphasis is on symmetry.

The way the three snakes are entangled in the woodcut *Snakes* (Figure 4.37) and the nature of the grid, itself made with entangled circles, on which these snakes are living, is a perfect example of rotational symmetry (of order 3).

The following four illustrations involve wallpapers. A recurrent goal in Escher is to design wallpapers with a figurative motif which, nevertheless, would fill the plane. The fundamental region in *Horsemen* (Figure 4.38) is a horseman with his horse. The motif[5] is obtained by combining two copies of such a horseman, which are glidal images of each other. A quick

---

[5] We are, just as with hexagonal unit cells, accepting as a motif a figure that is not a parallelogram but which nevertheless spans the whole wallpaper by combining copies of it obtained by translations from the symmetry group of the wallpaper.

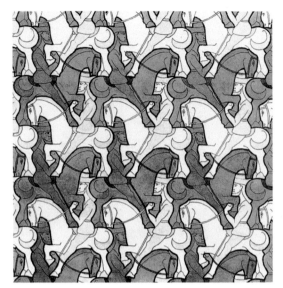

Figure 4.38 **M.C. Escher's** *Horsemen*, **woodcut, 1946 (left) and** *Tessellation 85*, **ink, pencil and watercolour (right)** Ⓒ **2012 The M.C. Escher Company-Holland. All rights reserved. www.mcescher.com.**

use of the flowchart in §3.6.3 shows that this wallpaper is of type *pg*. Similarly, the fundamental region of *Tessellation 85* (Figure 4.38) is composed of a lizard, a fish and a bat tightly fitted together. Another visit to our flowchart yields type *p3m1* for this wallpaper.

The wallpaper group of *Eight Heads* is of type *p1* (Figure 4.39). There are no symmetries besides the translational. It is remarkable, however, the way these heads accommodate to fill the plane. Speaking properly, *Sky and Water* (Figure 4.39) is not a wallpaper. Its would-be motif is fading in both up and down directions. But at the centre of the image one perceives this motif as formed by a fish and a bird. Up in the sky, fish dissolve into air to let birds gain freedom and definition. Low in the sea, it is the other way around with birds becoming dark water.

Another example of Escher's playfulness around ideas of symmetry is in the woodcut *Day and Night* (Figure 4.40). There is an overall bilateral symmetry here which is well defined on the sides (left and right) but vanishes in the middle of the scene. This bilateral symmetry, precise to the extreme in the two villages and the rivers they are respectively perched on, is emphasized by the inversion of black and white on these extremes that makes the left-hand side "day" and the right-hand side "night". Some attention to the middle part suggests a wallpaper pattern. The pattern is, however, diluting. Cells are light or dark and we see the former becoming birds towards the right and background towards the left and vice versa.

Figure 4.39 **M.C. Escher's** *Eight Heads*, **woodcut, 1922 (left) and** *Sky and Water*, **woodcut, 1938 (right)** © **2012 The M.C. Escher Company-Holland. All rights reserved. www.mcescher.com.**

Figure 4.40 **M.C. Escher's** *Day and Night*, **woodcut, 1938** © **2012 The M.C. Escher Company-Holland. All rights reserved. www.mcescher.com.**

In addition, both light and dark cells become land plots when moving towards the bottom. In the upper part of the middle section, where this wallpaper is perhaps clearest, the unit cells are birds and the wallpaper group is of type *pg*. D. Hofstadter (2002) describes his first viewing of this work in a most expressive way.

I saw white birds flying one way, black birds flying the other way, the two flocks meshing perfectly together to fill up space. As my gaze drifted downwards, I saw the bird-shapes distorting and turning into a diamond-like grid of black and white fields. To the left of the fields, I saw a peaceful village by a river, basking in bright sunlight, while to their right, I saw a mirror-image village by a mirror-image river, calmed by soft starlight.

I [. . .] was charmed by the idea of walking back and forth on the little roads linking these two villages, thus easily sliding, in a mere five minutes, between noon and midnight. [. . .] I wondered, "How could two flocks of birds fly right *through* each other, without even the tiniest space? For that matter, how could they even breathe, with no air between them? And how could three-dimensional birds, roughly half a meter in length, turn into two-dimensional fields, roughly 100 meters on a side?" None of this symmetric picture made any sense, but at some other level, it made *perfect* sense.

5 Reflections on the mirror

> Beauty! Where shall one begin on a subject so vast, one that has stirred so many celebrated writers? What a theme! An endless one; and let it be said in passing, just such questions as these are the most interesting, because with them, the last word is never said and because everyone can have a different opinion.
>
> E. Delacroix (1938: 705)

Why so many examples of symmetry in man-made objects? Does symmetry add beauty? Certainly, the implied consequence of an affirmative answer to the latter, namely the fact that the more symmetries an object possesses the more beautiful the object is, cannot be true. For the plane itself is the most symmetrical of all objects and yet it is wanting life. But then how does symmetry relate to beauty?

These questions do not have a definite answer. Yet, such a lack of definition appears not to have hindered attempts at formulating answers, and it could be argued that it has, in fact, provided a reason for multiple efforts to do so. For beauty itself has been the subject of many philosophical reflections and its perception studied as a branch of psychology. It would perhaps be as foolish to endeavour a description of all this pondering as it would be blind to ignore it. A possible compromise consists of selecting those reflections whose nature best suits the purpose of this book and whose contents will, therefore, throw light on our subsequent exposition. In doing so we will be oblivious to most of the discussion around beauty, including arguments which are close to our theme.

5.1 Aesthetic order

In his introduction to *The Blackwell Guide to Aesthetics*, Peter Kivy describes the regard of aesthetics in the eyes of the leading philosophers at the middle of the twentieth century as a relegation "to the category of emotive grunts and groans" (Kivy, 2004: 2). Such a state of affairs is in contrast with the situation of aesthetics and the philosophy of art nowadays, which, according to Kivy, "flourish as never before".

A central ingredient in this revival is the pre-eminence gained in recent decades by the so-called "analytical philosophy" in English-speaking academia together with the prominence in this brand of thought for focus on the details as opposed to all-encompassing philosophical systems.[1] Such a prominence allows one to ponder on the aesthetic phenomenon per se, without servitudes to philosophical dwellings on other issues.

A recent example of a study on aesthetics within this stream is the book *Aesthetic Order* by Ruth Lorand (2000), in which she states the existence of two orders: discursive and aesthetic. The former is what traditionally is understood as order, an arrangement of the elements of a set which follows a family of rules external to the set, pre-existent and independent of it. Aesthetic order, in contrast, does not suppose compliance with any a priori laws. Rather, the arrangement of the elements making such an order is unique to the instance at hand. To explain the nature of this arrangement Lorand makes use of the notion of *interpretation*, a notion that, she claims, is presented in the literature in a way which is "either too wide or to narrow". Too wide, because it includes activities which are not proper interpretations. Too narrow, because there is a tendency to limit interpretations to verbal activity. In addition, interpretation is commonly associated with a search for some kind of truth. In contrast, Lorand writes

The gist of my suggestion is that interpretation should be viewed not via the concept of truth but rather via the concept of problem solving. A proposed solution to a problem is neither true nor false; it is rather appreciated as a better or worse solution than its alternative.

To make this idea clearer Lorand takes a negative approach and enumerates some other cognitive activities – with their associated problem-solving features – against which that of interpretation could distinguish itself. Those are description, explanation, clarification and decoding.

A *description* is "a detailed verbal account of the object's qualities, situation and context". It is meant to be independent of the particular beholder's views and supposes these "qualities, situation and context" as prior to the description. The account's adequacy to this prior makes the description liable to judgements of truth.

An *explanation* is "a verbal manifestation of discursive order. It regards the explanandum as a particular case of a general principle (a law, a rule or a theory)". Unlike the purely enumerative character of the description, there is a logic linking the different parts of an explanation which,

---

[1] "There is, I think, a widespread presumption within the tradition that it is often possible to make philosophical progress by intensively investigating a small, circumscribed range of philosophical issues while holding broader, systematic questions in abeyance. What distinguishes twentieth-century analytical philosophy from at least some philosophy in other traditions, or at other times, is not a categorical rejection of philosophical systems, but rather the acceptance of a wealth of smaller, more thorough and more rigorous, investigations that need not be tied to any overarching philosophical view" (Soames, 2003: xv).

usually, fits the thing to be explained as a particular case of an existing theory or framework. In this sense, an explanation is good in the measure that correctly fits this framework and true in the measure that the framework is so.

In contrast with the previous two activities, a *clarification* does not intend to transmit new information. It rather transmits some information already known by means different from those used in the original transmission. The new means do not need to be verbal; gestures, drawings, or other non-verbal ways of expression may be used. Lorand mentions the distinction made by Wittgenstein between clarifications and propositions: "A proposition is informative and subject to truth conditions; a clarification is not. It either achieves its goal by evoking understanding or it fails".

The object of a *decoding*, like that of an explanation, can be regarded as a set of signals, each with an individual meaning but with the set itself lacking a global meaning. The difference is that the solution proposed by an explanation is subject to a theory and different theories will yield different explanations, whereas the solution of a decoding is unique. The wealth (or lack of it) of different countries will be explained in different ways as a function of the economic theories that sustain the explanations. But there is only one possible decoding of, say, a ciphered message or the solution of a murder investigation. In a decoding, the basic assumption is the existence of a hidden meaning and the decoding succeeds when this meaning is revealed. Truth, understood as the coincidence between this hidden meaning and the proposed decoding, is of the essence.

A first characteristic of *interpretations* pointed out by Lorand is the kind of problem they solve.

Interpretation seeks to solve a specific kind of problem, not just any problem. Only the *apparent incompleteness* calls for interpretation; other problems call for different activities such as clarifications or explanations [...] We interpret when we believe in the genuine coherence of the object and wish to demonstrate this coherence by struggling with difficulties on the object's "surface".

Consequently, not all objects are objects of interpretation. To be one an object needs to appear incomplete.

An interpretation shares some of the characteristics of a decoding and some of an explanation. Like an explanation (and unlike a decoding) there may be several interpretations of an object. Unlike an explanation (but possibly like a decoding) an interpretation does not attempt to solve its problem within a theory or framework. Interpretations may be tailor-made for the object at hand. These distinctions point us to the logical status of interpretations. A decoding is either true or false (according to whether it reveals the hidden meaning). An explanation may be true or false if the theory underlying it is so (and the explanation is correct within this theory). But an interpretation can be considered neither true nor false. It can only be considered as possible and the facts that support this possibility can justify the proposed solution but not validate it.

Interpretations present possibilities that are, by definition, neither true nor false, but differ in their values as better or worse solutions to their problems.

Interpretation as defined above is, however, too encompassing. Aesthetic order consists of one particular kind of interpretation, which Lorand distinguishes from other kinds and refers to under the name of complementary interpretation. Again, she best illustrates what complementary interpretation is by first describing what it is not. The argument goes by classifying interpretations according to "How does the interpretation in question relate to the pre-interpreted object?" Without going into detail, Lorand lists interpretations of several types. First, *signs*, in which X (the pre-interpreted object) is best completed when viewed as Y (the proposed interpretation). We may say that X is best viewed as a sign of Y. Second, *substitutions*, in which Y is a replacement for X. A paradigm here is a translation of a poem: a literal (mechanical) translation will be incomplete. Therefore, the translator attempts to complete this attempt by searching for expressions that best replace the original ones. Third, the *metonymy*,[2] in which the general is interpreted by the particular or vice versa. Fourth, the *analysis*, in which X "is best demonstrated when disassembled and reconstructed in a proposed way – Y". Lastly, *complementary interpretation*:

Complementary interpretation is concerned with the relation between the raw material and the final product. It opposes interpretative analysis in the sense that it is the construction of the whole out of segregated materials, rather than the disassembly of the whole. The pre-interpreted object serves as the raw material; the proposed interpretation is the final product. [...] Personal events, feelings and observations often serve as raw materials for a novel, and body gestures in various situations supply materials for creating a new dance.

This form of interpretation entangles materials with products. "The product interprets the material by demonstrating what can be done with them [...]. In this sense, the product is not only an end but also a means for a new and better understanding of the materials."

Aesthetic order is, in Lorand's elaboration, complementary interpretation, and beauty is "a high degree of aesthetic order". Once this is set, Lorand proceeds in her analysis to derive a number of opposites of beauty either as a consequence of a low aesthetic order (boredom, insignificance), as a form of disorder (ugliness, meaninglessness), or because of other possible reasons (irrelevance, kitsch). We will not attempt to summarize here how this is done.

---

[2] The naming is mine.

A point made clear by Lorand is that discursive and aesthetic orders coexist in every object and are mutually dependent.[3] The issue of symmetry in art is, I believe, at the heart of this coexistence.

Lorand describes symmetry as a state of disorder, a statement, she acknowledges, that "seems to run counter to common sense." She even includes "concepts such as balance [and] equilibrium" as expressing disorder. The basic justification for this – which we will not develop here – leans on the homogeneity of the different copies of the repeated motif (or fundamental region) in the symmetric object and its predictability once the pattern is grasped. I disagree with this reasoning. For, on the one hand, this homogeneity is common to repetitions which do not possess global symmetry (e.g. a number of identical objects placed at random) and are actually better examples of disorder, and, on the other hand, and as we shall see in Section 6.1, symmetry does not need to (and maybe cannot) be perfect. The dismissal of symmetry does not end there, since Lorand writes that "even if one does not accept that symmetry is, in some sense, a form of disorder [...], there are strong reasons why one must accept that a rigid symmetrical set exhibits poor aesthetic order, in spite of a common inclination to praise symmetry". I lean again towards disagreement, and only an extreme acception of the word "rigid" in the sentence above prevents me from fully doing so.

I view symmetry, using Lorand's terms, in a way that runs along with common sense. That is, as a form of discursive order. But the distinction between discursive and aesthetic orders not only allows for their coexistence, as Lorand points out, but does not prevent one of them from becoming an object for the other. To see why, consider first the distinction between description and interpretation. This distinction does not impede one of these activities from becoming the material of the other. I can describe an interpretation (of a performance I saw yesterday). Conversely, a description can be the basis for an interpretation. A person going blind can describe their ailment to a physician in very simple (and accurate) terms (I am going blind, I am losing my eyesight). But this description may be felt as incomplete in some ways (e.g. how this condition affects his life) and may be the object of an interpretation as in the poems by Milton or Borges in which they describe their own blindness (see Section A.2). In the same manner, discursive and aesthetic orders can serve as each other's objects. Aesthetic order becomes the object of discursive order in Lorand's book. Conversely, elements of discursive order can be given an aesthetic character. For instance, logic itself – arguably

---

[3] One of the two, however, dominates the perception of the object allowing for its categorization. This dominant character may change over time and an object once considered as discursive may with time be regarded as aesthetic (e.g. a text once seen as an explanation of the world's laws may now be read for purely literary reasons).

the pillar of discursive order – becomes a raw material for complementary interpretation in mystery novels. I hold that such is the case for the occurrence of symmetries in art.

5.2 The aesthetic measure of Birkhoff

In 1933, George David Birkhoff published *Aesthetic Measure* (Birkhoff, 1933). One of the leading American mathematicians during the first half of the twentieth century, his contribution to the discussion on aesthetics is doubtless singular.

As its name indicates, the purpose of Birkhoff (1933) is to describe a quantitative measure for aesthetic quality. From the outset, Birkhoff indicates that the existence of such a measure is not out of the question due to "the fact that the objects belonging to a definite class admit of direct intuitive comparison with respect to aesthetic value. The artist and the connoisseur excel in their power to make discriminations of this kind". The implicit reasoning is clear. If artists and connoisseurs can compare and rank (agreeing in most cases) the merits of art pieces belonging to a definite class, the reason is that there is a subjacent measure that makes these comparisons possible.

Nonetheless, Birkhoff limits the description of his aesthetic measure to a specific framework. He notes that what he calls the associations entering the aesthetic experience divide into two kinds. On the one hand, *formal* associations which can be unambiguously measured or verified. The metric of a given line in a poem or the symmetry about a given axis on a figure are instances of this kind of association. On the other hand, *connotative* associations whose nature defies (or, at any rate, discourages) measurement. In the examples just mentioned, the meaning of the line in the poem or the contents of the figure are obvious cases of connotative associations. Given a class of comparable objects, Birkhoff's measure will be defined in terms of the formal associations only; connotative associations are excluded in this measure.

The property of a formal association which can be measured (or whose existence can be verified) is called an *element of order* of the object. These elements of order are not unspeakable entities. The way Birkhoff gives substance to them is of sufficient importance to us to be quoted in full.

The actual types of formal elements of order which will be met with are mainly such obvious positive ones as repetition, similarity, contrast, equality, symmetry, balance, and sequence, each of which takes many forms [...] Furthermore there is a somewhat less obvious positive element of order, due to suitable centers of interest or repose, which plays a rôle. For example, a painting should have one predominant center of interest on which the eye can rest; similarly in Western music it is desirable to commence in the central tonic chord and to return to this center at the end.

On the other hand, ambiguity, undue repetition, and unnecessary imperfection are formal elements of order which are of strongly negative type. A rectangle nearly but not quite a square is unpleasantly ambiguous; a poem overburdened by alliteration and assonance fatigues by undue repetition; a musical performance in which a single wrong note is heard is marred by the unnecessary imperfection.

Once the (formal) elements of order in a class are identified, a measure O of order for the members of this class naturally follows if we assume that the amount of each of these elements a member of the class possesses can be quantitatively estimated. If this is the case, there one simply takes O to be the sum of these amounts over all the elements of order in the class. To illustrate this measure of order, I summarize (briefly; detailed definitions are in (Birkhoff, 1933: §21–25)) the simplest case dealt with by Birkhoff, the class of plane polygonal figures. For this class Birkhoff identifies five elements of order:

V The existence of bilateral symmetry with a vertical axis. The amount of such an element can only be 0 (if there is not such a symmetry) or 1 (if there is).

E The existence of equilibrium. Birkhoff distinguishes between "complete optical equilibrium" and equilibrium "in the ordinary mechanical sense" (as in the polygon ⮎). He assigns value 1 to the polygons having the former, value 0 to those having only the latter and value -1 to those having none.

R The existence of rotational symmetries satisfying certain additional conditions. The value obtained in this case is in the set $\left\{0, 1, \frac{3}{2}, 2, \frac{5}{2}, 3\right\}$.

HV Certain relations of the polygon to a wallpaper created using it as unit. This element of order can take values in $\{0, 1, 2\}$.

F The existence of "unsatisfactory" features (e.g. having two vertices too close to each other). A list consisting of a few of them is given and polygons are assigned value 0, -1, or -2 according to whether the polygon exhibits none, one, or more than one of them.

Then, for a given polygon, its measure of order is

$$O = V + E + R + \text{HV} + F.$$

Note that this definition implies that $-3 \leq O \leq 7$. The intended meaning of O is that of a pleasing perception when $O > 0$ (the larger O the more pleasing), of neutrality when $O = 0$ and of a certain displeasure when $O < 0$. It constitutes half of Birkhoff's aesthetic measure.

The other half is a measure of complexity C purported to quantify the "effort of attention [...] necessary for the act of perception". In most cases, this is a simple count of certain constitutive elements of the object. In our example above C is defined to be the number of different straight lines containing at least one side of the polygon. For the class of musical melodies it is the number of notes. In all cases studied by Birkhoff one has $C > 0$. The larger C is, the greater is the required effort of attention.

For a class where O and C have been made explicit, Birkhoff defines its *aesthetic measure* to be

$$M = \frac{O}{C}.$$

The character of this quotient is immediate. The aesthetic measure of an object is a balance (given mathematical form as a ratio) between the feeling of pleasure provided by the object's perception and the effort required by this perception.

A justification of this aesthetic measure will have to consider two different issues: on the one hand, a general argument as to why the quotient O/C captures a measure of beauty; on the other hand, the precise setting of O and C needs to be accounted for each time a class of objects is selected.

The latter task is not easy, since both the elements of order associated with a class and their numerical values are dependent on the historical context. And likewise for the elements of complexity. Once in the possession of a set of elements of order (and of complexity), though, the values associated with the different degrees of compliance with them may be determined (as is usual for most mathematical models) by calibration. That is, a family of objects in the class is ranked by "artists and connoisseurs" and, subsequently, a collection of values for these degrees of compliance is calculated to produce a measure M which best fits the expert ranking within this family. For instance, for the class of polygons described above one could perfectly well give the value 8, instead of 1, to the existence of bilateral symmetry with vertical axis. Similarly, one could take $\{0, \frac{1}{5}, \frac{2}{5}, \frac{3}{5}, \frac{4}{5}, 1\}$ as the possible values for R instead of $\{0, 1, \frac{3}{2}, 2, \frac{5}{2}, 3\}$. The first change would increase the influence, or importance, of vertical bilateral symmetry in the aesthetic measure. The second would decrease that of rotational symmetries. Birkhoff calibrated his measure for a number of classes. Probably the more illustrating (and the least controversial) of them is the class of vases (Birkhoff, 1933: Chapter IV).

To make good the ratio O/C as a measure of beauty Birkhoff has recourse to the physiology of perception. For the effort associated with perception Birkhoff notes that:

> [...] in order that the act of perception be successfully performed, there is also required the appropriate field of attention in consciousness. The attentive attitude [...] ensures that the motor adjustments requisite to the act of perception are effected when required. These adjustments are usually made without the intervention of motor ideas [...] and in this sense are "automatic".

The measure of order is independent of these adjustments:

This feeling of effort is correlated with the efferent part of the nerve current which gives rise to the required automatic motor adjustments, and has no direct reference to aesthetic feeling.

For the cause (physiologically speaking) of aesthetic feeling, we must look to that complementary part of the nerve current which, impinging on the auditory and visual centers, gives rise to sensations derived from the object, and, spreading from thence, calls various associated ideas and their attendant feelings into play.

The aesthetic measure M thus results in a "ratio of return" quantifying the yield in aesthetic pleasure of an "investment" in effort.

Note that in this quantitative measure of beauty the numbers themselves are irrelevant. The only purpose of the exercise is to compare objects of the same kind (that is, to be able to rank them). Precise numbers depend on the choice of unit. Thus, the situation is akin to comparing distances between cities in a map. The scale of the map is irrelevant to check that, say, Rennes is closer to Paris than Toulouse is. Not even the ratios between numbers are important. Any increasing transformation $f : \mathbb{R} \to \mathbb{R}$ satisfying $f(0) = 0$ would put forth the same ranking and the same basic splitting between pleasant and unpleasant objects.

There is an aspect in Birkhoff's formulation that appears to me to be ill-conceived. When $O < 0$, we have that the larger C is the smaller is M. For instance, the classification of polygons shown in Birkhoff (1933: Chapter II plates II to VII) exhibits 90 polygons ranked by aesthetic measure. The last three of them have negative measure: -0.1, -0.11 and -0.17. It appears, however, that their complexities are 10, 9 and 6 respectively, and that these complexities are the cause of their relative positions in the ranking. Their measures are obtained by dividing a shared measure of order of -1 by the complexities above. In cases such as this we face a situation that seems to run counter to Birkhoff's view of M as a ratio of return on an investement in effort; namely, that when the object is displeasing, the more effort its perception requires the less displeasing the result is (since the bigger its aesthetic measure is).

Birkhoff's ideas, needless to say, have been criticized in many ways. Lorand (2000: 67–68) offers a summary of these critics in her discussion of Birkhoff. She uses the expression "the formula for beauty" and states that such formulae may not be found since "the idea that beauty can be decoded clashes with the intuition that beauty is unexpected [. . .]". While I completely agree with Lorand's statement, I believe that, as a judgement of Birkhoff's ideas, it is not fair. Birkhoff never claimed to have found a formula for beauty or a way of decoding its many manifestations. He simply proposed a way of ranking objects which, on the one hand, required these objects to be in very restrictive classes (so that no symphony could be compared with a sculpture) and, on the other hand, restricted this ranking to the formal aspects of the object. This is by and large a modest exercise, and much less ambitious than Lorand's implicit goals in the sentences above. Furthermore, there is no claim to universality in Birkhoff's writings, neither in time nor in space. Both the elements of order and the complexity measure for a given class may change: chords once considered dissonant may become consonant in time in the same manner that intervals found pleasant in India may not be so in Europe. Birkhoff himself fully acknowledges this limitation. For instance, while he proposes a measure for diatonic chords, he notes that "since the appreciation of relationships of order among musical notes is continually expanding and developing, it would be highly absurd to try to formulate a definite theory of aesthetic measure, valid for the music of the future as well as of the past" (Birkhoff, 1933: 88). Similarly, after having developed the aesthetic measure for the musical quality in poetry, he warns "I would like to emphasize once more that musical quality is only one of the essential elements in poetry, and that even this quality cannot be measured in its more delicate *nuances* by any mechanical method, such as that given above" (Birkhoff, 1933: 189).

The calibration of Birkhoff's measure – and the evaluation of its ranking against accepted judgements of taste – depends on "the artist and the connoisseur", and these live in a given place at a given time. This is patent in the analysis of the diatonic chords[4] and the pairs of consecutive such chords performed by Birkhoff. He first formulates the aesthetic measure for individual chords and notes that this is done in a way which is consistent with the taste of his time as displayed in the book of Ebenezer Prout (1903), a book that was then in its 31st edition and was "thoroughly conventional in its treatment, and representative of a crystallized classical point of view". Then, he defines the aesthetic measure of a sequence of two chords (with the measure of the individual chords at hand as a constituent part) and compares his ranking of the 144 essentially different chords with a classification for them in Prout (1890) to find that the ranking derived from the aesthetic measure matches (and actually refines) the classification of such pairs (as "good", "possible" and "bad") proposed by Prout.

5.3 Gombrich and the sense of order

Although he was Austrian-born, the work of Ernst Gombrich was almost entirely developed in England. This work studies different aspects of art and includes a highly popular introduction to the visual arts (Gombrich,

---

[4] A brief explanation of some musical terms, such as "diatonic" and "chord", can be found in Section 8.1.

1989) which has enjoyed a long string of editions. It also includes a book, *The Sense of Order* (Gombrich, 1984), whose core on the perceptual aspects of visual arts fully matches the focus of this chapter.

A starting point for Gombrich, for which he refers to Hogarth (1753), is the ascertaining that vision varies between the two extremes of global perception and localized detail. There is a region, the *fovea centralis*, where the eye focuses. This region is small (less than 1°) and around it vision loses definition or clarity. Detailed pictures of objects can be obtained by displacing the fovea around the object and reconstructing the object from the resulting "mosaic of small snapshots".

Departing from Hogarth, however, Gombrich distinguishes between "focus" and "attention" by noting that while we *look at* the fovea we nevertheless *see* more than it. This *peripheral vision* may not be at the service of our conscious attention (except for detectives, Gombrich remarks), but we seem to attend to it notwithstanding. "We may see a newspaper lying somewhere and become aware of the fact that a given word such as our own name was printed somewhere on the page though we may have subsequently to search for it with our conscious attention." The loss of definition characteristic to peripheral vision should not be put on the same level with that produced by other causes, such as distance. The fact that the former can be confirmed by a momentary shift of vision and the latter cannot is of the essence. Furthermore, this confirmation does not necessarily require a detailed inspection and, Gombrich claims, generally does not receive one. The reason, he argues, is the large amount of redundancy surrounding us. Gombrich uses the term *redundancy* with the meaning given to it in information theory; that is, as additional information protecting the meaning against noise. The simplest form of redundancy is, quite appropriately in the context of our discussion, repetition. A more elaborate form is context. In the case of perception, Gombrich continues, it is the matching of our surroundings with our expectations which can be considered as redundancy. This matching allows for "[...] tremendous economies in reducing the amount of information to which we have specially to attend. These economies increase progressively as we are confronted with a familiar environment, where the objects we encounter conform to expectations".

The question is posed, What is the nature of these expectations? Perception starts with a few observations and, most likely based on them, some assumptions. We operate from this couple (observations plus assumptions) until, if ever, they are refuted.[5] It is this set of assumptions that gives substance to our expectations. Furthermore, Gombrich claims,

[5] Here, Gombrich acknowledges influence from Popper's ideas, which point to the asymmetry in scientific discovery between the confirmation of a hypothesis, for which accumulating positive instances only increase evidence, and its refutation, for which a single negative instance suffices.

these assumptions have a particular structure in the sense that are biased towards continuity rather than change. He calls this bias the *assumption of continuity* and points to the necessity of this assumption for our survival since "our senses could not cope with the task of mapping the environment afresh every moment".

Thus, we expect continuity and take it for granted. Refutation occurs when this continuity stops.

Our whole sensory apparatus is basically tuned to the monitoring of unexpected change. Continuity fails to register after a time, and this is true both on the physiological and the psychological level. [...] Any continuous body feeling, sight or sound, will sink below the threshold of attention. The rush of wind and water, the rustling of leaves, the ticking of the clock and even the roar of traffic outside the window become mere background and will be ignored unless they interfere with other sounds. The best proof we have, however, that we still hear what we no longer notice comes from the well-known observation that we realize it when the sound changes or stops. We must have been "monitoring" it unconsciously all the time.

At this point it is important to notice that the word "continuity" can apply at different levels. The occurrence of a tree at the roadside represents a discontinuity in the landscape as I drive. But when other trees follow to form a sequence of them, separated by a roughly constant distance, this (local) discontinuity becomes (globally) continuous and I come to expect it. After a while I will no longer register these trees and it is not until the regularity of the sequence is abruptly changed that my attention is called back to it.

Gombrich calls these changes in continuity *breaks* or *accents* and we can summarize his thesis in the statement that our sensory system is attuned to the detection of these accents. In addition, the continuity-probing at the centre of this detection proceeds "from the general to the particular, from the large to the small". It is graded to register changes on many different levels.

How does this defining feature of our sensory system relate to geometry? What is the final effect produced by the geometry of the perceived object on our break-spotter? These questions are probably too broad and Gombrich points out that "what we wish to investigate are not the inherent qualities of form [...] but rather the potentiality of design to arouse certain responses".

A starting point for this investigation is the contrast between restlessness and repose. Our eyes focus only on a small area (the fovea). If many signals compete for attention then the focus needs to shift and this produces restlessness (this observation is akin to Birkhoff's justification of his measure of complexity). But if these signals are ordered within a pattern there is repose as there are "more continuities and fewer breaks". Because of our break-spotting mechanism "our eyes are attracted to points of maximal information content. [...] too many such magnets competing with each other in an unexpected way would result in uncoordinated movement". But too few result in dullness and boredom. A balance is needed; a trade-off must be struck:

> The realm of decoration extends between these extremes of confusion and monotony: the one will send the "break-spotter" spinning, the other will perhaps cause it to switch off.

The first geometric accent Gombrich then considers is bilateral symmetry. This property has long been associated with a feeling of balance.[6] In addition, it is easily detected. Gombrich gives the example of the following sequences of characters

$$((((O(((O((O(O((((O(((O((O(O$$

and

$$((((O(((O((O(O(OO)O))O)))O))))$$

observing that we see the bilateral symmetry of the second with less effort than the translational symmetry[7] of the first. The explanation is that the axis of reflection acts as a break attracting the eye. A small neighbourhood around the axis is then enough to guess the bilateral symmetry and peripheral vision will only have to confirm it. A side effect of this mechanism, highly relevant in our future discussions, is that

> the role of peripheral vision in the global impression of bilateral redundancy also helps to explain why we rarely notice minor deviations from symmetry [...].

We may add a quantitative aspect to this explanation. Even taking as given that symmetry is assumed, the confirmation of this assumption involves different mechanisms depending on the nature of the symmetry. In bilateral symmetry, some immediate confirmation is already in the fovea as the axis is thin and, therefore, the fovea contains a repetition of part of the unit. If peripheral vision confirms that bilateral symmetry holds in a larger neighbourhood then we are most likely to accept it. In contrast, confirmation of translational symmetry will require (besides the guessing of the magnitude of the translation vector) either a memory stack to register a copy of the motif or a back-and-forth between the original motif and its translated copy, or, most likely in my opinion, a

---

[6] There is a short discussion about this association in Weyl's book on symmetry (Weyl, 1952: 16).

[7] Strictly speaking, we should talk about a translational correspondence where a single copy of the unit is considered. Translational symmetry involves the repetition of infinitely many such copies.

combination of both processes. Whatever the process, peripheral vision seems to be insufficient for even a first verification of translational symmetry, a task for which the bigger the motif the more effort there is required to be carried out.

Gombrich also mentions rotational symmetry to point out that, in the absence of bilateral symmetries, some motifs with rotational symmetry (like the whirl in Figure 3.4) produce a feeling of movement:

I would suggest that this dynamic derives from the way we test our primary impression for redundancies [...] In the vortex motif we have at least to carry out an imaginary rotation to see that [its arms] are invariant [...] Rotational symmetry represents an order which is visually less easy to grasp. I hope it is a pardonable exaggeration to say that it is not the motif which is unbalanced but that it upsets the balance of our mind.

All the preceding discussion deals with what Gombrich calls the *sense of order*, which he opposes to the *sense of meaning*. We have already found a version of this distinction in the writings of Lorand opposing discursive order to aesthetic order and a different version in those of Birkhoff referring to the formal and connotative associations of the aesthetic experience. Gombrich defines these "senses" in evolutionary terms by saying that the former "enables us to locate a stimulus" while the latter "enables us to respond to it in the interest of our survival". Leaving aside the way this distinction is made precise, a difference between Birkhoff on the one hand and Gombrich and Lorand on the other is that the analysis of the first, being of a quantitative nature, focuses on formal associations and excludes the connotative ones. In contrast, Gombrich attempts to give some account of "the complex forces that the interaction of order and meaning bring into play" as this interaction "constitutes the warp and woof of the decorative arts" (and, similarly, Lorand writes of the mutual dependency of discursive and aesthetic orders).

To do so, Gombrich focuses on (symmetrical) repetition as a paradigm of order and claims that repetition removes meaning (from the motif) because it annihilates (its) individuality. But the removed meaning is replaced by a new one arising from repetition. "It is precisely by draining the individual elements of their identity that the overall order makes them fuse into a larger unit which tends to be perceived as an object in its own right." This is why Rorschach ink-blots are symmetrical. They are still random, but bilateral symmetry "invites us to search for a meaningful description".

This intrusion of order in the realm of meaning is fought back. This is why "nonfigurative artists fight the tendency of looking for representational elements in their shapes and colours, for such projection can have the most disruptive effect on the intended dynamics of form. Meaning can subvert order, just as order can subvert meaning".

5.4 Between boredom and confusion

Not chaos-like together crushed and bruised,
But, as the world, harmoniously confused:
Where order in variety we see,
And where, though all things differ, all agree.

A. Pope (1966: Windsor Forest)

Look at the following drawing:

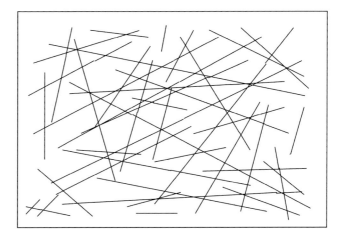

It contains a number of lines (or, to be precise, line segments) which, on first inspection, appear to be placed at random, to follow no pattern. In opposition to this disarray, consider instead this arrangement of lines:

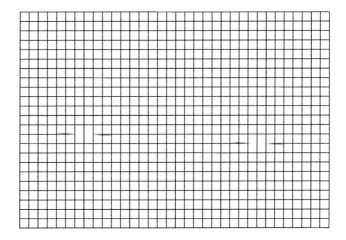

They are drawn to follow a law which is only too obvious: one begins with two lines crossing at a right angle and then adds parallel copies of each of these lines at regular (i.e. equally spaced) intervals. The lack of order[8] in the first drawing may produce confusion. But in just the same measure, the excess of order in the second may make it boring. The painting *Composition 10 in black and white, 1915* by Piet Mondrian in Figure 5.1 may be seen as searching a middle point in this tension between boredom and confusion.

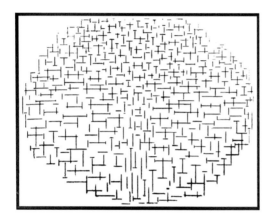

Figure 5.1 *Composition 10 in black and white, 1915* **by Piet Mondrian. (Collection Kröller-Müller Museum Otterlo, The Netherlands, ⓒ 2013 Mondrian/ Holtzman Trust c/o HCR International, USA.)**

It follows no law (as the grid does), but it is not completely lawless either (as the random pattern appears to be). For instance, all segments are either horizontal or vertical. Also, their lengths vary within a somehow limited range. Furthermore, Mondrian's painting has no symmetries (unlike the grid, which has many) but it contains several "local" symmetries: translational copies of small crosses like + occurring all around as a result of the lines being either horizontal or vertical. These copies, being all oriented in the same manner, are easily verified. In Gombrich's terms, once the break + is identified, peripheral vision confirms the occurrence of many copies of it, a matching of the painting with our expectations which allows for "tremendous economies in reducing the amount of information to which we have specially to attend."

The middle point mentioned above can be seen as a solution to a trade-off mentioned by Gombrich: "The realm of decoration extends between these extremes of confusion and monotony: the one will send the "break-spotter" spinning, the other will perhaps cause it to switch off".

It also relates to Birkhoff's aesthetic measure. Birkhoff suggests the presence of such a trade-off already in his description of elements of

---

[8] To be precise we should write "lack of deterministic order". Random events follow laws of their own which allow one to predict patterns in certain situations.

order by mentioning "such obvious positive ones as repetition" just to balance out this choice by remarking that "undue repetition [is a formal element] of strongly negative type". The way the aesthetic measure is defined, in addition, also appears to evidence this trade-off. To see why, recall the aesthetic measure for polygons. The presence of symmetries increases order, but in a bounded way. The maximum possible order is 7 and adding more symmetries to the polygon will not increase O. But too many symmetries will increase, now in an unbounded way, the complexity of the polygon (since the only way to get more symmetries is by having more sides). Hence, after a certain point the addition of symmetries increases complexity without increasing order and, consequently, reduces the aesthetic measure. This property (boundedness of O, unboundedness of C) extends to other classes analysed by Birkhoff. It must be noted, however, that these classes are somehow "simple enough" or, at least, their formal aspects are. It is this simplicity which allows for Birkhoff's analysis. To define order or complexity for objects such as the above painting by Mondrian would require careful thought. For the number of line segments in the painting would suggest a high complexity and yet, as we argued above, we apprehend the painting with "tremendous economies" in perceptual effort.

The woodcut *Day and Night* we encountered in Figure 4.40 is another example of a search of equilibrium between boredom and confusion. The way birds intermingle between them and dilute into a grid of paddy fields is prone to confusion. But the bilateral symmetry and the wallpaper pattern attenuate this tendency to suggest an order, and they do so without the boredom of undue repetition. Using Pope's words, the elements of *Day and Night* are "harmoniously confused".

The discussion above has focused in the purely formal side of the tension between boredom and confusion. It would be a mistake to infer that these formal aspects capture the whole of it. Even in contexts where formal associations play a distinguished role, the solution to this tension will partially lie in connotative associations. One cannot say that a particular group of wallpaper symmetries is boring or confusing: it is the coupling of a particular unit with this group which will result in one or the other. Formal features are raw materials for complementary interpretations, not interpretations on themselves. It is at the meeting point of formal and connotational associations that the following words from Mozart, taken from a letter to his father about three piano concertos he had just finished, can be best understood. Mozart writes (I am quoting from Gombrich (1984: 303))

They are exactly between too hard and too easy. Very brilliant, pleasing to the ear, naturally without lapsing into emptiness, here and there only connoisseurs will find satisfaction but in such a way that even non-connoisseurs must feel content without knowing why.

A raw material

6

That is, of you, who are a firmament
Of virtues, where no one is grown, nor spent;
They are your materials, not your ornament.

J. Donne (Carey, 2000: 232)

Birkhoff, Gombrich and Lorand purport different, and at times antithetical, views on art and beauty. The three coincide, however, in taking for granted a basic dichotomy in its analysis.

On the one hand, the *formal aspects* of an artwork (which roughly coincide with the formal associations in Birkhoff, the sense of order in Gombrich and the discursive order in Lorand) and, on the other hand, the *semantic aspects* (connotational associations, sense of meaning and aesthetic order respectively). Not that these are watertight compartments. On the contrary, the interplay between them is a defining feature of many artforms (poetry being a notable example).

Unlike Lorand, but in agreement with Birkhoff and Gombrich, I see symmetry – in the various forms we have seen in Chapter 3 and in other forms we will meet – as a formal aspect, and a notable one for that matter. Furthermore, I see discursive and aesthetic orders as possible raw materials for each other (see the end of Section 5.1). In particular, I see symmetry as a raw material in the interplay mentioned above. This view is the theme of the present chapter and a starting point to develop it is the realization that, as a raw material, symmetry does not need to be exact. Not only will a number of imperfections not be an obstacle for a sense of symmetry, they may even be creatively used to avoid boredom.

6.1 The veiled mirror

In the Anatolian village and nomadic rugs there appear to be considerably less concern with the exact accuracy of the mirror image than in most other rug-producing areas. The detail of an Anatolian symmetrical image was never mechanical, as I had expected, but idiomatically drawn. Even the classical Turkish carpet was not as particular with perfect border solutions as was its Persian counterpart. M. Feldman (1985: 124)

There is a gate in Japan, a gate in Neiko, which is sometimes called by the Japanese the most beautiful gate in Japan; it was built in a time when there was great influence from Chinese art. [. . .] But when one looks closely he sees that in the elaborate and complex design along one of the pillars, one of the small design elements is carved upside down; otherwise the thing is completely symmetrical. If one asks why this is, the story is that it was carved upside down so that the gods will not be jealous of the perfection of man.   R. Feynman (Feynman *et al.*, 1963–1965: vol. 1, 52-12)

Symmetry is rarely perfect. Objects we perceive as symmetric deviate from this perfection in two common ways. First, and independent of the accuracy in the different repetitions of the fundamental region, the number and placement of these repetitions may be faulty: a frieze never contains all the (infinitely many) copies of its motif; sometimes the two halves of a bilaterally symmetric figure are slightly tilted.[1] Second, the copies of the fundamental regions may differ in small details. The quotes by Feldman (a musician) and Feynman (an eminent physicist) give evidence on the indifference with respect to, and sometimes purposeful departure from, this faithfulness in the repetition of the fundamental region. Moving from particular instances to general remarks, Hermann Weyl (1952: 13) notes that "occidental art, like life itself, is inclined to mitigate, to loosen, to modify, even to break strict symmetry. But seldom is asymmetry merely the absence of symmetry. Even in asymmetric designs one feels symmetry as the norm from which one deviates under the influence of forces of non-formal character".

Weyl's words move us to the focus of this section which we can summarize in the following three points:

1. We rarely notice minor deviations from symmetry.
2. There are a number of elementary feelings associated with the various forms of symmetry which are perceived by the observer even in the presence of deviations, as long as these are noticed as such.
3. Artists made (and still make) use of the fact that these feelings are perceived by the observer even in the presence of noticeable deviations from strict symmetry to convey them without fully subjecting to the boredom of an excessive order.

To argue for points (1) and (2) we begin considering the case of bilateral symmetry, a case that has already been dealt with by Gombrich. Indeed, for point (1) we already quoted Gombrich stating that, for bilateral symmetry "we rarely notice minor deviations from symmetry" (Gombrich even explains that this is partially explained by "the role of peripheral vision in the global impression of bilateral redundancy"). Concerning point (2), we also remarked elsewhere that bilateral symmetry has long been associated with a feeling of balance. Other impressions commonly associated with bilateral symmetry are "proportion", "repose" and "equilibrium". The latter is doubtless related to physical equilibrium,

---

[1] There is a difference in these two deviations: the first is a necessity, the second is not.

a relationship which appears stronger when the symmetry axis is vertical, since, in this case, assuming the figure is made of some homogeneous material, bilateral symmetry entails physical equilibrium.

Although the mechanisms may differ, we rarely notice minor deviations from other forms of symmetry (rotational, translational, glidal) as well. And these other forms also have some elementary feelings associated with them. Thus, Gombrich also pointed out that, in the absence of bilateral symmetry, rotational symmetry produces a feeling of movement. That is, we perceive a tendency to spin on whirls but not on rosettes, and Gombrich suggests that this is because "in the vortex motif we have at least to carry out an imaginary rotation to see that [its arms] are invariant". In the special case of central symmetry (without a bilateral one) we mentioned, additionally, a feeling of compenetration and interdependence.

Some of these ideas extend to translational symmetry, but the situation now seems to be more involved. Consider first the case of friezes, which to fix ideas we assume have a horizontal basic vector. In the absence of vertical bilateral symmetry, a reasoning similar to the one above would associate a feeling of linear movement. The sense of balance in the presence of bilateral symmetry with a vertical axis may (and in general does) annihilate this feeling (think, for instance, on the barcode in §3.5.1 or in a pattern like ▲ ▲ ▲ ▲ ▲). A possible explanation of this paralysis could lie in an indifference of our sensory apparatus toward left or right. In contrast, for friezes without (vertical) bilateral symmetry, the shape of the motif can suggest a direction of movement, as in

▶ ▶ ▶ ▶ ▶

or

◀ ◀ ◀ ◀ ◀

Furthermore, if the frieze contains non-induced glides (i.e. for friezes of type *p1a1* or *pma2*) this linear movement is compatible with a feeling of alternation:

The feelings associated with wallpapers seem more difficult to isolate. Indeed, in such a pattern there are movements in many possible directions, and to verify the wallpaper character one should follow two such non-parallel directions. Yet, we can carry out only one such imaginary movement, and the choice of a particular direction for it will depend heavily on the motif. For instance, in the pattern on the right this choice may be toward up-and-right (i.e. moving as ↗) or toward down-and-left (i.e. as ↙). It is also logically possible – and actually necessary to confirm the wallpaper structure – to move purely rightward (→) or

leftward (←) or up (↑) or down (↓), or in directions obtained as combinations of the ones already mentioned. But it would seem that the tilt of the motif prevents our sensory apparatus from following these possibilities.

It is worth noting at this juncture that the motion direction in the reading of an image may have consequences in the way we perceive the image. The psychologist Joseph Jastrow (1900) elaborated on this issue more than a century ago using as an example the drawing on the left, originally published in the German magazine *Fliegende Blätter* (23 October, 1892, p. 147). If you read the image from left to right you see a rabbit, whereas if you do it from right to left you see a duck. We will not pursue Jastrow's explanations here. It suffices for our purposes to have pointed out the phenomenon.

Having thus argued for points (1) and (2) above we can now proceed with (3). A first example of this claim may be appreciated in the paintings by Piero della Francesca of Federigo da Montefeltro and his wife shown in Figure 6.1. These portraits illustrate the interaction between order and

Figure 6.1 **Piero della Francesca, *Federigo da Montefeltro* (right) and his wife *Battista Sforza* (left) oil on panels, *c.* 1472. (Galleria degli Uffizi, Florence, Italy/Giraudon/The Bridgeman Art Library.)**

meaning which, quoting Gombrich, "constitutes the warp and woof of the decorative arts". For, if repetition removes meaning because it annihilates the individuality of the subject, a measure of imperfection may bring it back. The double portrait by Piero keeps the equilibrium associated with bilateral symmetry, but in giving two differentiated copies of the motif it allows each of them to keep an individual character. Looking for the "unity in diversity" in this couple, the painting balances out the order of perfect bilateral symmetry with the variety in the models' differences.

A second example is the painting of the Silvaplana Lake done by Ferdinand Hodler during his holidays in 1907 (Figure 6.2). There are two approximate bilateral symmetries: one which (literally) reflects the mountains on the surface of the lake and another which is present on the shape of the mountains themselves. As a consequence (recall Proposition 3.11) there is also central symmetry. The feeling of repose derived from the painting is a consequence of these two bilateral symmetries and is doubtless akin to the repose (again literally) that Hodler was taking beside the lake.

Figure 6.2 *The Lake of Silvaplana*, oil on canvas, 1907, by F. Hodler.

The Ecuadorian painter Osvaldo Guayasamín provides us with an instance of a creative use of central symmetry. His canvas *Los Amantes* (Figure 6.3) shows the heads of a pair of lovers. In this sense, it may be said to share subject matter with the above work of Piero della Francesca. Affinities, however, stop here. Not only because 500 years separate the styles of both paintings, but also due to the difference in the ways we are

Figure 6.3 **Osvaldo Guayasamín, *Los Amantes (The Lovers)*, oil on canvas, 1983, Fundación Guayasamín, Quito.**

intended to feel the relations behind these couples. Federigo and his wife are separated by the reflection axis (a feature emphasized by the physical separation of the canvas in two pieces). There is an equilibrium in the relation of the two bodies, but there is also a detachment between them. The choice of central symmetry (without any bilateral one) in Guayasamín's painting emphasizes, in contrast, a feeling of compenetration, a complicity between these lovers expressing a balance which is unrelated to repose. One is drawn to an empathy between these lovers which is distant from the passionless relation we forebode in the Montefeltro–Sforza couple.

For a last example of imperfect symmetry we return to Escher, more precisely to his lithograph *Drawing Hands* (Figure 6.4). Even though the central symmetry is broken all over the engraving, the feeling of such a symmetry is strong. The interdependence associated with central symmetry becomes corporeal in its subject: the very existence of each hand is a consequence of the other one's drawing, an activity emphasized as ongoing in the unfinished quality around the cuffs. Even the work's name suggests this interdependence in its ambiguity, for the word "drawing"

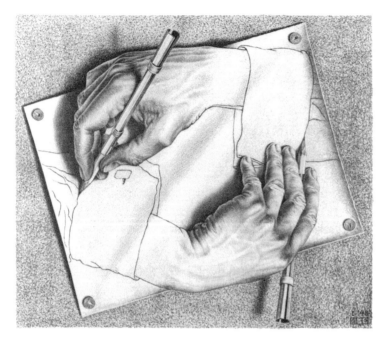

Figure 6.4 **M.C. Escher's** *Drawing Hands,* **lithograph, 1948** © **2012 The M.C. Escher Company-Holland. All rights reserved. www.mcescher.com.**

may be an adjective (meaning that the hands are drawing) or a verb (meaning that the hands are being drawn).[2] These twin roles become apparent when one makes a (verbal) frieze with this name "... hands drawing hands drawing hands drawing hands drawing hands ...".

6.2 Between detachment and dilution

The nature of friezes and wallpapers poses a particular question in relation to the departure from strict symmetry. How many copies of the motif should one include? The choice of an answer to this question is of no small consequence, since "repetition devaluates the motif while isolation enhances its potential meaning" (Gombrich, 1984: 152). An extreme choice, displaying only one copy of the motif, will fully focus on the latter at the expense of the pattern. But in the other extreme, if the motif is infinitely repeated and consequently shrunk, it will no longer be noticed: the wallpaper becomes texture. There is a trade-off

---

[2] The Dutch version of the official M.C. Escher website (http://www.mcescher.nl) lists this piece as *Tekenende Handen*. This means "drawing hands" as in "hands that are drawing", not "hands that are drawn". I do not know whether the original title as set by Escher was in Dutch or in English. If it was the former, the English version would be an example in which, for once, *traduttori* are not *tradittori*.

between detachment and dilution whose compromise solutions determine the relative importance of the motif's meaning and the pattern's structure.

In addition to the meaning/structure aspect of this trade-off there is a relationship with another tension we are familiar with, namely that between boredom and confusion. For the excessive repetition of the motif also risks creating boredom. In many decorative arts, and when the situation at hand limits the space to be filled with a wallpaper or a frieze to a predetermined enclosure, it is the size of the motif (as well as its particular shape) that provides a solution to these trade-offs, since this size determines the number of times the motif is repeated within the enclosure.[3] There is the need to choose an appropriate scale.

In other contexts, a different approach may be put to profit which leans on the possible imperfections of symmetry we discussed in Section 6.1. This approach involves counterbalancing the order of repetition with a measured amount of variation. Such an idea has illustrious roots. Edmund Burke, whose *Philosophical Enquiry into the Origin of our Ideas of the Sublime and Beautiful* is one of the earliest works in the Anglosaxon thinking on aesthetics, devoted two chapters to "gradual variation" as a cause of beauty (see (Burke, 1990: III.XV and IV.XXIII)). Burke's main focus was on the shape of some contours. Nevertheless, he stated the idea in quite general form (Burke, 1990: 140):

Nothing long continued in the same manner, nothing very suddenly varied can be beautiful; because both are opposite to that agreeable relaxation, which is the characteristic effect of beauty. It is thus in all the senses.

One needs not to agree with the entirety of Burke's position (I, for one, do appreciate sudden variations in some artworks) notwithstanding the acceptance of the idea of gradual variation as a counterbalance of an excess of order. The work of Andy Warhol makes a systematic use of this idea. Warhol began painting in the early 1960s and made his first solo exhibition as a painter in July 1962, at the Ferus Gallery in Los Angeles. The exposition consisted of 32 canvases of 20 inches × 16 inches, each depicting one of the varieties of Campbell's soup that were marketed at the time (Figure 6.5). The level of variation in this painting is low, the difference between soup cans being reduced to their variety names. This level is further reduced in a different version of the same theme – *200 Campbell's Soup Cans* – where the 32 varieties are now repeated to fill a 72 inches × 100 inches canvas displaying 200 cans. A steering towards dilution where (Bourdon, 1989: 96)

---

[3] Gombrich refers to this issue when he talks about the double task of "filling" and "framing" (Gombrich, 1984: 75 and ff.).

Figure 6.5 **Andy Warhol, *Campbell's Soup Cans*, Synthetic polymer paint on canvas, 1962, MOMA, New York. (© The Andy Warhol Foundation for the Visual Arts, Inc. /DACS, London, 2007. Trademarks Licensed by Campbell Soup Company. All Rights Reserved.)**

The repetition resulted in an allover pattern that encourages viewers to scan the work as a whole, almost as if it were an abstraction [. . .] One's gaze may occasionally linger on a particular unit, but it is more likely to keep sweeping around the entire field in a vain quest for a focal point.

Warhol also made use of more dramatic forms of variation. The day after the closure of the exhibition at the Ferus Gallery, Marilyn Monroe died. Warhol went on to purchase a photo of Marilyn which he would later use as a motif for a number of paintings, some of which made the nucleus of his first solo exhibition in New York, also in 1962. The set of 10 screen-prints in Figure 6.6 shows the use of colour as a means for variation.[4]

Other means of variation used by Warhol included the emphasizing and de-emphasizing of different parts of the motif (as in the series on Liz Taylor in Figure 6.7) and even the use of randomness. Bourdon (1989: 126) mentions that Warhol worked in a rushed manner that was prone to imperfections in the different screenings. Warhol's response to this was "I like it that way." These imperfections introduced variations whose characteristics were unintended and random and did not go unnoticed

---

[4] This is a common resource. As Gombrich (1984: 72) notes, "The pattern-maker often enjoys creating classes of motifs which are like in one respect and different in another. Colouring is his most elementary device in achieving this end".

Figure 6.6   **Andy Warhol, *Marilyn Monroe*, screenprints on paper, 1967. (© The Andy Warhol Foundation for the Visual Arts/Artists Rights Society (ARS), New York/DACS, London 2012. Photo: Hamburger Kunsthalle, Hamburg, Germany/The Bridgeman Art Library.)**

Figure 6.7   **Andy Warhol, *Ten Lizes*, oil and lacquer over silkscreen printing, 1963. (© The Andy Warhol Foundation for the Visual Arts/Artists Rights Society (ARS), New York/DACS, London 2012.)**

by the public. Bourdon (1989: 130) quotes Lawrence Alloway[5] on stating that "these failures of identical repetition take on gestural vitality; they become the reassurance of human touch or, at least, casualness".

A different use of gradual variation occurs in the work of Escher. We have already encountered an example in *Sky and Water* (Figure 4.39). Here, variation affects the shape of the various copies of the motif and it does so in a carefully calculated manner, a manner that allows birds to become background sea and fish background air. His woodcuts *Metamorphosis* (there are three versions of it) further push this idea (see Figure 6.8). Bees, fishes, horses, lizards, birds, ships and even winged envelopes smoothly transform into each other in a sequence whose physical length makes it difficult to display. One may nonetheless zoom in on particular sectors to convey an idea as shown in Figure 6.9.

---

[5] The art critic credited with the coinage of the expression "Pop Art".

Figure 6.8 M.C. Escher's *Metamorphosis III*, woodcut, 1967–1968 © 2012 The M.C. Escher Company-Holland. All rights reserved. www.mcescher.com.

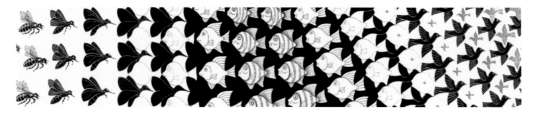

Figure 6.9 M.C. Escher's *Metamorphosis III* detail © 2012 The M.C. Escher Company-Holland. All rights reserved. www.mcescher.com.

6.3 A blurred boundary: I

> The trend toward pure art betrays not arrogance, as is often thought, but modesty. Art that has rid itself of human pathos is a thing without consequence – just art with no other pretenses.
>
> J. Ortega y Gasset (1968: 52)

Brickwork, tiling, tapestries and other textiles, ceramics, furniture, glassware, jewellery and metalwork, and, of course, friezes (those which grace buildings) and wallpapers (those which cover actual walls). The list could doubtless go on to eventually provide an extensional definition of the decorative (or applied) arts, but it should suffice for our purposes, along with the following two remarks. First, that decorative arts have often been looked down upon when compared with their nobler siblings, the fine arts. Second, that symmetry is common in them, to say the least.

It is probably this association of symmetry with decorative arts that occasionally grants the former, within the context of art theory, the low esteem afforded to the latter. We have already encountered this dismissal in Lorand's writings. She is not alone. Rudold Arnheim, one of the major characters in the efforts to explain visual arts in terms of perceptual psychology, also expressed such views in his best-known work, even though he was careful to limit the extent of his comments to "strict symmetry". The latter, he writes, "is as rare in painting and sculpture as it is frequent in decorations and the applied arts, such as ceramics or architecture" (Arnheim, 1984: 149).[6] The cause of this rarity is made patent in

---

[6] It is worth noting that Arnheim, contrary to the established view of architecture as one of the fine arts, does not hesitate to list it as an applied art, most likely because of its adherence to strict symmetry.

Arnheim's description of the painting of the Silvaplana Lake by Hodler (Figure 6.2): "by turning nature into ornament, the artist has obtained a chilly preponderance of order". In the tension between boredom and confusion an excess of order will favour the former and Arnheim is quick to note that this is not unknown to the artist, since "even in works in which an overall symmetry is appropriate to the subject, its severity is always mitigated by enlivening deviations". We have already discussed this issue in Section 6.2.

Leaving unanswered the question of whether the repose at Silvaplana ultimately turned into a bore for Hodler, we may return to the relations between symmetry, decoration, and fine arts. In doing so, we first note that the distinction between decorative and fine arts is, at times, vague. Herbert Read traces back the origin of this distinction to the Renaissance and relates it (not exactly in these terms) to the "purposeless purposivity" described by Kant (1952).[7] Read (1961: 12) states that

a complete distinction was henceforth to be made between the artist who made things to satisfy a practical purpose, such as the builder and the architect, and the artist who made things (essentially non-utilitarian) for the delectation of individuals.

This distinction did not grow overnight. To give an idea of how it could have developed, Read proceeds

Originally the illuminations in a manuscript were conceived strictly as decorations subsidiary to the text [. . .] They were incomplete and meaningless when divorced from the book. But gradually [. . .] the illuminator began to conceive his decoration *as a page*, complete as a page. It was then but a logical step to divorce the illumination from the book [. . .] In this manner, toward the end of the fourteenth century and during the fifteenth century, the panel picture came into existence.

These pictures would grace the rooms of wealthy bourgeois fulfilling no practical purpose:

We might speak of chamber painting just as we speak of chamber music; and just as such music is music divorced from its original purpose as an accompaniment to dance, the march, or the religious service, to become a self-consistent unity, the satisfaction of an appetite for sweet sound, so painting [. . .] becomes an art aspiring to the same freedom from purpose, the same purity of delectation.

The possession (or lack of it) of a practical pursuit is not, however, a sharp attribute and may change with time when the original purpose of an artpiece is no longer relevant (carpets are not primarily used nowadays to attenuate variations in temperature, nor are ceramic vases used

---

[7] A detailed analysis of how the modern system of the arts came to be (in the eighteenth century) is given by Kristeller (1965).

to store oil[8]). This removes precision in the boundaries between fine and decorative arts. Furthermore, these boundaries are not uniquely determined by the presence or not of functionality. Size seems to play a role as well in this distinction, making, for instance, large sculptures belong to the realm of fine arts and small sculptures to that of decoration. But a boundary determined by size is, again, not sharp.[9]

We may still mention a contrast between fine and decorative arts which, notwithstanding being the last we list, is of the essence in what follows. Susan Sontag (1967: 4) holds that the conception of art that the West inherited from the Greeks, which makes art a mimesis of an outer reality, created the need of a defence of art per se, beyond individual artpieces.

And it is the defense of art which gives birth to the odd vision by which something we have learned to call "form" is separated off from something we have learned to call "content," and to the well-intentioned move which makes content essential and form accessory.

From these premises Sontag unfolds a troubled marriage bewteen art and interpretation:[10]

What the overemphasis on the idea of content entails is the perennial, never consummated project of *interpretation*. And, conversely, it is the habit of approaching works of art in order to *interpret* them that sustains the fancy that there really is such a thing as the content of a work of art.

We will not pursue Sontag's arguments any further, as it is sufficient for our purposes to note that the attribution of content has been, chiefly, a privilege (or a burden, as Sontag would have it) of the fine arts. Jewels, textiles or ceramics have assuredly inspired critical writings, but a cursory inspection of them reveals that they almost invariably deal with issues of materials, techniques, or formal design.

We may pinpoint the origin in time for this attribution to the Renaissance, evolving *pari passu* with the divide between decorative and fine arts. More importantly, perhaps, we can trace early signals of doubts being cast on it back to the nineteenth century and we can observe

---

[8] See also Footnote 3 in Chapter 5.

[9] The distinction between fine and decorative arts is, in addition, frequently enhanced by the circumstances in the display of the artpiece. Typically, a painting will be framed or a big sculpture mounted on a pedestal. In contrast, rugs, vases or jewels are rarely displayed in such a way. There is, in this framing, a demand for attention on the framed object. Decoration does not effect such a demand; it is just there.

[10] The word "interpretation" in Sontag's writings is closest to what Lorand calls "signs" (cf. Section 5.1). It is certainly not the "complementary interpretation" that underlies aesthetic order. Also, an interpretation (sign) assumes the existence of a "hidden" meaning as opposed to the "apparent" meaning, if any, of the artpiece.

Figure 6.10 **Reginald Neal, *Square of Three*, litho and paint on canvas, 1964. (From (Gombrich, 1984).)**

these doubts develop in tandem with the twentieth century. The chamber music mentioned by Read in the passage quoted above is a case at hand. "Divorced from its original purpose as an accompaniment to dance, the march, or the religious service, to become a self-consistent unity", it would seem that the presence of meaning is excluded by this self-consistency. But we have to wait until very recently for this conclusion to be upheld (a theme we will return to in Section 8.7). Meaning also began to be questioned in the visual arts. We have already quoted Sontag on the issue at large and we may add critics writing on particular artists or works such as, for instance, Bourdon (1989: 92) stating that "it is seldom a good idea to plumb deeply for meaning in Warhol's art". But it is perhaps equally revealing to pay attention to the artists' intentions as well, and a case illustrating the point is that of the Op Art which bloomed in the 1960s and of which the works of Reginald Neal (Figure 6.10) and Bridget Riley (Figure 6.11) are representative.

The defining characteristic (not necessarily declared but always explicit on the artpiece) of paintings within this current was to create

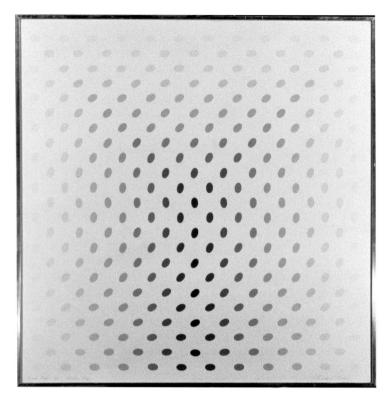

Figure 6.11 **Bridget Riley, *Nineteen Greys C*, screenprint on card, 1968. (Private Collection/The Bridgeman Art Library.)**

a certain flicker of the eye.[11] As Jasia Reichardt (1974: 238) puts it, "the term Opt Art [...] is applied to that type of illusion where the normal processes of seeing are brought into doubt, mainly through the optical phenomena of the work". Furthermore, Reichardt adds

[...] all Op Art is abstract, essentially formal and exact and [...] it could be seen as [...] the essence of Malevich's aim to achieve "the supremacy of pure sensibility in art".

In this triumph of pure sensibility, meaning is conspicuously absent.

Reichardt's use of the word "abstract" in the quote above brings us back to our discussion. Because if the nature of Op Art lies on

[11] Although not of the essence for our present arguments, it is not without interest to ponder on the source of this flicker. To this effect, we return to Gombrich's (1984: 134) description of our sensory apparatus which "is good in recognizing continuities and redundancies, but bad [...] in 'locking in' on a particular feature of repeated elements". It is because of this badness that, when there is a multiplicity of possible readings – of comparable size – the "extrapolator" is lost and oscillates between these readings.

some perceptual features rather than on possible meanings, the very essence of abstraction is the stress on the work per se as opposed to possible references to an outer reality. Even though there is no consensus on the use of the word "abstract" in art, the prevailing view is that abstract art is non-figurative, with the latter defined by the group Abstraction-Création (Harrison and Wood, 1992: 374) as follows:

non-figuration, that's to say a purely plastic culture which excludes every element of explication, anecdote, literature, naturalism, etc. . . .

A similar tenet was expressed by Alfred Barr Jr[12] when writing (Harrison and Wood, 1992: 382) that the idea of abstract painting and sculpture

is based upon the assumption that a work of art, a painting for example, is worth looking at primarily because it presents a composition or organization of color, line, light and shade [. . .] since resemblance to nature is at best superfluous and at worst distracting, it might as well be eliminated.

We also find it implicit in the claim of Edward Lucie-Smith (1995: 63) that "it is not going too far to say that abstract painting has been one of the main victims of the return to content which is one of the most striking features of the contemporary art in the 1990s". These views are confirmed by the artist's production. The work of painters such as Agnes Martin (Figure 6.12) or Eva Hesse (Figures 6.13) has little apparent content (and to look for a hidden one is an exercise in intellectual acrobacy).[13]

One is tempted to conclude that, with the abandonment of a pre-eminence for content, abstract art has further blurred the boundaries between fine and decorative arts. We have not found such a contention fully developed even though there is a substantial body of opinions, from both artists and art theorists, supporting it. For instance, the following comments of El Lissitzky (Harrison and Wood, 1992: 319) on Mondrian's painting:

Mondrian accomplished the ultimate solution in the development of Western painting. He reduces surface to its primeval state, namely surface *only*, in the sense that there is no longer any spatial in or out of a given surface. Whenever Mondrian's principle is transposed by fashionable A[rts] onto the three surfaces of a room, it turns into decoration.

---

[12] The first Director of the Museum of Modern Art in New York.

[13] A similar thought is expressed by Kivy (2009: 196) regarding a search for content in absolute (i.e. purely instrumental) music: "True, some of us [. . .] do, by what I would call laborious digging and tortuous interpretation, come up with representational stories for works in the absolute music repertoire. But many of us who read these stories fail to hear them in the music".

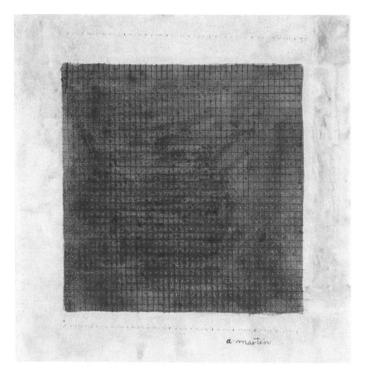

Figure 6.12 **Agnes Martin, *Stars*, ink and watercolour on paper, 1963. (© 2012. Agnes Martin/DACS.)**

Or the standpoint of Hans Hofmann (Harrison and Wood, 1992: 373):

We must discriminate between the word "decorative" in a pictorial sense and decorative in a naturalistic sense. The pictorially decorative effect is achieved through musical contrasts and rhythmic relations conditioned by space. This leads to abstraction [. . .] The word "decorative" used in a naturalistic sense implies the representation of objects [. . .] subordinated to some literary meaning foreign to the art of painting for itself.

Similar views are also present in the work of Herbert Read we have already cited. Read (1961: 36) is explicit when writing "my contention is then that the utilitarian arts – that is to say, objects designed primarily for use – appeal to the aesthetic sensibility *as abstract art*" a term he defined, consistently with the descriptions above, as being "nonfigurative art, which has no concern beyond making objects whose plastic form appeals to the aesthetic sensibility".

We cannot close this section without a reference to industrial design. In a peculiar twist of language[14] the word "design" has come to be

---

[14] The Italian form of the word "design" was used by Vasari in the expression *Arti del disegno* from which the expression "Beaux Arts" (the French equivalent of "Fine Arts") probably evolved and gave name to the first Art Academy, the *Academia del Disegno*, founded in Florence in 1563 (Kristeller, 1965: 182).

Figure 6.13 **Eva Hesse, *Untitled*, ink on paper, 1967, Tate Gallery. (©) Tate, London 2012.)**

associated with a variety of industrially produced objects for which a care for appearance is patent. The undeniable practical purpose in these objects would fit design within the decorative arts, as noted early in the twentieth century by Walter Gropius (Harrison and Wood, 1992: 311):

The Bauhaus strives to coordinate all creative effort, to achieve, in a new architecture, *the unification of all training in art and design*. The ultimate, if distant goal of the Bauhaus is the *collective work of art* – the Building – in which no barriers exist between the structural and the decorative arts.

If our claim that abstraction has blurred the boundaries between decorative and fine arts holds some truth one would expect abstraction and design to cross paths. And indeed, this is precisely maintained by Ben Nicholson (Harrison and Wood, 1992: 399),

I think, too, that so far from abstract art being the withdrawal of the artist from reality (into an "ivory tower") it has brought art once again into common every-day life – there is evidence of this in its common spirit with and influence on many things like contemporary architecture, aeroplanes, cars, refrigerators, typography, publicity, electric torches, lipstick holders, etc.

The blurring of boundaries between fine arts and decoration has provided the two with mutual characteristics. For instance, at a sociological level, the worship of individuality, long associated with the successful fine artist, seems to have reached the applied artist as well. Philippe Starck, whose designs comprise inter alia motorbikes and interior designs, optical mouses and citrus squeezers (Figure 6.14), has a room devoted to his work at the Musée National d'Art Moderne in Paris. Conversely, the words of Ortega y Gasset opening this section suggest that twentieth-century fine arts may have acquired some of the characteristics of its humbler sibling.

Figure 6.14 **"Juicy Salif" design, Philippe Starck, 1990.
Citrus-squeezer in aluminium casting. (Alessi S.p.A.,
Crusinallo, Italy.)**

6.4 The amazing kaleidoscope

> Let them praise His name with dancing;
> Let them sing praises to Him with timbrel and lyre.
>
> Psalm 149:3, *The Bible*

In 1819 Sir David Brewster published his *Treatise on the Kaleido-scope* (Brewster, 1819) describing a device he had patented 2 years earlier.

The name Brewster gave to his invention (from the Greek καλός (beautiful), εἶδος (form) and σκοπεῖν (to look at)) betrays the perception he had of his invention as a generator of beautiful images. Brewster actually had great expectations for the kaleidoscope. He saw it as a help for all sorts of artists, from architects to bookbinders. The fate of the kaleidoscope, despite these expectations, would prove more modest. An accidental viewer would, after a brief moment of curiosity, put the kaleidoscope down and abandon its use. Gombrich (1984: 151) explains this invariable erosion of the first interest by the lack of meaning in the images generated by the kaleidoscope:

There must be a conflict, or at least a tension, between the two functions of perception to which we referred at the outset, the perception of things and the perception of order. By things I here mean elements in our environment which have a meaning for us in terms of our survival and our interests.

While there is some truth in Gombrich's explanation, I do not think that it captures the whole of the problem. It is true that the successive images displayed by the kaleidoscope are randomly generated, a feature that may aggravate their lack of meaning and separate them from, say, an abstract painting. I contend, however, that the weakness of the kaleidoscope lies elsewhere.

It is not relevant for our purposes if the images created by a kaleidoscope are a discrete sequence (say $S_1, S_2, S_3, \ldots$) or vary continuously with time. It suffices to say that in both cases the symmetry group of the images S_t is the same for all time t. They are actually of the form D_n (with the value of n depending on the kaleidoscope used). That is, to the repetition present in each image due to a symmetry of the kind D_n, we add the repetition derived by the persistence on time of this group structure. Under such circumstances boredom is ensured.

A kaleidoscope (literally) of a different nature is possible. A starting point in arguing this tenet is in the article "On style" by Susan Sontag (1967: 21), where she mentions, en passant, "those arts which are abstract or have largely gone abstract, like music and painting and the dance". Of the three, the one which, like the kaleidoscope, displays a sequence of images in time is the dance. These images are not necessarily symmetric; actually, most are not. But in the merging of those which are with those which are not, and in the changing symmetry structure of the former, lies an amazing kaleidoscope.

Sontag's claim that dance has "gone abstract" points to the emphasis on formal aspects – at the cost of content – which made the substance of Section 6.3 and is mirrored in the development of a philosophy of dance, a summary of which is given by Noël Carroll (2003). Carroll refers to the question of "What makes dance an artform?" as one becoming pressing in the eighteenth century and points to John Weaver and Jean-George Noverre as the most notable providers of early answers to this question. Both of them choreographers, they maintain that what makes dance an artform is its representational, imitative character. If this character is absent from a dance (as occurs in most social dances) then we cannot consider it dance *qua* art. Dissent with this view did not wait long to manifest itself and is clear in the opinions of Théophile Gautier, Stéphane Mallarmé or Paul Valéry, with the first of these stating in 1837 (quoted in (Carroll, 2003: 587)) that

"After all, dancing consists of nothing more than the art of displaying beautiful shapes in graceful positions and the development of them in lines agreeable to the eye; it is mute rhythm, music that is seen. Dancing is little adapted to render metaphysical themes."

Carroll puts forward André Levinson as the author having given "the most powerful formulation" of this theory (known as *formalism*) and his article "The spirit of the classic dance" (Levinson, 1974) as a reference for this formulation. He then describes a third answer to the nature of dance as an artform, *expressivism*, which makes this nature the expression of feelings, particularly emotions, and is best described by John Martin (1972) and Susan Langer (1953). Both expressivism and formalism are consistent with dance "going abstract" and each has actually served as theoretical background for several choreographies doing so.

Arguably the major contemporary figure in the tendency towards abstraction in dance is George Balanchine, the founder of the New York City Ballet. His plotless choreographies emphasize geometry, and symmetric configurations are a preponderant means to do it. This should come as no surprise. After all, the spatial repetition present in the synchronized motion of the troupe (choreographers talk about moving in unison) is one of the most noticeable features of dance. Balanchine (1968: 558) explains why this repetition is so pervasive as follows:

The eye can focus perfectly only on objects which are in the center of its field of vision. Those objects which are not head-on are seen clearly only because the observer knows and imagines what they are, while he focuses on the center object. If some new or different form is placed in the secondary part of a composition, the eye instinctively changes its focus and convinces itself of the identity of each individual form. And as vision is the channel through which the art of choreography reaches its audience, this inevitably results in confusion and a loss of attention to the main theme. But the eye can follow the movements of a large group of dancers if these form a harmonious pattern within its central field of vision.

Thus, Balanchine suggests "a harmonious pattern" (i.e. translational repetition) as a way to facilitate the perception of the stage's content in a reference to ease of perception that brings to mind the investment in

Figure 6.15 *Symphony in C* by Georges Bizet, choreographed by George Balanchine performed by the San Francisco Ballet.

Birkhoff's "rate of return". Also, the perceptual mechanisms underlying Balanchine's explanation are precisely the *fovea centralis* and the peripheral vision doing so in Gombrich's thought. The following extract from the Dance Review by Alastair Macaulay – published in the *New York Times* on 28 January, 2009 – on some of his choreographies shows how Balanchine's kaleidoscope works.

All three Balanchine ballets are fascinating to watch in terms of combined stage arithmetic and geometry. "Monumentum pro Gesualdo" – for a corps of six male-female couples, with a lead couple at the center – starts completely symmetrically. Its patterns – subtly shifting two- and three-part visual harmonies – keep rearranging symmetrically just long enough for us to sense the major shift when suddenly all the dancers make one diagonal line across the stage.

As new patterns follow, we realize that they, too, are symmetrical, but no longer facing us: the axis of symmetry has become the stage's diagonal. Different geometries follow. These features – though they are just one layer of the choreography – are so timed as to make us listen with a rare attentiveness to the music.

The constitution of most dances makes them, by the reasons described above by Ballanchine, suitable for translational repetition. This is the form of symmetry displayed in Figures 6.15 and 6.16.

Other forms of symmetry are, the above notwithstanding, also possible. For instance, bilateral symmetry is at the heart of the *Thousand-Hand Avalokitesvara Bodhisattva*, a choreography performed by the China Disabled People's Performing Art Troupe (Figure 6.17), on which the extensive use of this form of symmetry makes up for the limited mobility of a dance company whose performers are blind.

Figure 6.16 **Alvin Ailey American Dance Theater's M. Rushing, K. Boyd, G. Allen Sims and C. Brown. Photo by Andrew Eccles.**

Figure 6.17 **China Disabled People's Performing Art Troupe, directed by Tai Lihua.**

The feeling of alternation proper of glidal symmetry is common in motions such as a line splitting with alternate dancers moving in opposite directions. It is also apparent in Figure 6.18.

Central symmetry, with its associated feelings of compenetration and interdependence, is frequently featured but not obvious to perceive. The reason is that this form of symmetry is mostly displayed in patterns

Figure 6.18 **Alvin Ailey American Dance Theater in Alvin Ailey's** *Revelations*. **Photo by Andrew Eccles.**

Figure 6.19 *Concordia* **of Matjash Mrozewski.**

that appear as centrally symmetric when seen from above (an unlikely viewpoint for most spectators). Figure 6.19 shows Muriel Maffre and Pierre-François Vilanova performing a *pas de deux* in the ballet *Concordia* by Matthew Hindson, choreographed by Matjash Mrozewski.

Figure 6.20 **The Pilobolus Dance Theater.**

The central symmetry from a zenithal viewpoint is unmistakable. Just as its associated feelings are hard to avoid.

Other forms of rotational symmetry have been choreographed as well, an example being the arrangement[15] in Figure 6.20 performed by the Pilobolus Dance Theater.

6.5 The strictures of verse

> For rhyme the rudder is of verses,
> With which, like ships, they steer their courses.   S. Butler (1967: 14)

> It is the scandal of poetry that "contingent" features of sound and rhythm
> systematically infect and affect thought.   J. Culler (1997: 80)

We close this chapter by discussing a subject seemingly departing from our line of thought. Indeed, by entering into the domain of poetry we abandon the realm of the immediately visual. Images can be evoked but cannot be exhibited and, as a consequence, the notion of symmetry does not directly apply. The passage of a poem through time, however, allows

---

[15] This image is reminiscent of the Hollywood musical movies choreographed by Busby Berkeley in the 1930s and 1940s which featured many such mandalas.

one (and we will not be remiss) to find symmetry in the various types of repetitions that occur in this passage. It is true that most forms in our catalogue of symmetries have no place in this context. Yet, in return for this modesty, poetry offers an outstanding example of interaction between sense of order and sense of meaning.

A recurrent view of poetry centres around the idea of "semantic density". Poems carry considerable amounts of meaning in just a few lines. An equally recurrent analysis of poetry explains the possibility of doing so by a number of resources at the poet's hand. Most of these resources are grouped in what is known as *figures of speech*, a set which, in turn, is commonly divided into *schemes* and *tropes*. Even though there is no complete agreement on the definitions, the general idea is that tropes are expressions or phrases a part of which is used in a way that departs from their face-value meaning. In this sense, they rely on semantic content. In contrast, schemes are groups of (closely placed) words that follow some acoustic arrangement.[16] The best known examples of tropes are the *metaphor* (where two things are compared by replacing the word for one by the word for the other), the *simile* (as before, but explicitly stating the comparison by using words such as "like" or "as"), or the *metonymy* (the replacement of a word denoting an object or action by another word denoting a property of the former). Equally well-known examples of schemes are the *alliteration* (repetition of the first consonant sound in words within a short passage of verse) or the *assonance* (ditto with repetition of vowel sounds). A detailed list of figures of speech would extend beyond the reasonable limits of this section (the interested reader can find one in Lanham (1991)).[17] It should suffice for our purposes to note that schemes, being in principle of a purely acoustical nature, belong to the group of features of an art which we may call "formal" and that most, if not all, of the critical literature on poetry emphasizes the mutual influence between its formal and semantic aspects. Thus, Mary Oliver (1998: 6):

The metrical poem is a pattern made with sound just as much as it is a statement made through sound.

Or Jonathan Culler (1997: 80):

The poem is a structure of signifiers that absorbs and reconstitutes the signifieds in that its formal patterns have effects on its semantic structures [. . .]

---

[16] This division is too general and, therefore, bound to meet exceptions. For instance, *homonymy* or *homophony*, by their very definition, are neither purely semantic nor purely acoustical.

[17] Figures of speech rely on the underlying language and, hence, so does their use in poetry. For the sake of clarity we will restrict the remainder of this section to English poetry, say of the period extending between the sixteenth and early twentieth centuries.

Schemes are not the only formal resources of the poet. They are not even those claiming the defining character of poetry. That place is taken by *rhyme* and *rhythm*.[18] Words in English are composed of syllables which, when sounded, are either stressed (uttered with emphasis) or unstressed. Consider, for instance, the sentence opening this section. Using a stroke ´ placed at the end of a syllable to indicate that it is stressed and a breve ˘ on top of one to indicate it is unstressed, we obtain

For rhyme the rudder is of verses

There is a pattern which repeats four times in this phrase, namely a pair of syllables composed of an unstressed syllable followed by a stressed one. Such a pair is called an *iamb* and we may draw it as ˘´. Other metrical patterns used in English poems are the *trochee* ´˘, the *dactyl* ´˘˘ and the *anapest* ˘˘´. These patterns are commonly referred to as *feet*.

Classical English poems are organized as a sequence of *lines*, each of them having a certain number of feet. The line above contains four feet; it is said to be a *tetrameter*. Names for other line lengths include *trimeter* (three feet), *pentameter* (five feet), *hexameter* (six feet) and *octameter* (eight feet). The choice of the feet and line length determines the rhythm of a poem,[19] the latter being responsible for the local variations in stress and the former for the rests that normally accompany a line end. Among these choices, the one favoured by the greatest English poets (e.g. Shakespeare, Milton, Wordsworth) is the iambic pentameter. The poem *On His Blindness* by John Milton, which we quoted in Section A.2, is a superb example of such metre.

Milton's poem also forcefully brings to us the second formal resource mentioned above: rhyme. This is no more than a structured repetition of the sound at the ends of the lines in a poem. In Milton's poem, this repetition follows a pattern that, using the same letter to denote the same sound, identifies the endings of the 14 lines with the letters in ABBA ABBA CDE CDE. There are, needless to say, names to distinguish different sorts of rhyme, such as *masculine* (when the rhyming syllables are stressed), *feminine* (when they are unstressed), *imperfect* (when small phonetic differences are allowed), *perfect* (when they are not), and so on. But again, we will not deepen on this direction.

We will, however, now point to the obvious: the fact that both rhythm and rhyme rely on repetition and, actually, on a very structured form of

---

[18] There is no purposeful alliteration in this pair. Actually, both words come from the Latin *rithmus*, but the first through a detour via French. For a short explanation of how this process took place, see (McArthur, 1995: 29).

[19] Even after a choice of metre and length has been made – say dactylic hexameter – it is common to deviate from it. These *accommodations* may be necessary for a variety of reasons. When the metre is regular we say it is *pure*, otherwise we say it is *impure*.

repetition. A form which, even though we are no longer in the realm of the visual, we can refer to as symmetry. There is certainly bilateral symmetry on the pattern ABBA, as there is a translational correspondence in both the repetition of this pattern and the repetition of the CDE one in the last two tercets. Also, there is glidal symmetry in the alternation of unstressed and stressed syllables. Indeed, fix a certain level of stress (phonetic emphasis) for the unstressed syllables and another, higher, for the stressed ones. Then, since syllables have roughly the same length, we may depict a line in an iambic pentameter (say the opening line of *On His Blindness*) by a saw-like curve as in the following graph:

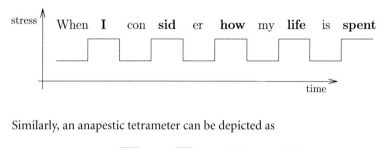

Similarly, an anapestic tetrameter can be depicted as

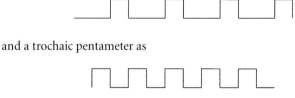

and a trochaic pentameter as

The possible symmetries for rhyme and rhythm, one soon realizes, are not many. Furthermore, some combinations have earned a position of favour among poets, maybe not without reason.[20] Notwithstanding this want, the role of formal features in poetry allows us to revisit a number of the issues that occupied us in this and the preceding chapter and, hopefully, to shed light on them from a different context. We close this section by doing so.

A first remark is that poetry relies on anticipation. Patterns are set and we expect the poem to follow them. We have already seen these expectations at work in the visual experience as described by Gombrich. Although differing in the details, a similar mechanism is in place with literature, as pointed out by Culler (1997: 73):

---

[20] As Mary Oliver (1998: 29) writes (referring to line lengths but holding more generally): "Poets are free to select their own designs (within the rules, I mean) and, among all the options, to determine the brevity or the extravagance of their lines. But they rarely do – they rarely go beyond the usual. In truth, much has been tried, and much has failed. What works, works for profound and understandable reasons. And so the workable formulae have become the formal structures that are used and used again".

For readers, genres are sets of conventions and expectations: knowing whether we are reading a detective story or a romance, a lyric poem or a tragedy, we are on the lookout for different things and make assumptions about what will be significant.

In the case of a poem, we will be on the lookout (even if not necessarily consciously) for patterns of rhythm and rhyme. Any such pattern plays an essential role in our reception of the poem. Indeed, in a stanza (Oliver, 1998: 46) "the initial group of lines sets the design to be followed, establishing the metrical pattern, the line length or lengths, and the system of rhyme." After which (Oliver, 1998: 31)

we see how forceful is the pattern's work upon us, for we quickly, within a line or two, believe the pattern that is given; we put on its "atmosphere" and adjust our mood according to the length of the line. It is amazing. And it is reliable.

And there is pleasure in this fulfilment (Oliver, 1998: 40): "Rhyme – that decisive repetition of sound at the end of lines – gives the kind of pleasure felt with any anticipation and arrival". Hence, we find in poetry an example of symmetry (bilateral or translational, depending on the rhyme pattern) which appears to contradict Lorand's statement that "symmetrical sets exhibit poor aesthetic order" because the symmetrical occurrences of an already perceived motif are to be expected. It turns out that the satisfaction of this expectation may be of the essence.

The above being as it may, excessive repetition – in either rhythm or rhyme – toys with boredom (Richards, 1995: 108): "Verse in which we constantly get exactly what we are ready for and no more, instead of something which we can and must take up and incorporate as another stage in a total developing response is merely toilsome and tedious". The use of variations in the pattern, as discussed in Section 6.2, is a standard solution to keep boredom at bay. The prevailing metrical pattern may be occasionally substituted by another one, and ditto for the rhyme. How much or how often to do so? Oliver (1998: 14) is quick to note that

The question of how much substitution is allowed is unanswerable. [. . .] – if substitution is used too much, the pleasing sense of a pattern will be lost, at which point all is lost. If no substitution is used, the poem may take on a doggerel quality, and a feeling that complexity has been sacrificed for regularity.

Oliver's words are forcefully bringing to us the trade-off between boredom and confusion – this time in the context of poetry – and the need for a solution for it.

A last remark, regarding the discussion on feelings associated with symmetric forms in Section 6.1 applied to a visual context is in order. In the aural context of poetry other associations will hold. Mary Oliver points to some of them, related to rhythm, which we summarize to conclude this section. She first remarks (Oliver, 1998: 19) that "Iambic meter gives a graceful, motionful sense of balance and unexcited progress". When sounded in pentameters (Oliver, 1998: 30) it delivers "a message of capability, aptitude, and easy fulfillment". Oliver suggests that this is due to the fact that the pentameter length roughly coincides with our lung capacity. Furthermore (Oliver, 1998: 31), "lines shorter than pentameter leave the reader feeling slightly hurried and, thus, agitated" while "the longer line gives a feeling of abundance – sometimes with a sense of energy and brimming over, at other times with a feeling of extraordinary authority and power" (Oliver, 1998: 33).

The agitation of shorter lines can be fully appreciated, for instance, in the poem *The Destruction of Sennacherib* by Lord Byron. Written with anapestic tetrameters, it recounts the attempt by the Assyrian king Sennacherib to conquer Jerusalem, an attempt that failed (according to the Bible (2 Kings 19:35) because the angel of Jehovah struck down the Assyrian forces). The anapestic tetrameter captures not just the agitation of the battle; it even has a galloping sonority, very much appropriate to the description of the Assyrian hordes, as you can check in the first quatrain of the poem:

> The Assyrian came down like a wolf on the fold,
>
> And his cohorts were gleaming in purple and gold;
>
> And the sheen of their spears was like stars on the sea,
>
> When the blue wave rolls nightly on deep Galilee.

Stretching the plane

The drawing on the left of Figure 7.1 shows a fish (known as *Argyropele-cus olfersi* among zoologists) over a grid with square units. The drawing on the right shows another fish (known as *Sternoptyx diaphana*), also over a grid whose cells are now parallelograms.[1]

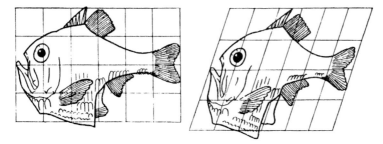

Figure 7.1 *Argyropelecus olfersi* (left) and *Sternoptyx diaphana* (right). (From (Thompson, 1961: Figures 146 and 147).)

Their original purpose was to illustrate a way of comparing the forms of these two species which would be simpler and more precise than that favoured by morphologists at the beginning of the twentieth century. For the morphologist, Thompson writes (1961: 274), "when comparing one organism with another, describes the differences between them point by point, and 'character' by 'character'" even though "he is from time to time constrained to admit the existence of 'correlation' between characters [...]." Thompson is thus finding fault in the "local" nature of the morphologist's comparisons, the fact that they rely on an accumulation of details lacking a "global", all encompassing, capacity of explanation. In contrast to this approach, Thompson suggests that a correlation in the forms of the two species above can be found of such a particular nature that it would explain all the "point by point differences". More precisely, he maintains that the form on the right is the image of that on the left by a transformation of the plane.

A cursory glance at Figure 7.1 shows, however, that the required transformation is not an isometry: squares cannot become (non-square) parallelograms under the action of an isometry. The goal of this chapter is to describe the kinds of transformation subjacent in producing the shape of *Sternoptyx diaphana* from that of *Argyropelecus olfersi*. In doing so, we will see a new form of symmetry as well as some style innovations during the Mannerist period, and we will get the occasion to revisit

---

[1] A parallelogram is a plane quadrilateral figure whose opposite sides are parallel.

our discussion on shapes and geometries in Section 1.2 within a more rewarding context.

7.1 Homothecies and similarities

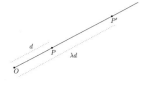

Let O be a point and $\lambda > 0$ be a real number. A *homothecy* with centre O and coefficient λ is a transformation $\mathrm{homot}_{O,\lambda} : \mathbb{E} \to \mathbb{E}$ which associates to any point $P \in \mathbb{E}$ the point P' obtained as follows. If $P = O$ then $P' = O$. Else, let r be the half-line with origin at O and passing through P. The point P' is the only point in r satisfying $\mathrm{dist}(P', O) = \lambda \, \mathrm{dist}(P, O)$.

When $\lambda > 1$ the homothecy is *expansive*, whereas when $\lambda < 1$ we say it is *contractive*. Note that if $\lambda = 1$ then $\mathrm{homot}_{O,\lambda} = \mathrm{Id}$; that is, the homothecy is the identity transformation we defined in Section 3.2. We can follow the footprints of Chapter 2 to visualize the effect of a homothecy (in the instance below, with coefficient 1.5 and centre at the origin of coordinates).

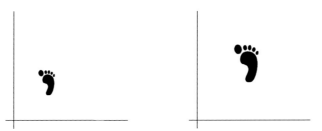

We next focus on the properties preserved by homothecies.

Proposition 7.1 *Homothecies preserve lines, directions and angles. That is, if h is a homothecy and ℓ is a line then $h(\ell)$ is a line as well and it is parallel to h. Furthermore, if ℓ_1, ℓ_2 are lines then the angle between $h(\ell_1)$ and $h(\ell_2)$ equals the angle between ℓ_1 and ℓ_2.*

Proof We proceed slowly to describe all the details of an argument we will frequently use. The idea is to use analytical methods. More precisely, we choose the x-axis parallel to ℓ and passing through the centre O of h. We next choose the y-axis passing through O as well. Then the equation of ℓ is

$$y = b$$

for a certain $b \in \mathbb{R}$. Furthermore, this choice of coordinate axes allows for a simple analytical description of h. Indeed, if λ is the coefficient of h then

$$h(x, y) = (\lambda x, \lambda y).$$

The idea is to find the equation of $h(\ell)$ from the description of h and the equation of ℓ. To do so, consider an arbitrary point $(x, y) \in h(\ell)$. This is equivalent to saying that $h^{-1}(x,y) \in \ell$. And this, in turn, amounts to saying that $h^{-1}(x,y)$ satisfies the equation of ℓ. In our case h^{-1} admits a simple description (which is deduced from that of h), namely

$$h^{-1}(x,y) = \left(\frac{x}{\lambda}, \frac{y}{\lambda}\right).$$

Therefore, a point (x,y) is in $h(\ell)$ if and only if $(\frac{x}{\lambda}, \frac{y}{\lambda})$ satisfies the equation of ℓ; that is, if and only if

$$y = \lambda b.$$

We have thus shown that $h(\ell)$ is also a line and that, in addition, it is parallel to ℓ. If we now consider two lines ℓ_1 and ℓ_2 we have that $h(\ell_1)$ and $h(\ell_2)$ are parallel to ℓ_1 and ℓ_2 respectively. Therefore, the angle between $h(\ell_1)$ and $h(\ell_2)$ coincides with that between ℓ_1 and ℓ_2. □

Homothecies do not preserve distances, but they do change them in a precise manner.

Proposition 7.2 *Let P, Q be two points and h a homothecy of coefficient λ. Then*

$$\mathrm{dist}(h(P), h(Q)) = \lambda \, \mathrm{dist}(P, Q).$$

Proof We choose the coordinate axes so that the origin of coordinates is the centre of h. Let (x_P, y_P) be the coordinates of P in this system and (x_Q, y_Q) those of Q. Then

$$h(P) = (\lambda x_P, \lambda y_P) \quad \text{and} \quad h(P) = (\lambda x_Q, \lambda y_Q)$$

and therefore

$$
\begin{aligned}
\mathrm{dist}(h(P), h(Q))^2 &= (\lambda x_P - \lambda x_Q)^2 + (\lambda y_P - \lambda y_Q)^2 \\
&= \lambda^2 (x_P - x_Q)^2 + \lambda^2 (y_P - y_Q)^2 \\
&= \lambda^2 \left[(x_P - x_Q)^2 + (y_P - y_Q)^2\right] = \lambda^2 \, \mathrm{dist}(P, Q)^2.
\end{aligned}
$$

Taking square roots on both sides, the statement follows. □

Proposition 7.2 allows us to establish other features preserved by homothecies.

Proposition 7.3 *Homothecies preserve proportions. That is, for any homothecy h and any three points P, Q, R we have*

$$\frac{\mathrm{dist}(h(P), h(Q))}{\mathrm{dist}(h(P), h(R))} = \frac{\mathrm{dist}(P, Q)}{\mathrm{dist}(P, R)}.$$

Proof Let λ be the coefficient of h. Then, using Proposition 7.2,

$$\frac{\mathrm{dist}(h(P), h(Q))}{\mathrm{dist}(h(P), h(R))} = \frac{\lambda \, \mathrm{dist}(P, Q)}{\lambda \, \mathrm{dist}(P, R)} = \frac{\mathrm{dist}(P, Q)}{\mathrm{dist}(P, R)}.$$

  □

Proposition 7.4 *Homothecies preserve circles. That is, for any homothecy h and any circle C, h(C) is a circle as well. Moreover, if C has radius R then h(C) has radius λR, where λ is the coefficient of h.*

Proof Let O be the centre of C. For any point P, we have[2]

$$P \in h(C) \iff h^{-1}(P) \in C \iff \text{dist}(h^{-1}(P), O) = R$$
$$\iff \text{dist}(h(h^{-1}(P)), h(O)) = \lambda R \iff \text{dist}(P, h(O)) = \lambda R.$$

We deduce that $h(C)$ is the circle of radius λR with centre $h(O)$. □

In Section 3.2 we showed that we can endow the set of all isometries on the plane with an "arithmetic" given by the operation of composition. This operation allows us to apply one isometry to an object, then another, and another, ...; and no matter how many such isometries we apply nor they nature (reflection, rotation, etc.), their composition is an isometry as well. Were we willing to do the same with homothecies we would meet an obstruction: the composition of two homothecies is not necessarily a homothecy. For instance, the composition of $\text{homot}_{O,2}$ with $\text{homot}_{P,\frac{1}{2}}$ yields a translation with vector $\overrightarrow{OQ}$ with Q the midpoint of the segment $\overline{OP}$.

To check this, note that this composition preserves distances (apply Proposition 7.2 twice together with the fact that $2 \times \frac{1}{2} = 1$), leaves the line ℓ through O and P invariant, and takes O to Q and Q to P.

The simplest way to remove this obstruction is to consider the family of all transformations of the plane that can be obtained by composing (a finite number of) homothecies. In doing so, we would obtain a class of transformations containing at least all the homothecies and all the translations. Other isometries, however, such as reflections, would not be in this class. If we want to have all isometries (if not for any reason other than for the sake of completeness) the simplest course of action is to allow them in the compositions. We thus obtain the class of similarities.

Definition 7.5 A *similarity* is a composition of an arbitrary number of isometries and homothecies. That is, a transformation $\varphi : \mathbb{E} \to \mathbb{E}$ of the form

$$\varphi = \psi_n \circ \psi_{n-1} \circ \cdots \circ \psi_2 \circ \psi_1$$

where, for $i = 1, \ldots, n$, ψ_i is either an isometry or a homothecy. In particular, any isometry is a similarity. We denote by $\text{Sim}(\mathbb{E})$ the set of all similarities.

---

[2] We use the symbol $\iff$ as an abreviation of "if and only if".

We next show that the class of similarities, together with composition, forms a group (in the sense of Definition 3.6). We will repeatedly use the argument of the proof in several contexts.

Theorem 7.6 *The set* $\mathrm{Sim}(\mathbb{E})$ *is a group (of transformations on* $\mathbb{E}$*).*

Proof It is clear that if φ and ψ are similarities then so is the composition $\varphi \circ \psi$. The neutral element of $\mathrm{Sim}(\mathbb{E})$ with respect to composition is, as usual, the identity function $\mathrm{Id} : \mathbb{E} \to \mathbb{E}$ defined by $\mathrm{Id}(x) = x$. Associativity is a trivial consequence of this property for the composition of functions. We are thus left with the task of showing that every element $\varphi \in \mathrm{Sim}(\mathbb{E})$ has an inverse. And this is an easy consequence of the definition of similarity. Indeed, if $\varphi = \psi_n \circ \psi_{n-1} \circ \cdots \circ \psi_2 \circ \psi_1$ then each ψ_i has an inverse ψ^{-1} and these inverses are also homothecies and isometries. Define $\varphi^{-1} = \psi_1^{-1} \circ \psi_2^{-1} \circ \cdots \circ \psi_{n-1}^{-1} \circ \psi_n^{-1}$. Then, by construction, $\varphi^{-1} \in \mathrm{Sim}(\mathbb{E})$ and one has

$$\varphi \circ \varphi^{-1} = \psi_n \circ \psi_{n-1} \circ \cdots \circ \psi_2 \circ \psi_1 \circ \psi_1^{-1} \circ \psi_2^{-1} \circ \cdots \circ \psi_{n-1}^{-1} \circ \psi_n^{-1}$$

$$= \psi_n \circ \psi_{n-1} \circ \cdots \circ \psi_2 \circ \mathrm{Id} \circ \psi_2^{-1} \circ \cdots \circ \psi_{n-1}^{-1} \circ \psi_n^{-1}$$

$$= \psi_n \circ \psi_{n-1} \circ \cdots \circ \psi_2 \circ \psi_2^{-1} \circ \cdots \circ \psi_{n-1}^{-1} \circ \psi_n^{-1}$$

$$\vdots$$

$$= \psi_n \circ \psi_n^{-1}$$

$$= \mathrm{Id}$$

and, similarly, $\varphi^{-1} \circ \varphi = \mathrm{Id}$. ☐

The next, easy, result gives a summary of the properties preserved by elements in the group $\mathrm{Sim}(\mathbb{E})$.

Theorem 7.7 *Similarities preserve straight lines, angles, proportions and circles.*

Proof It is enough to see that both isometries and homothecies preserve these objects. If this is so, then each of the transformations that composes to yield a similarity φ will in turn take, say, a straight line and yield a new one, and the whole composition will do so.

We already know that isometries preserve straight lines, angles, proportions and circles and have just seen that homothecies do so as well (Propositions 7.1, 7.3 and 7.4). ☐

7.2 Similarities and symmetry

The introduction of the group of similarities naturally raises the question of what are the possible symmetries defined by invariance under similarities. We will abstain from developing a catalogue, since such an

Figure 7.2 **M.C. Escher's** *Smaller and Smaller*, **woodcut, 1956** ©️ **2012 The M.C. Escher Company-Holland. All rights reserved. www.mcescher.com.**

endeavour would take us far astray, and limit our exposition to some instances of invariance under similarities. The first such instance is the woodcut *Smaller and Smaller* by Escher (Figure 7.2). Its central part has a pattern exhibiting invariance under a homothecy (of coefficient approximately 2). It is also invariant under a rotation of 45° (with the same centre as the homothecy's[3]). To fit the pattern into a square, Escher modified the shape of the lizards in the outer group but otherwise the design shows an unusual kind of wallpaper. Its title alludes to the infinite sequence of copies of the motif that decrease towards the centre, a sequence whose perception is only limited by the artist's capacity to miniaturize and the viewer's perceptual finesse, in opposition to a "larger and larger" sequence of motifs that would require, in addition, the infinite extent of the Euclidean plane.

Another example of invariance under a similarity is also worth mentioning. In his analysis of perceptual effects Gombrich (1984: 134) mentions the "baffling trick [...] played on our 'extrapolator' by the

---

[3] The fact that the centres of the homothecy and the rotation are the same implies that the composition of these two transformations is commutative; we obtain the same transformation by performing first the homothecy and then the rotation than by proceeding the other way round.

Figure 7.3 **The Fraser spiral.**

so-called Fraser Spiral [...] which is not a spiral at all, but a series of concentric circles superimposed on vortex lines".[4]

Gombrich explains the "trick" as follows: "These lines, it turns out, tend to deflect our searching gaze so that we always lose our place and settle for the most plausible "templet", the continuous spiral". We can add to this explanation by noting that the Fraser spiral (Figure 7.3) is invariant under a very precise similarity, namely the composition of a homothecy of coefficient $\lambda \approx 1.5$ with a rotation of angle $\theta = 22.5°$, both with the same centre (as in Escher's woodcut). Figure 7.4 shows, framed in red, a fundamental region for the Fraser spiral. Rotations of $22.5°$, 16 of them to be precise, generate a circular crown. And successive homothecies of coefficient 1.5 (as well as $\frac{2}{3}$) of this crown generate increasingly larger (respectively, smaller) copies of it. The figure is, as usual, the union of all these copies.

Now, recall the way Gombrich explains the feeling of movement in a whirl (which we discussed near the end of Section 5.3). He claimed that "we have at least to carry out an imaginary rotation to see that [the arms

---

[4] This optical illusion owes its name to the British psychologist James Fraser who first described it in 1908.

Figure 7.4 **The Fraser spiral (fundamental region).**

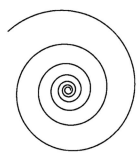

of the whirl] are invariant". It is perhaps not by chance that the typical invariant figure for the similarity ψ described above is the logarithmic spiral (see the upper drawing on the left).

A piecewise linear version of such a kind of spiral can actually be obtained by joining with line segments the successive images (for both the similarity ψ and its inverse ψ^{-1}) of a given point. The central drawing shows a few steps of this process with $\lambda \approx 1.082$ and $\theta = 22.5°$. The dotted lines are concentric and equiseparated by an angle θ. The initial point P is on one of them. Its successive images under ψ describe the branch of the spiral that recedes from the centre (of both the rotation and the homothecy) moving counterclockwise. The corresponding images under ψ^{-1} produce the branch approaching this centre.

While I agree, in general terms, with the explanation given by Gombrich, I find a detail to be both missing and crucial. Gombrich claims that it is the vortex lines which "deflect our searching gaze". This cannot be true. The reason is that these vortex lines are both centrally and bilaterally symmetric. Yet, our searching gaze is deflected in a way that leaves no doubts about the "direction" of the spiral. As in the two upper drawings, the spiral we perceive in Figure 7.3 is receding from its centre counterclockwise. An unhurried glance at this figure (or at its magnified version in Figure 7.4) shows that the "concentric lines" referred to by Gombrich are not bilaterally symmetric. They are made of a number of fitting black and white segments which – and I claim this to be at the heart of the spiral's direction – when traversed counterclockwise slightly move away from the centre. The lower drawing shows a piece of a concentric line as a circular crown between two dotted arcs of a circle with centre at O together with the boundary of one of these segments.

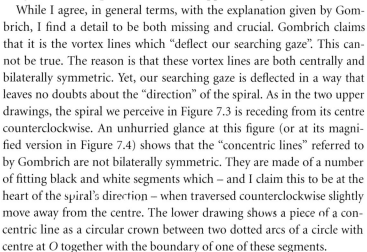

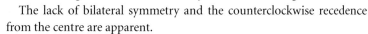

The lack of bilateral symmetry and the counterclockwise recedence from the centre are apparent.

A number of Op-artists made use of this sort of symmetry. An example from Spyros Horemis is shown in Figure 7.5. The chequered crown is invariant, again, under a similarity obtained by composing a rotation with a homothecy with the same centre. Unlike the two previous examples, however, the crown displays additional bilateral symmetries.

The sequence of iterates of, say, black "squares" builds spirals receding both clockwise and counterclockwise and the bilateral symmetry of these spirals gives no clue to the eye as to which sequence to follow. The resulting alternance between the two possible templets is the cause of the "flicker of the eye" so characteristic in Op Art.

7.3 Shears, strains and affinities

> In the shape of the labellum we see the affinity of Goodyera to Epipactis.
>
> C.R. Darwin (1962: 115)

Affinities preserve proportions; they only change the scale of objects. We next turn our attention to transformations not preserving proportions.

Figure 7.5 **Design by Spyros Horemis. (From (Horemis, 1970) with permission.)**

Given an oriented line ℓ and a number $\sigma > 0$, the *shear* with axis ℓ and coefficient σ is the transformation $\text{shear}_{\ell,\sigma} : \mathbb{E} \to \mathbb{E}$ associating to any point P in the plane the point P' obtained as follows. If $P \in \ell$ then $P' = P$. Otherwise, consider the half-line q passing through P, parallel to ℓ and with the direction of ℓ.

We take P' to be the only point in q, whose distance to P is σ times the distance (say d) from P to ℓ. A shear with coefficient $\sigma < 0$ is defined similarly by taking the half-line in the direction opposite to that of ℓ (and $\text{dist}(P, P') = -\sigma \text{dist}(P, \ell)$). A shear with coefficient $\sigma = 0$ is the identity Id. An equivalent, analytical, definition might be simpler to grasp. Take ℓ to be the y-axis. Then, for any σ and any point (x, y) on the plane, $\text{shear}_{\ell,\sigma}(x, y) = (x, y + \sigma x)$.

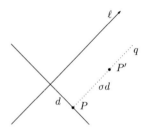

To keep with our habits, we visualize the effect of a shear with a footprint, which we now enclose in a rectangle to further convey the effect of shearing. The axis of the shear is the bottom side of the rectangle and its coefficient is 1.

We see in the picture that the four sides of the rectangle enclosing the footprint remain straight lines after the transformation and that opposite sides remain parallel. The rectangle, however, became a parallelogram. We next have a closer look at the properties preserved by shears.

Proposition 7.8 *Shears preserve lines. That is, for any shear φ and any line r, $\varphi(r)$ is a line. Furthermore, φ preserve parallelism.*

Proof Let ℓ be a line and σ a number such that $\varphi = \text{shear}_{\ell,\sigma}$. Choose a system of coordinates with ℓ as y-axis so that for all points $(x,y) \in \mathbb{E}$ we have $\varphi(x,y) = (x, y + \sigma x)$. Let a, b, c be real numbers such that, in this coordinate system, $ax + by + c = 0$ is an equation for r.

For any point $(x,y) \in \mathbb{E}$ we have that $(x,y) \in \varphi(r)$ if and only if $\varphi^{-1}(x,y) \in r$. Since $\varphi^{-1} = \text{shear}_{\ell,-\sigma}$ we have $\varphi^{-1}(x,y) = (x, y - \sigma x)$. Hence, $(x,y) \in \varphi(r)$ if and only if

$$ax + b(y - \sigma x) + c = 0.$$

It follows that $\varphi(r)$ has equation $(a - b\sigma)x + by + c = 0$, which is the equation of a line.

Let now p and r be two parallel lines given by the equations $y = mx + q$ and $y = mx + s$. The coefficients in x are the same because of Proposition 1.3. The images of these lines under φ have equations $y = (m-\sigma)x + q$ and $y = (m-\sigma)x + s$ respectively. Using Proposition 1.3 again, we deduce that these images are parallel as well. $\square$

Shears do not, in general, preserve proportions. The drawing on the left shows the shear (with horizontal axis) of an isosceles triangle which is no longer so after being sheared.

Nonetheless, shears do preserve proportions when restricted to lines. Such proportions are called *simple ratios*.

Proposition 7.9 *Shears preserve simple ratios. That is, for any shear φ and any three aligned points P, Q, R we have*

$$\frac{\text{dist}(\varphi(P), \varphi(Q))}{\text{dist}(\varphi(Q), \varphi(R))} = \frac{\text{dist}(P, Q)}{\text{dist}(Q, R)}.$$

Proof Let ℓ be a line and σ a number such that $\varphi = \text{shear}_{\ell,\sigma}$. Also, let r be the line where P, Q and R lie.

Because of Proposition 7.8 the image $\varphi(r)$ of r is a line. Also, the three lines passing through P and $\varphi(P)$, Q and $\varphi(Q)$, and R and $\varphi(R)$ are each all parallel to ℓ and hence parallel to each other.

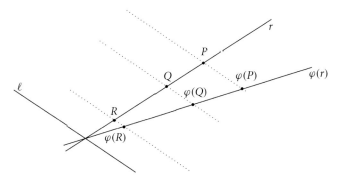

The desired equality is given by Thales' theorem (Theorem 1.2). $\square$

Remark 7.10 Another property preserved by shears is *area*. We have not provided a definition of what the area of a plane figure is and we will not do so now, since this definition is too technical[5] and, anyway, we will not use the area-preserving character of shears. In the absence of a formal definition we cannot prove this character. Nevertheless, we can briefly argue why this holds true.

To begin, we consider a rectangle having one side parallel to the axis of the shear. The picture below needs no words to convince the reader that the image of such a rectangle has the same area as the rectangle:

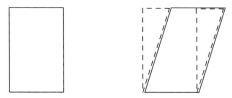

Now consider a figure F on the plane. We can place within the figure a number of non-overlapping rectangles (with one side parallel to the axis of the shear). Then the area of the union of these rectangles is at most the area of F. Furthermore, the area of the union of the images of the rectangles is at most the area of the image of F. And, as noted above, the former equals the area of the original rectangles.

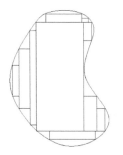

The remaining part of the argument uses the idea of "limit". We can keep adding (always non-overlapping) rectangles within F so that the difference between the area of F and the total area of the rectangles keeps decreasing. In the limit – and this means that the number of rectangles may grow indefinitely – this difference becomes zero, and the same happens with the difference of the areas of the images under the shear. Thus, in the limit,

---

[5] The reader is doubtless familiar with formulae computing the area of a rectangle or a triangle. The computation of the area for less-regular planar objects is considerably more involved.

$$\text{area}(F) = \text{area}(\text{rectangles}) = \text{area}(\text{shear}(\text{rectangles})) = \text{area}(\text{shear}(F)).$$

This finishes the argument.

The transformations we will next consider are more general than shears. As usual, we begin with the formal definition and then show an illustration with our familiar footprint.

Given a line ℓ and a real number $\lambda > 0$, the *strain* with axis ℓ and coefficient λ is the transformation $\text{strain}_{\ell,\lambda} : \mathbb{E} \to \mathbb{E}$ associating to a point P in the plane the point P' defined as follows. If $P \in \ell$ then $P' = P$. Otherwise we let r be the line passing through P and perpendicular to ℓ and Q be its intersection with ℓ. Then P' is the only point on r such that P and P' are on the same side with respect to ℓ and $\text{dist}(P', Q) = \lambda\,\text{dist}(P, Q)$.

If $\lambda > 1$ the strain is *expansive* and if $\lambda < 1$ it is *contractive*. Note that if $\lambda = 1$ the strain is the identity.

Strains with coefficient $\lambda < 0$ are similarly defined, the only difference being that P' is chosen on the other side of ℓ. Again, if $\lambda < -1$ the strain is *expansive* and if $\lambda > -1$ it is *contractive*. The case $\lambda = -1$ recovers an old friend of us, as stated in the next, trivial to prove, proposition.

Proposition 7.11 *A strain with axis ℓ and coefficient -1 is a reflection with axis ℓ.*

Strains produce a change of scaling in one direction. When this direction coincides with a natural direction of the strained figure the transformation has a different feeling than the one produced by a shear. The following figure gives an idea of the effect of a strain (with axis the dotted line and coefficient 1.5):

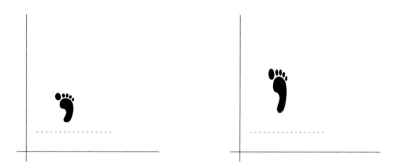

In general, however, the transformation produced by a strain is reminiscent of that produced by a shear (without the area-preserving feature). The picture on the left shows the straining of a rectangle with axis one of the rectangle diagonals.

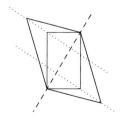

The corresponding straining of the footprint would look as follows:

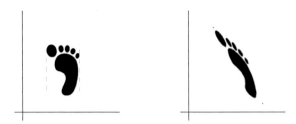

We now turn, in the next two propositions, to properties preserved by strains.

Proposition 7.12 *Strains preserve lines. That is, for any strain φ and any line r, $\varphi(r)$ is a line. Furthermore, strains preserve parallelism.*

Proof The proof is similar to that of Proposition 7.8. We need to use the property that, if $\varphi = \text{strain}_{\ell,\lambda}$, we may choose a system of coordinates with ℓ as y-axis so that for all points $(x, y) \in \mathbb{E}$ we have $\varphi(x, y) = (\lambda x, y)$ and, hence, $\varphi^{-1}(x, y) = (\frac{1}{\lambda}x, y)$. Using this we show that the image of the line with equation $ax+by+c = 0$ is the line with equation $\frac{a}{\lambda}x+by+c = 0$. ☐

Proposition 7.13 *Strains preserve simple ratios.*

Proof Again, the proof is similar to that of Proposition 7.9, the only difference being that the lines passing through P and $\varphi(P)$, Q and $\varphi(Q)$, and R and $\varphi(R)$ are each now perpendicular to the axis of the strain (instead of parallel to the axis of the shear). ☐

The composition of an arbitrary number of shears and strains gives a group of transformations that contains the group of similarities. These transformations, as we have seen, do not necessarily preserve proportions. We next formally define this group and then show that it contains the group of similarities.

Definition 7.14 An *affinity* is a composition of an arbitrary number of shears and strains. That is, a transformation $\varphi : \mathbb{E} \to \mathbb{E}$ of the form

$$\varphi = \psi_n \circ \psi_{n-1} \circ \cdots \circ \psi_2 \circ \psi_1,$$

where, for $i = 1, \ldots, n$, ψ_i is either a shear or a strain. We denote by $\text{Aff}(\mathbb{E})$ the set of all affinities.

The proof of Theorem 7.6 carries over, *mutatis mutandis*, to prove the following result.

Theorem 7.15 *The set $\text{Aff}(\mathbb{E})$ is a group (of transformations on $\mathbb{E}$).*

Our next goal is to show, in Theorem 7.17, that any similarity is an affinity. The next proposition will be used towards this goal.

Proposition 7.16 *Any homothecy can be written as the composition of two strains with perpendicular axes and the same coefficient.*

Proof Consider any homothecy $\mathrm{homot}_{O,\lambda}$ with centre O and coefficient λ. Take two perpendicular lines, say X and Y, passing through O and the strains $\mathrm{strain}_{X,\lambda}$ and $\mathrm{strain}_{Y,\lambda}$. Then, taking coordinates with X and Y as axes, we have, for all points (x, y),

$$(\mathrm{strain}_{Y,\lambda} \circ \mathrm{strain}_{X,\lambda})(x, y) = \mathrm{strain}_{Y,\lambda}(\lambda x, y)$$

$$= (\lambda x, \lambda y) = \mathrm{homot}_{O,\lambda}(x, y).$$

That is, we can represent $\mathrm{homot}_{O,\lambda}$ as the composition of $\mathrm{strain}_{X,\lambda}$ and $\mathrm{strain}_{Y,\lambda}$. ☐

Theorem 7.17 *Any similarity is an affinity.*

Proof It suffices to show that both isometries and homothecies can be written as a composition of strains. The case of isometries is an immediate consequence of the facts that any isometry can be written as the composition of at most three reflections (Theorem 3.9) and that any reflection can be written as a particular strain (Proposition 7.11). The case of homothecies has been dealt with in Proposition 7.16. ☐

Theorem 7.18 *Affinities preserve straight lines, parallelism and simple ratios.*

Proof As in the proof of Theorem 7.7, it is enough to see that both shears and strains preserve these properties. And this is true by Propositions 7.8, 7.9, 7.12 and 7.13. ☐

The next result extends to affinities a property we already know for isometries (cf. Proposition 2.5). We will use it to prove Theorem 7.20 below.

Proposition 7.19 *Any affinity is determined by the images of three non-collinear points.*

Proof Let P, Q, R be three non-collinear points and P', Q', R' be their images under an affinity φ. Note that the last three points cannot be collinear since the first three are not and affinities preserve alignment (Theorem 7.18).

Now let S be any other point on the plane. We want to show that the image $S' = \varphi(S)$ of S is determined by P, Q, R, P', Q', R' and S. To do so, consider the lines q and r determined respectively by P and Q and by P and R. And similarly for p' and r'.

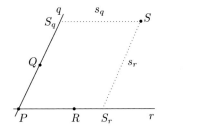
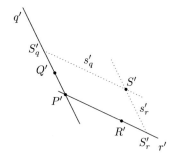

Consider the line s_q passing through S and parallel to r and let S_q be the intersection of this line with q. Let s_r and S_r be defined in the same manner. Then the image S_q' of S_q is determined by P, Q, S_q, P' and Q' because affinities preserve simple ratios (Theorem 7.18), therefore, S_q' must be the only point in q' satisfying that

$$\frac{\text{dist}(S_q', P')}{\text{dist}(Q', P')} = \frac{\text{dist}(S_q, P)}{\text{dist}(Q, P)}.$$

The point S_r' is similarly determined. Finally, consider the images s_q' and s_r' of s_q and s_r respectively. Since S is in both s_q and s_r, S' must be in the intersection of their images. But these images are determined by q, r, S_q, S_r and their images. Indeed, s_q is the only line through S_q parallel to r. Then, since affinities preserve parallelism (Theorem 7.18 once again), s_q' must be the only line through S_q' parallel to r'. And similarly for s_r'. We can thus conclude our proof, since S' is determined as the only intersection of s_q' and s_r'. □

The last result in this section shows that any triangle can be transformed into any other triangle by an affinity. In some sense, this describes precisely what affinities can do.

Theorem 7.20 *Given any pair A, B, C and P, Q, R of triples of non-collinear points there exists a unique affinity $\varphi : \mathbb{E} \to \mathbb{E}$ mapping A to P, B to Q, and C to R. Furthermore, φ can be written as the composition of a direct (orientation-preserving) isometry, a homothecy, a strain and a shear.*

Proof The idea of the proof (we will not present all the minute details) is well expressed in the following picture. On this picture we marked the positions of P, Q and R as small triangles and the points A, B and C as red dots. For each of the four transformations we will consider, we take the images of the red dots but leave the triangles fixed. We want to show that after applying the four transformations the red dots have been mapped onto the triangles.

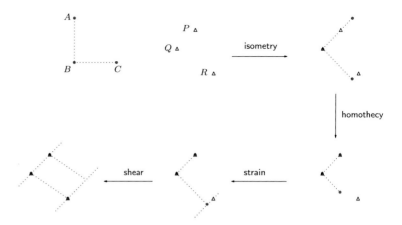

We first find an isometry mapping B into Q (a suitable translation, for instance) and the line $\overline{AB}$ into the line $\overline{PQ}$ (composed with a suitable rotation with centre Q). We then apply a homothecy with centre Q and coefficient chosen such that the image of A is P. Note that this homothecy leaves Q (and hence the image of B up to now) fixed. We thirdly apply a strain with axis $\overline{PQ}$ and coefficient chosen such that the image of C lies on the line passing through R and parallel to $\overline{PQ}$. Again, this strain leaves the line $\overline{PQ}$ (and hence the images of A and B up to now) fixed. We finally apply a shear with axis $\overline{PQ}$ (so that the images of A and B are always fixed) whose orientation and coefficient are such that the image of C is R.

The composition of these transformations is an affinity and maps A, B, C to P, Q, R. It is the only affinity doing so by Proposition 7.19. □

Remark 7.21 Strains may be seen to play the same role within affinities that reflections play within isometries. We have already seen that isometries and homothecies (and, hence, any similarity) can be written as a composition of strains. One can also prove, but we will refrain from doing so, that any shear can be written as a composition of three strains. Using Theorem 7.20, this implies that any affinity can be written as the composition of nine strains.[6]

Affinities do not preserve circles as similarities do, as shown in the drawing on the left depicting the effect of a strain, with horizontal axis and coefficient 2, on a circle.

When understanding the behaviour of affinities on circles a new class of objects naturally comes into play.

---

[6] The number nine is surely not optimal. However, I have been unable to find any reference in the literature to the minimal number of strains needed to represent an affinity.

7.4 Conics

> We are apt to think of mathematical definitions as too strict and rigid for common use, but their rigour is combined with endless freedom. The precise definition of an ellipse introduces us to all the ellipses in the world [...]
> D.W. Thompson (1961: 269)

The simplest curves (after straight lines and circles) are the conics. They can be characterized in several different ways, of which we will see three in this section. The following definition gives the first of these three.

$x^2 + 4y^2 - 4 = 0$

Definition 7.22 We call *plane conic* (or simply *conic*) any figure in $\mathbb{E}$ which, in a given system of coordinates, is the set of points satisfying an equation of the form

$$ax^2 + bxy + cy^2 + dx + ey + f = 0$$

for some constants $a, b, c, d, e, f \in \mathbb{R}$. If $b^2 - 4ac < 0$ we say that the conic is an *ellipse*, if $b^2 - 4ac = 0$ we say that it is a *parabola*, and if $b^2 - 4ac > 0$ that it is a *hyperbola*. The number $b^2 - 4ac$ is called the *discriminant* of the conic.

$x^2 - y = 0$

The pictures on the right show one conic in each class (from top to bottom, ellipse, parabola and hyperbola), together with their equations.

The dashed line in the parabola is its *axis*. One sees that the two "branches" of the parabola, as they separate from the axis, become increasingly aligned with it. In contrast, the branches of the hyperbola tend to align with the two dotted lines, called *asymptotes*.

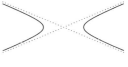

$x^2 - y^2 - 1 = 0$

Remark 7.23

(i) An important fact, that we will not prove here since any possible proof would deviate too much from our main line of thought, is that the property of a figure being a conic (actually, being an ellipse, parabola or hyperbola) is independent of the coordinate system chosen. If it is so for one coordinate system, it is so for any.

(ii) Let C be a circle of radius R centred at a point O with coordinates (α, β). A point P (with coordinates (x, y)) is in C if and only if $\text{dist}(P, O) = R$ or, equivalently,

$$\sqrt{(\alpha - x)^2 + (\beta - y)^2} = R.$$

This gives the equivalent equation

$$x^2 + y^2 - 2\alpha x - 2\beta y + (\alpha^2 + \beta^2 - R^2) = 0.$$

This shows that the circle is a particular case of an ellipse (the coefficients a and c in Definition 7.22 are equal and the coefficient b is zero).

(iii) The definition above counts two non-parallel lines as a hyperbola (for instance, $xy = 0$ is the equation of the figure formed by the two coordinate axes) and two parallel lines as a parabola (for instance, $x^2 - 1 = 0$ is the equation of the two lines $x = 1$ and $x = -1$). These conics (actually, any conic containing a line) are said to be *degenerate*. In the following we will not consider these degenerate cases.

Proposition 7.24 *A conic intersects a line in at most two points.*

Proof Choose the coordinate axes so that the line has the equation $y = 0$ (i.e. the line is the x-axis). The conic C has equation $ax^2 + bxy + cy^2 + dx + ey + f = 0$ for some constants a, b, c, d, e, f. A point in the line has coordinates $(x, 0)$ for some x. If it also belongs to C it must satisfy $ax^2 + dx + f = 0$. If $a = d = f = 0$ then all points $(x, 0)$ are in C. This means that the whole line $y = 0$ is included in C and, therefore, that the conic is degenerate, which we assumed it was not. Otherwise, the equation $ax^2 + dx + f = 0$ has at most two solutions. □

Conics had been already studied by the Greeks (though not by Euclid in the *Elements*) and a number of their properties were known to them. The Greeks did not define conics using equations (Descartes would be the first to do so) but as sections of a cone; that is, as intersections of a cone and a plane. The cone had to be symmetric around its vertex so that the hyperbola would have two components. Figure 7.6 shows these sections. On the left we see a parabola, in the middle are two sections (both yielding an ellipse – the lower one actually begin a circle), and on the right a hyperbola.

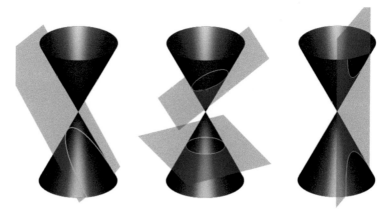

Figure 7.6 **Conic sections. (Pbroks13/CC-BY-3.0.)**

Conics can also be defined without recourse to either analytic geometry or three-dimensional geometry. Thus, ellipses turn out to be characterized as sets of points for which the sum of the distances to two fixed points (called the *foci* of the ellipse) is constant.

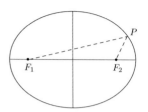

The line segment within the ellipse passing through the foci is its *major axis* and the line segment passing through the midpoint of the major axis and perpendicular to it is its *minor axis*.

Similarly, hyperbolas can be characterized as sets of points for which the absolute value of the difference of the distances to two fixed points (called the *foci* of the hyperbola) is constant.

The line passing through the foci (F_1 and F_2 in the drawing) is called the *transverse axis* and the line passing through the midpoint of the foci and perpendicular to the transverse axis is called the *conjugate axis*. Note that for all points P in the hyperbolas the value $|\mathrm{dist}(P, F_1) - \mathrm{dist}(P, F_2)|$, which we know is constant, is the distance between the *vertices* of the hyperbolas, the latter being the points where the hyperbolas cut their transverse axes.

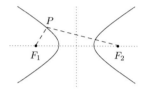

Finally, a parabola can be described as the set of points equidistant to a fixed point (the *focus* of the parabola) and a fixed line (its *directrix*).

The fact that conics can be defined in terms of distances implies that they are preserved under isometries. One can also prove this fact from Definition 7.22 (together with Remark 7.23(i)). Or derive it from our next result together with Theorem 3.9 and Proposition 7.11.

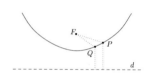

Theorem 7.25 *Affinities preserve ellipses, parabolas, and hyperbolas.*

Proof It is enough to see that both shears and strains preserve ellipses, parabolas, and hyperbolas.

Let ℓ be a line and $\lambda > 0$ be a real number. Also, let C be a conic. We want to show that $C' = \mathrm{strain}_{\ell,\lambda}$ is also a conic and of the same type C is. To do so, choose a coordinate system with ℓ as y-axis. Let a, b, c, d, e, f be numbers such that, in this coordinate system, C has equation $ax^2 + bxy + cy^2 + dx + ey + f = 0$.

For any point (x, y) on $\mathbb{E}$, $(x, y) \in C'$ if and only if $\mathrm{strain}_{\ell,\frac{1}{\lambda}}(x, y) \in C$. That is, if and only if

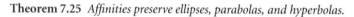

$$a\left(\frac{x}{\lambda}\right)^2 + b\frac{x}{\lambda}y + cy^2 + d\frac{x}{\lambda} + ey + f = 0.$$

That is, C' is a conic with equation $\frac{a}{\lambda^2}x^2 + \frac{b}{\lambda}xy + cy^2 + \frac{d}{\lambda}x + ey + f = 0.$ Furthermore, its discriminant is

$$\left(\frac{b}{\lambda}\right)^2 - 4\frac{a}{\lambda^2}c = \frac{1}{\lambda^2}\left(b^2 - 4ac\right).$$

Since $1/\lambda^2 > 0$ this discriminant has the same sign as the discriminant of C, which implies that C and C' are of the same type.

The argument for a shear is similar, the equation of C' now being

$$ax^2 + bx(y - \sigma x) + c(y - \sigma x)^2 + dx + e(y - \sigma x) + f$$
$$= (a + c\sigma^2 - b\sigma)x^2 + (b - 2c\sigma)xy + cy^2 + (d - e\sigma)x + ey + f = 0.$$

Its discriminant is

$$(b - 2c\sigma)^2 - 4(a + c\sigma^2 - b\sigma)c = b^2 - 4ac;$$

that is, the discriminant of C. Hence, C and C' are, again, of the same type. $\qquad\square$

7.5 The eclosion of ellipses

During the first half of the sixteenth century, a number of Italian artists reacted against the naturalistic style associated with the height of the Renaissance as exemplified by Leonardo or Raphael. To the poised, restrained manners and ordered compositions prevalent in Renaissance paintings they opposed a style (eventually known as *Mannerism*) which has been described as follows (Cheney, 1997: 7):[7]

Form is characterized by an emphasis on the exaggerated expression of the human body – serpentine or twisted poses by a concern to elongate the human body, thus distorting the figure and limiting its spatial relation – by using bright, acid, sharply contrasting colors for the purpose of emotionality, by abruptly heightening or diminishing light effects, thus creating visual disturbances, by grouping the figures in a shallow or deep illusionistic space, thus creating a disjointed relationship between the space of the canvas and the painting, and by depicting rich, elaborate, and exotic textures and creating a highly polished and decorative surface. The composition is central, unbalanced, over-rhythmical, and devoid of physical harmony. The organic unity of the form is separated by the strong emphasis on creating tension and paradoxical relationships through use of the elements of design.

Among the many characteristics in the lengthy description above, it is the elongation of the human body – which Arnold Hauser refers to as "the slender forms of the Mannerist style" (Hauser, 1962: volume II, 96) – that we wish to focus on. Two examples of it, representing early and late Mannerism, are shown in Figure 7.7.

---

[7] This passage appears to be lacking punctuation marks. I have added some with the only goal of improving legibility.

Figure 7.7 **Parmigianino,** *Madonna with the Long Neck*, **oil on panel, 1534–40 (Galleria degli Uffizi, Florence, Italy/Alinari/The Bridgeman Art Library) (left) and El Greco,** *Saint Martin and the Beggar*, **oil on canvas, 1597–1600 (National Gallery of Art, Washington, DC, USA/Index/The Bridgeman Art Library) (right).**

The delicate hands of the Madonna in Parmigianino's painting or the torso of the beggar in El Greco's are representations of the human body possessing a strange eloquence, a beauty that, incidentally, cannot fail to evoke what Umberto Eco (2004: 428) described as "the anorexic charms of the latest fashion models".

It is perhaps interesting to ponder the occurrence of similar elongations in Mannerist architecture, not on the human figures possibly displayed in ornamental work (either painted or carved), but on the more structural elements of design. These elements were supported by shapes drawn from a small collection of geometric objects such as rectangles, triangles and circles. The elongation of these objects would amount to applying to them a strain (with an axis parallel to a rectangle side in the case of this figure). And it is at this point that we note that, no matter the position of the axis, the strain of a circle produces an ellipse not a circle. If elements of design that up to the sixteenth century had a circular shape (windows, dome sections, enclosed space layouts) were going to be elongated, we should witness the emergence of elliptical designs for such elements. As indeed we do.

Figure 7.8 **Francesco Borromini, *San Carlo alle Quattro Fontane* (frontal view).**

The introduction of elliptical designs in architecture is commonly exemplified in the work of the Baroque arquitect Francesco Borromini, notably by his church at San Carlo alle Quattro Fontane. Anthony Blunt (1979: 67) writes, however, that:

In using a basically oval plan [for San Carlo alle Quattro Fontane] Borromini was following a tradition which had been growing in importance since the mid-sixteenth century. Peruzzi and Serlio had invented several ingenious designs based on a simple oval, but they were not carried out. Da Vignola had used an oval dome over a rectangular substructure at S. Andrea sulla via Flaminia (1550–53), and in his later S. Anna dei Palafrenieri (begun 1565) he made the body of the church itself oval. In S. Giacomo degli Incurabili (begun just before 1590) Francesco da Volterra followed his example, but on a much larger scale.

Figures 7.8 and 7.9 (right) allow one to appreciate some of the ellipses in San Carlo alle Quattro Fontane. The most celebrated of these ellipses is, however, not visible in these figures. It consists of the section of the church's dome, which can be seen in Figure 7.10.

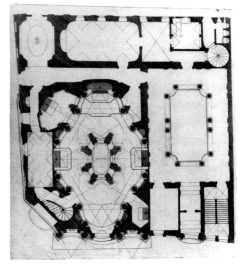

Figure 7.9 **Francesco Borromini, *San Carlo alle Quattro Fontane* (elevation and groundplan).**

Figure 7.10 **Francesco Borromini, *San Carlo alle Quattro Fontane* (dome).**

Another well-known elliptical structure from the seventeenth century is the shape of St Peter's square, in front of St Peter's Basilica in Vatican City, designed by Gian Lorenzo Bernini.

The reflection of Mannerist painting in architectural design is returned in kind to Baroque painting. Here, elongation occurs not as a feature of the depicted bodies but as a compositional device. This is apparent in several of Rubens' works (Belkin, 1998). The canvas shown in Figure 7.11 was painted by Rubens in an early period of his career when he worked in Mantua at the court of the Duke Vincenzo Gonzaga. It is divided into an upper section – showing a representation of the Holy Trinity on a tapestry held by angels – and a lower section – showing the Duke Vincenzo and his father Guglielmo on the left along with their respective wives, Eleonora de Medici and Eleonora of Austria, on the right, in adoration. The painting was eventually cut up and some of its fragments lost. The main parts, nevertheless, remain in Mantua, and these remains, however patchy, are enough to make our point.

A quick glance at the painting reveals that in each of the two sections the characters are positioned in a way that exhibits bilateral symmetry with respect to a central vertical axis. In addition, they suggest an elliptic shape. This is more clearly visible in the upper section with the ellipse roughly drawn by the fabric held by the angels as a background and by the angels themselves. We have overdrawn such an ellipse in the left part of Figure 7.13. In this figure we also show a copy of the ellipse translated to the lower section of the painting. The shape of this lower ellipse is again suggested by fabric, more precisely by the red cloaks held by the Gonzaga couple. Observing the upper and lower compositions, a difference between them stands out: the foci of the upper ellipse are situated at the figures of the Father and Son (and its centre slightly below the Holy Spirit) whereas the Gonzaga couple are more eccentric. Rubens is thus putting an emphasis in the Holy Trinity that is not put on his benefactor. The positions of the four figures within the lower ellipse also suggests a triangle, with base at the feet of these figures and vertex at the upper angel, whose sides pass through Father and Son. Such a triangular composition contributes to the emphasis mentioned above by directing the eye upwards.

Another example of elliptical design in Rubens' paintings occurs in his well known *The Exchange of Princesses* (Figure 7.12) produced between 1622 and 1625 as part of a cycle commissioned to Rubens by Marie de' Medici, wife of King Henry IV of France, themed around Marie's life. The picture shows the meeting of Princess Anna of Austria and Princess Isabella of France, the daughter of Marie, on 9 November 1616, before their weddings to each other's brothers, namely the Prince of Spain (later to become Philip IV) and that of France, hence son of Marie (later to become Louis XIII). On the left of the painting

Figure 7.11 **Peter Paul Rubens,** *The Gonzaga Family Worshipping the Holy Trinity,* **oil on canvas, 1604–05. (Palazzo Ducale, Mantua, Italy/Alinari/The Bridgeman Art Library.)**

we see Isabella, then aged 13, being taken by (a personification of) Spain, while on the right we see Anna being gently pulled by France. Above them, on the sides, two putti wave hymeneal torches, while, on top, a group of winged putti surrounds Felicitas Publica who showers the couple with gold. Below the main group, the river-god Andaye rests on his urn and a nereid presents a piece of coral and a strand of pearls to the princesses, while a triton blasts a conch to trumpet the occasion.

The elliptical structures in the exchange are, again, apparent. In the lower section, an ellipse is suggested by the extended arms of Spain and France (as well as that of Isabella). This ellipse confines the princesses, Spain, France, and the youth between Anna and France, usually thought to be Hymen. We have overdrawn this ellipse at the right of Figure 7.13. Another ellipse, which we have not made explicit in this figure, is nevertheless made clear in the arms of the winged putti in the upper part of

Figure 7.12 **Peter Paul Rubens, *The Exchange of Princesses*, oil on canvas, 1622–25. (Louvre, Paris, France/Giraudon/The Bridgeman Art Library.)**

the painting. This celestial ellipse brings prominence to Felicitas Publica by having her at its centre. Down at earth, as Severo Sarduy (1974: III.3) notes, an asymmetry occurs. In this terrestrial ellipse, a reflection of the celestial one, one of the foci is occupied by Anna, who shines like the sun. Isabella, instead, is placed below the other focus in a display of dullness.

The two paintings discussed above are not isolated examples of elliptical structure in the work of Rubens. This kind of structure can also be seen, for instance, in his 1608 *The Virgin and Child Adored by Angels*.

The mention by Sarduy of the sun as the luminous focus of an ellipse refers to the first of the three laws of planetary motion postulated by Kepler: the orbit of every planet is an ellipse with the Sun at

Figure 7.13 **Ellipses in the compositions of** *The Gonzaga Family Worshipping the Holy Trinity* **(left) and** *The Exchange of Princesses* **(right).**

one of the two foci. This law – bringing the eclosion of ellipses to other arenas – was published by Kepler in 1609, after the painting of *The Gonzaga Family*, where no marked asymmetries occur, but well before that of *The Exchange of Princesses*. I assume that Rubens must have known of it, but I have read no evidence to sustain this assumption.

Also, the view of ellipses as elongated circles and the suggested connection of painting and architectural design are, as described in the discussion above, a mere analogy. The possible existence of a causal relationship is left for the art historians to decide.

7.6 Klein (aber nur der name)

At this point we can return to the discussion we initiated in Section 1.2. We claimed there that perception takes place within a context which, so to speak, chooses a geometry and we illustrated this claim by elaborating on the notion of shape, a notion which we informally related with the features that are essential to a geometry. We may now become more precise.

In the previous chapters we have considered a fixed two-dimensional space, the Euclidean plane $\mathbb{E}$, over which we studied the action of three groups of transformations: isometries, similarities and affinities. For each of these classes of transformations, we have shown a number of features that are preserved and a number which are not. Furthermore, since isometries are similarities and similarities are affinities, the features preserved by affinities are preserved as well by similarities, and those preserved by similarities are preserved by isometries.

In 1872, Felix Klein linked geometry and algebra by defining a geometry as a space together with a group of transformations acting on this space. This view allowed him to see a geometry as an abstract object and to classify geometries based on purely group-theoretical considerations. The details of Klein's viewpoint (known as the *Erlangen program*) are, needless to say, too technical to be described here. They are also not too relevant to our discussion. Suffice it to say that the endowing of space $\mathbb{E}$ with the groups of isometries, similarities and affinities yields, in Klein's terms, three different geometries, which we will call *metric*, *Euclidean* and *affine* respectively. The fact that, for instance, every isometry is a similarity makes metric geometry more restrictive than Euclidean geometry. Indeed, every theorem in the latter is a theorem as well in the former. But the converse is not true since statements involving the distance between two points will be true, or false in metric geometry but are meaningless in Euclidean geometry (since similarities may change this distance). We can say that Euclidean geometry has less theorems than metric geometry but they are more general. Also, adding transformations to a geometry results in the possible addition of new symmetric objects. An example at hand is the logarithmic spiral, which is not symmetric within metric geometry but does possess symmetries within Euclidean geometry. Finally, we can precisely specify the features we consider as essential to each of these three geometries. They are, respectively, those preserved by isometries, similarities and affinities.

Although not commonly considered, a fourth geometry on the space $\mathbb{E}$ can be defined by endowing it with the group of *area-preserving* transformations. This is the subgroup of the set of affinites generated by shears and isometries. That is, the set of all transformations which can be written as a composition of shears and isometries. The resulting geometry, which we will call *authalic*, has as essential features alignment, simple ratios and area. It is more general than metric geometry and more restrictive than affine geometry. It does not compare with Euclidean geometry. Representing "A is more general than B" with an arrow from B to A we can summarize the relations between these four geometries as follows:

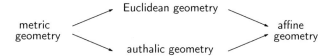

We can revisit the notion of shape – or, to be more precise, of two objects having the same shape – under the light of the ideas above and become more specific: within a given geometry, two objects have the same shape when there is a transformation mapping one into the other.

Consequently, with this definition, the two tiles in Section 1.2 have the same shape within affine geometry (it is possible to prove that in this geometry every parallelogram can be mapped into any other) but not within Euclidean geometry (since the latter preserves angles).

Our claim that the context within which perception takes place selects a geometry can also now be given a more specific meaning. It is this context which classifies geometric features as relevant or otherwise. For instance, when travelling by car with the help of a map, distances between points in the map are not relevant but proportions are. These proportions allow me to estimate driving times between different points on my route independently of the scale of the map (within, of course, a range of map scales allowing for a comfortable readability). Euclidean geometry appears to be appropriate in this context. In contrast, when finding my way in London's underground system by using the system's map, proportions are not important. Actually, even the notion of being a straight line is not relevant in this context (and, indeed, diagrams for the underground systems in many cities systematically disregard the curvature of their different lines and represent them by piecewise linear segments joined by rounded elbows). The relevant features here are related to such things as connectedness (whether I can reach a station from another station), intersection of lines (in particular, how many line changes a given path has) and number of stations (usually drawn as ▬○▬) met along a path (a number naturally related to the time it takes to travel the path). Any system map respecting these features (again, within some limits ensuring comfortable readability) is appropriate; hence, while designing such maps, it is common to overlook proportions, angles or curvatures and to choose a representation that allows for a clear view of details such as names of stations, intersections of lines, and so on. The geometry having such essential features is called *topology* and the transformations defining it in the sense of Klein are known as *homeomorphisms*. It will not occupy us in the rest of this book.

D'Arcy Thompson (1961: 326), whom we met at the beginning of this chapter, avowed he was "no skilled mathematician" but that "something of the use and beauty of mathematics I think I am able to understand". It is apparent that he did indeed, in so far as the role of transformations in geometry (and the relation of this role with the notion of shape) is concerned. An affinity mapping any of the drawings in Figure 7.1 into the other expresses a very definite relation for the shapes of the two fish, namely that they have the same shape under an affine geometry viewpoint. Affinities of this sort are much more precise than the one mentioned by Darwin at the opening of Section 7.3. In the last chapter of his book (Chapter IX, "On the theory of Transformations, or the comparison of related forms") Thompson produces a number of

Figure 7.14 **Section of the shell of *Nautilus pompilius*. (Chris73/Wikimedia Commons.)**

comparisons based on geometric transformations (affinities and more general). In other chapters he exhibits instances in nature where the form of an organism can be both described mathematically and explained in mechanical terms. A case bearing relation with our previous discussion in this chapter is provided by the form of the *Nautilus pompilius*, whose central shell section draws a beautiful logarithmic spiral (Figure 7.14). Another example of the logarithmic spiral mentioned by Thompson is the trajectories that certain insects follow towards a candle. Further examples of spirals in nature can be found in Cook (1979).

Aural wallpaper

8

All art constantly aspires to the condition of music.

W. Pater (1986: 86)

[...] why do so many of us constantly try to explain the beauty of music, thus apparently depriving it of its mystery? The fact is that music is not only a mysterious and metaphorical art; it is also born of science. It is made of mathematically measurable elements [...] And so any explication of music must combine mathematics with aesthetics [...]

L. Bernstein (1976: 9)
Reprinted by permission of the publisher.
Copyright © 1976 by Leonard Bernstein.

In 1893 Erik Satie composed *Vexations*, a musical piece consisting of a short theme – of about a minute and twenty seconds – which had to be played "840 times in succession" with the player(s) prepared beforehand "in the deepest silence, by serious immobilities". A possible goal of this piece was, apparently, to bore his critics.[1] In the trade-off between boredom and confusion Satie was structuring his piece to achieve the former.

Yet repetition is, in many forms and in smaller doses, an essential constituent of music, allowing themes to be re-expressed and patterns to be displayed. What is more, its many forms correspond to different geoemetric transformations whose use is neatly exemplified in the various kinds of canon, a form of imitative counterpoint common in the Baroque. They are also the building bricks which, via the notion of invariance, are used to obtain the different symmetries occurring in music.

To describe the above we first need to provide some basics of music theory. Such a provision, even if highly simplified, should endow us with a common language and allow us to embed the space where music lives into the Euclidean plane.

---

[1] Of the Chorale opening *Sports et Divertissements*, which Satie was going to compose in 1914, he said "I have put all I know about Boredom. I dedicate it to those who don't like me". *Vexations* was only published in 1949, after Satie's death, and was first performed on 9 September 1963, in New York, under the initiative of John Cage. By then, most (if not all) of Satie's critics, as well as Satie himself, were no longer alive to be bored.

8.1 Elements of music

We will not attempt to define what music is. It is less controversial (and sufficient for our needs) to observe that music involves a combination of different "structured" *sounds* during a period of time. Each of these musical sounds is called a *note* and it has associated with it a certain definite *pitch*.[2] Pitch is easy to measure since it corresponds to different vibration frequencies. Since these frequencies can take a continuum of values, it follows that there is a continuum of possible pitches. Yet, while some musical traditions, like the Indian, may claim to rely on a large family of musical pitches, Western tradition is based upon a limited number of them.

The basic set of notes, the *chromatic scale*, comprises 12 pitches. By doubling (or halving) the frequency associated with any of these pitches we obtain a pitch that, although new, sounds closely similar to the original. Musicians give to this new note the same name but say that it is an *octave* higher (or lower) than the original. The chromatic scale equally divides the interval – also called *octave* – between its first note (lowest pitch) and the same note one octave higher in 12 equal parts, which are called *half-steps*. We may obtain the set of all notes used in traditional Western music by concatenating several copies of the chromatic scale. A *diatonic scale* is an (arbitrary) selection of seven of these notes in some arrangement of whole or half-steps (a whole-step being, not surprisingly, two consecutive half-steps). We will return to the issue of diatonic scales later on this section and to the role of the remaining five pitches (called *accidentals*). For the time being, and for the purposes of illustration, we consider the diatonic scale (said to be in the key of C-major, we will soon enough see why) consisting of the notes C, D, E, F, G, A and B. This is a convenient choice, since these notes correspond to the white keys on a piano keyboard (the black keys being the accidentals of C-major). In Figure 8.1 we see a fragment of the piano keyboard corresponding to three octaves.

Musical works are sets of notes organized in particular ways. To understand this organization, two notions that stand out are *melody* and *harmony*. Basically, the first term refers to the way notes are sequenced in time and the second to the way they are simultaneously sounded at a given instant in time.

---

[2] What can be considered a "musical sound" is actually a matter of discussion. And in today's musical inventory one can find works for unpitched percussion, such as Edgar Varèse's *Ionisation*. It is in this sense important to bear in mind that the contents of this section apply to the *common practice* period spreading, in Western music, from about 1700 to 1900.

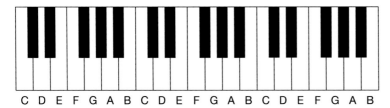

Figure 8.1 **A piano keyboard (fragment).**

To the uninitiated, melody is certainly the most basic, and the simplest to understand, of the above two notions. It is the melody of a song that we hum or whistle when we want to transmit that song to somebody else (or recreate it for ourselves). Humming or whistling, though, are not convenient as a means to transmit music among musicians. For that purpose, musicians have developed a notation (just as mathematicians have) that allows them to communicate smoothly. Take, for instance, the sequence of notes (a simple melody)

CDEFGABCBAGFEDC.

We could express this melody as shown above. Musicians, however, have agreed on a different system, shown in Figure 8.2. The set of five parallel lines is called a *staff* and the (vertical) position of the ovals on the staff determines which note each oval corresponds to. The accidentals are denoted by using, in addition, the symbols ♯ and ♭ (called *sharp* and *flat*), but we will not describe how here since this is not relevant to our discussion.

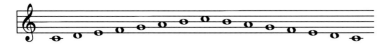

Figure 8.2 **The C-major scale ascending and descending.**

In addition to pitch, musical sounds have a *duration*. The duration of the notes in a melody is as essential as their pitch, and the way these durations are sequenced determines the *rhythm* of the melody. Again, one could in principle allow for a continuum of durations, but in Western tradition this has not been done (until recently). On the contrary, only a handful of durations are commonly used. The left of Figure 8.3 displays six common durations. They receive the names of breve, semibreve, minim, crotchet, quaver and semiquaver. At the right

Figure 8.3 **Breve, semibreve, minim, crotchet, quaver and semiquaver.**

of Figure 8.3 appear the symbols for a silence with the corresponding durations (silences that in music are, maybe inappropriately, called *rests*).

A less essential, but nonetheless important, element of melody is the *quality* of a sound. This includes features of the sound such as its loudness.

Before moving to harmony, a remark on contiguity is of the essence. When sounding two notes, one after the other, the difference in pitch between these notes is not irrelevant to the ear. As it happens (and this can be explained in terms of the quotient of the corresponding frequencies), some differences sound pleasant to the ear while others do not. A pair of consecutive (in time) notes is called a *melodic interval*. It receives a name derived from the number of notes between them. Thus, the interval B–C is a *second* and the interval C–F is a *fourth*. Intervals pleasant to the ear are called *consonant* and those which are not are *dissonant*. While the classification of intervals as consonant and dissonant has historically changed, a point worth emphasizing in our discussion is that arbitrary sequences of notes are very likely to contain a large number of dissonant intervals. This property highlights a complexity issue in the task of composing melodies. In the universe of possible melodies, coming up with a consonant one is not trivial, and coming up with a beautiful one is much less so.

We next focus on harmony. Since this deals with pitch simultaneity, a basic notion is that of a set of notes sounded together. Sets of two notes played together are called *harmonic intervals* and sets of three or more are called *chords*. Harmonic intervals are named just like the melodic ones (see Figure 8.4, where the U stands for *unison*).

U   2nd 3rd 4th 5th 6th 7th 8ve

Figure 8.4 **Harmonic intervals.**

Harmonic intervals and chords can also be consonant or dissonant. And both consonances and dissonances are used in musical works. But dissonances are felt as creating *tensions* that need to be *resolved* through the (subsequent[3]) use of consonances. We find again that arbitrary sequences of intervals or chords are likely to contain a large number of

---

[3] The resolution of a dissonance by a consonance (traditionally by the dissonant upper pitch descending by a half- or a whole-step to create a consonance) can be delayed by the insertion of combinations of pitches that do not specifically resolve the dissonance in question. This results in a *prolongation* of the dissonant function, a prevalent means of achieving a longer musical work.

unresolved tensions. Again, this points to a complexity issue in music composition.

It is not hard to imagine that, since music composition offers difficulties related to both melody and harmony, the composition of works involving simultaneous melodies must require the composer to possess considerable skill. It is therefore not surprising that early uses of several voices were *monophonic* in *texture*. That is, they had all the voices singing the same melody with a fixed interval between them. For instance, all male voices singing a given melody and all female voices singing the same melody one octave higher.[4] In the Middle Ages, monophony was extended to *homophony*, a texture in which one voice sings a melody and one or more other voices accompany it, mainly, to produce harmonic consonance. Eventually, homophony evolved into *polyphony*, a texture in which two or more voices sing independent melodies. The way in which two or more voices relate harmonically while keeping independent melodies is known as *counterpoint*. It captures a combination of the complexities of both melody and harmony. In the words of John Rahn (2000: 177):

It is hard to write a beautiful song. It is harder to write several individually beautiful songs that, when sung simultaneously, sound as a more beautiful polyphonic whole.

The study of counterpoint started during the Renaissance period and arguably reached its culmination during the Baroque period with an emphasis on what is known as *imitative counterpoint*, a practice in which two or more voices sing some version of a given melody. The clearest example of imitative counterpoint is the *canon* and its most elaborate exponent is the *fugue*. The latter is more complex, as it has more voices, but is less strict than the former. The canon will be the focus for the next two sections, where we will see how this form embodies the use of geometric transformations in music.

Before diving into the structure of canons, we look further into the structure of diatonic scales. We already mentioned that such a scale is a selection of seven pitches out of the 12 in the chromatic scale in some arrangement of half- and whole-steps. Over time, two *modes* of selection prevailed. Beginning at an arbitrary note, which we identify with the number 0, the *major* mode contains the notes

$$0\ 2\ 4\ 5\ 7\ 9\ 11$$

and the *natural minor* mode contains the notes

$$0\ 2\ 3\ 5\ 7\ 8\ 10.$$

---

[4] Despite this last example, it should be noted that, in music, the word "voice" is used for both vocal and instrumental melodic lines. We extend this abuse of notation by consequently using the verb "to sing" in both cases.

Thus, the diatonic scale C D E F G A B is called C-major. Let us denote the 12 tones beginning with C as follows:

$$\text{C C♯ D D♯ E F F♯ G G♯ A A♯ B.}$$

Then, for instance, D-major will be the scale

$$\text{D E F♯ G A B C♯ (D)}$$

(and note, some of the "white" notes in this scale are "black" in C-major). Similarly, B-minor will be

$$\text{B C♯ D E F♯ G A (B).}$$

One immediately remarks that, as a subset of the chromatic scale, D-major and B-minor are the same. Yet, there is a difference in the use of these scales when composing.[5] The first note of a scale is called its *tonic* and plays a central role in all compositions on this scale (or, as musicians say, in this *key*). In particular, the listener feels it as a final destination: there is a longing for the tonic. An expectation often taken care of by the composer as most compositions in a given key finish on the corresponding tonic (this is so engrained in musical tradition that in the Middle Ages the tonic was actually called the *final*). Other endings would not be so satisfying and would leave the listener with a feeling of unresolved tension.

What happens when, while playing in a given key, we play an accidental? Doing so creates a momentary replacement of the original tonic by a new one (to which the accidental belongs). This is felt by the listener and, therefore, it creates a tension that needs to be resolved, no matter whether in the short or long term.

8.2 The geometry of canons

We can now return to an issue mentioned at the beginning of this chapter, namely the idea of embedding "the space where music lives into the Euclidean plane". The way scores are written suggests how we may consider this embedding. In the two-dimensional picture of a staff, the x-axis represents time and the y-axis pitch. We can even choose unit measures for both axes (e.g. measuring time in breves[6]). And it is under this embedding that the use of affinities in the composition of canons becomes apparent.

[5] The minor scales impose different selections of pitches for its ascending and descending versions.

[6] It is important to bear in mind that musical time progresses at a pace independent of the clock. Instead, we sense a continuous series of beats, like clock ticks, which may seem identical to a division of real time but are of necessity (since music is performed by people and not by machines, and is responsive to the ebb and flow of musical tension) somewhat flexible. We can speak of "musical time" as opposed to "clock time".

The word *canon* comes from the Greek word for "rule" or "law". And its use in music could not be more accurate, since it refers to a form of imitative counterpoint, the strictest, in which a voice is bound to imitate another. The latter can be thought of as the *leader* and the former as the *follower*. In this sense, canonical composition is an excellent example of this book's theme of creation constrained by rules. It is also an excellent example of the occurrence of geometric transformations in music.

Canon at the unison   In the simplest example of a canon the follower sings the same melody as the leader delayed by a certain amount of time. Popular canons of this kind that the reader must have heard (and may have sung) are *Row, row, row your boat* or *Frère Jacques*. A scheme that may be useful to visualize the geometric nature of the different canonical forms is to draw the basic melody as a triangle[7] and triangles corresponding to the leader and follower drawn on two parallel lines, with the leader's on top. With this scheme, our canon looks as the picture on the right.

It is important to note that we draw the two voices at different heights for the sake of clarity (that is, to avoid an overlap). This difference does not imply any difference in pitch (there is none in a canon at the unison) and is akin to writing two staffs, corresponding to two musical instruments, one on top of the other. Note also that (leaving aside the fact that we have drawn the two voices at different heights) the second voice is, in geometric terms, a translation of the first (with a horizontal vector $\longrightarrow$). We have emphasized canons with two voices (the leader and the follower). Yet, other additional voices can sing as well, usually delayed by the same amount of time. That is, the third voice would begin with a delay after the second which equals the delay the second has with respect to the leader. This is common in the execution of the two popular canons mentioned above.

Perpetual Canon   Another feature of these canons is that they may be repeated indefinitely. Such canons are often called *perpetual canons* (or *rounds*) and forcefully bring the issue of repetition to our discussion. A round can be considered as a musical version of a frieze. The corresponding unit cell is in this case the part of the canon played between two instants of time separated by the original delay between voices. This becomes clear using our scheme:

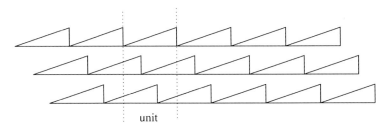

unit

---

[7] We have chosen a triangle because the effects of different transformations on this figure are apparent. This choice does not imply any kind of tone rising.

We remark that, just as a frieze in a building is limited in practice by the dimensions of the building, the theoretical property of a round to repeat indefinitely is limited in practice by a number of obvious conditions (including the patience of players and listeners!). We also note that the basic vector of the frieze is a multiple of the delay vector. In our scheme above, for instance, the unit cell is repeated after three such delays.

More sophisticated canons are obtained by allowing the follower variations other than a mere delay in time (or, in geometric terms, transformations other than a mere translation on the x-axis).

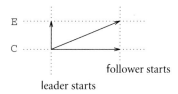

Canon at an interval This form requires the follower to begin on a pitch different from the one on which the leader began. For instance, the leader begins with C and the follower two notes higher (i.e. with E). Using our scheme this can be drawn as shown on the left, where we have indicated the position of the pitch C (for both leader and follower) by a dotted line. In geometric terms this also corresponds to a translation, but the translation vector is now more complex since it can be decomposed as a sum of a vertical component (corresponding to a displacement in pitch) and a horizontal one (corresponding to a displacement in time).

$$E \cdots\cdots$$
$$C \cdots\cdots$$
follower starts

leader starts

In the particular case that the displacement in pitch is one octave (fifth, second, etc.) the canon is said to be *at the octave* (or *fifth, second,* etc.).

Canon *contrario motu* A canon is said to be *contrario motu* (or *in contrary motion*) when the two voices evolve (leaving aside the difference in their starting moments) by following the same melodic intervals moving in opposite pitch directions. This corresponds to (leaving aside the translation on the x-axis) a reflection with respect to a horizontal line (in modern music terminology this is called an *inversion*). It can be depicted by the following picture:

where the dotted line indicates the (ideal) axis of the reflection. Taking the horizontal translation into account, the second voice in a canon in contrary motion is obtained by performing a glide on the first voice.

Using our simple scheme (with a dashed line for the axis of the glide) we obtain the picture on the right.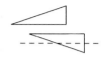

In the composition of canons in contrary motion, authors have often allowed for (minor) variations in the quality of the intervals in order to avoid dissonances or to preserve tonality. Exceptionally, a canon in contrary motion is composed such that the follower sings exactly the reflected image of the first voice. To emphasize this exactness such canons are called *mirror canons*.

Retrograde canons In this kind of canon, also called *crab canon* or *cancrizans*, the follower sings the leader's melody backwards. This corresponds to a reflection with respect to a vertical line. The idea is shown as follows:

where the dashed line indicates the axis of reflection. Usually, crab canons do not have horizontal translation. That is, the leader and follower begin simultaneously. Using our simple scheme, a crab canon looks like the picture on the right. In this particular case, it may be interesting to overlap the two voices to mimic the simultaneity of the two voices (see the second picture on the right). A sense of symmetry emerges: the crab canon has bilateral invariance. Yet, the symmetric character of the crab canon can rarely be heard.

Proportional canons With proportional canons we abandon the realm of isometries to allow for more general affinities. Indeed, in this kind of canon the follower progresses at a speed that is a fixed multiple of the speed of the leader. For instance, twice, or half, as fast. Geometrically (and leaving aside translation on the horizontal axis) this corresponds to the strain mapping $(x, y) \mapsto (\lambda x, y)$ where, for the speeds mentioned above, $\lambda = 1/2$ and $\lambda = 2$ respectively. A simple example is given in the following picture (with $\lambda = 1/2$):

Proportional canons are said to be *in augmentation* if $\lambda > 1$ (i.e. when the follower sings more slowly than the leader) and *in diminution* when $\lambda < 1$. The follower may, or may not, begin simultaneously with the leader. The drawing on the left schematizes a canon in diminution with a horizontal translation in addition to the strain (with coefficient $\lambda = 1/2$ and axis the dotted line).

Spiral canons In our discussion above, we have associated a frieze with a perpetual canon. The use of this simple scheme even allowed for a visual form of this frieze. We noted that the frieze's basic vector is a multiple of the time delay vector and we remark here that this implies that the basic vector is horizontal. In addition, although our discussion was for canons at the unison, the same holds true when the different voices sing at other intervals. The frieze in this case simply looks like this (the horizontal dotted lines indicating, say, the pitch C):

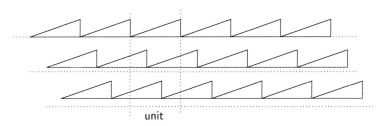

unit

Does this mean that all perpetual canons must have horizontal frieze vectors? Actually, it does not. In most canons the basic melody ends in the tonic of the key it is played in. This allows for each voice in the canon to be repeated without changing the key (neither its tonic nor its mode). In a *spiral canon* the basic melody ends on a note other than the tonic. Therefore, at the time of this ending, it is momentarily in another key. Which naturally affects the melody in two ways: first, it demands to be continued since there are unresolved tensions and, second, it forces the canon subject, when it starts again, to do so on a note different than the original starting one. In such a canon, each of the voices would contribute to the frieze with a pattern like

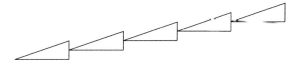

and the whole frieze would have a basic vector like ⟶ . The canon is said to be a *modulating spiral* (or a *tonal spiral*) if the scale is changed in such a way that the mode is preserved but the tonic is changed, and it is said to be a *modal spiral* if, in contrast, the tonic is preserved but not the mode.

The above list of canonical forms gives the basic ingredients that may be used in the composition of canons. These basic ingredients may be used together with a number of variants, such as whether the canon is *accompanied* (by additional voices that are not bound to the leader's melody), whether it is *double* (i.e. containing two leaders) or *triple* (ditto, with three leaders). They may also be combined to obtain, for instance, a canon in augmentation and contrary motion. Some of these combinations even have names of their own, such as the *table canon*: a canon that is both retrograde and in contrary motion (in geometric terms, a canon having a central symmetry). Its name derives from the fact that a score with a single melodic line was put on a table and shared by two performers who, sitting on opposite sides of the table, read the same line in opposite directions (see Figure 8.8).

8.3 The *Musical Offering* (revisited)

It may be that the canon is near to my (anti)symmetric plane-filling mania. Bach played with repetition, superposition, inversion, mirroring, acceleration and slowing down of his themes in a way which is in many regards comparable with my translation and glide-mirroring of my "themes" of recognizable figures.

M.C. Escher, quoted in (Schattschneider, 1990: 254)

A closer look at the painting of Bach by Haussmann in Figure A.2 would allow one to see that the paper Bach holds in his right hand contains three lines of musical notation with a title heading. The latter reads *Canon triplex à 6 Voc[ibus] per J.S. Bach*. This is certainly not by chance; there is no disagreement with the fact that with Johann Sebastian Bach the composition of canons reached its peak.

The *Musical Offering* contains two fugues (one for three voices and one for six), one trio sonata, and ten canons. It is nowadays believed that the fugue for three voices is the one improvised by Bach at the court. The fugue for six voices is certainly the one that Bach pledged himself "to work out [...] more fully" in Leipzig. The sonata features the "Royal theme" in the flute (Frederick's own instrument). But it is the 10 canons that get our attention. All of them feature (in some as the subject of the leader's melody; in others as an accompanying voice harmonizing with the leader and follower), the "Royal theme":

In addition, the 10 canons share the playful spirit of the inscribed dedica-tion in the measure that their scores are presented to the reader as riddles or puzzles. Bach did not completely specify the way the canons are to be played, but rather limited himself to providing some indications or clues to that effect, such as the title (which would instruct the type of canonical form to be used), the initial pitch and a starting point for the follower. In what follows we describe the 10 canons in the *Musical Offering*. For the first five we show the original score and a "solution" to the riddle written in a more contemporary fashion while accompanying the reader in the task of solving the riddle. With the remainder we become more succint.

Canon 1 (*a 2 cancrizans*)   The score as written by Bach is as follows:

For the "solution", we first look at the title Bach gave to this canon. Such titles give most of the essential information. The expression *a 2* indicates that there are two voices. In addition, Bach instructed that the second voice should be played backwards by the word *cancrizans*. The only ques-tions are when should the second voice start and on which pitch. For the latter, note that there is a C-clef (the symbol $\mathbb{B}$) at the beginning and a C-clef at the end of the score, and that the C-clef and the three flats at the end are facing backwards. They are actually the image of those at the beginning by a reflection with respect to an (ideal) vertical line at the middle of the score. This indicates that the second voice plays in the same key (and thus, both begin on the same pitch, namely C). As to where should the second voice start, the common practice was to signal the entry of the follower with the symbol 𝄪, but there is no such indica-tion in the score. One concludes that the two voices begin simultaneously and, finally, that the score can be written as follows:

We have framed in red the Royal theme, which occurs at the beginning of the leader's melody.

Canon 2 (*a 2 Violini in unisono*) The title here indicates two voices in unison (each beginning on the same pitch at the same octave). It also specifies the intruments (violins) for the two voices. Bach's score shows two melodic lines:

In the lower voice we recognize the Royal theme. Now look at the melodic line in the upper voice: it is not in any canonical relation with the lower one. It shows, however, at the beginning of the second measure, the symbol indicating the starting point for the follower. We conclude that there are three voices, two of which sing the canonic theme (the upper line) at the unison and with a delay of a measure (a semibreve in this case) and the third accompanying them both with the Royal theme. That is, the score looks as follows (again, we have framed in red the first notes of the Royal theme):

Canon 3 (*a 2 per Motum contrarium*) We see again the Royal theme, now in the upper voice, but played twice as fast as in the original score. And this time, the symbol 𝄋 is in the lower voice. Hence, it is this lower voice that is sung against itself to create the canon (and, as in Canon 2, the Royal theme is an accompanying voice).

The title of the canon indicates that the follower sings the leader's theme in contrary motion. The three flats (♭) following the second C-clef in the lower voice are a reflected version of the three flats following the first C-clef and provide the solution for the position of the axis of reflection with respect to which the contrary motion is executed. They are centred at the first and third lines in the staff. It follows that the second line is the axis of reflection. We thus obtain the following score:

There are no variations in the quality of the intervals in this canon.[8] It is actually a mirror canon at the fifth. Finally, note that the simultaneous presence of a time delay (i.e. a horizontal translation) and a contrary motion (a reflection with respect to a horizontal axis) results in a glide. This is apparent in the score above.

Canon 4 (*a 2 per Augmentationem contrario Motu*) With this canon we abandon the (geometric) realm of the isometries. Indeed, the title is telling us that the follower sings "per augmentationem," that is, slower than the leader. In fact, Bach used this augmentation to express his best wishes for the King by adding the inscription *Notulis crescentibus crescat Fortuna Regis* (As the notes increase may the fortunes of the King do likewise).

---

[8] Looking at the score it may seem that this is not so, since the first descending notes on the second line (the leader's) do not coincide with the first ascending notes on the third line (the follower's). This is due to the feature alluded to in Footnote 5 – the canon is played in a minor key – and does not alter the quality of the intervals.

Looking at Bach's score, we meet again two voices with the symbol 𝄎 in the lower one. There are also two clefs in the lower voice, a C-clef and an upside down G-clef (the symbol). It is therefore clear that the lower voice contains the leader's line as well as the indications for its canonical variation.

Note that the G-clef is reflected (with respect to a horizontal axis) in agreement with the fact that the follower sings in contrary motion. A similar agreement can be observed in the three flats following this clef which are similarly reflected. Bach does not give any indication of how much slower the follower should sing. But the most sensible choice seems to be twice as slow, because doing so results in counterpoint that follows the rules that were common practice in Bach's time and doing otherwise violates them. Putting all these remarks together, the score can be written as follows (where the leader is in the lower voice, the follower in the upper voice, and the middle one is an accompanying line):

Again, it is a canon at the fifth and, again, interval quality is strictly respected, so this is a mirror canon. The Royal theme is present in the accompanying voice, not in its pure form but modified by ornamentation (in this case by adding short bits of melody in between consecutive notes of the theme).

Canon 5 (*a 2 per Tonus*)   As in the previous canons, we notice that the upper voice is an accompanying voice and that the lower is the one with the lines for leader and follower:

There is an interval of a fifth between the leader and follower suggested by the positions of the F-clef (the symbol $\mathcal{9}\!\!:$) and the C-clef in the lower voice. Otherwise, both leader and follower move at the same rate and in the same (pitch) direction and it would seem that the riddle is solved. Yet, Bach titled the canon *per Tonus* and added to this title the wish *Ascendenteque Modulatione ascendat Gloria Regis* (As the keys ascend so may the glory of the King also ascend). And indeed, a closer look at both voices shows that the canon starts in C-minor but by the time it finishes it has moved one step higher, to D-minor. The canon is a modulating spiral. And Bach is wishing the King that his glory may spiral up as the key the canon is played in does.

In a contemporary rendition the (beginning of the) canon would look like the preceding score. (A red segment was added joining the initial note of the upper, accompanying, voice with the corresponding note in the second repetition of the melody which begins in the ninth measure. A quick glance at this repetition betrays the change of key.)

It is maybe interesting to note that while Bach doubtless wished the King a never-ending increase in his glory, the canon has a satisfying ending after six repetitions, a moment in which we reach again C-minor (but an octave higher).

Canon 6 (*Fuga canonica in Epidiapente*)   Of all the canons in the *Musical Offering* this is the one having the longest time delay between leader and follower: 10 measures. It is also one of the most complex, since, as its title indicates, it contains several expositions of a theme articulated as a fugue (a structure more complex than a canon). Yet, the canonical structure is simple. It is a canon at an interval and the Latin *in Epidiapente* states that this interval is a fifth. It is for three voices, the third one accompanying the two bound by the canonical structure and containing the Royal theme. Its score is quite long and it is better omitted here.

Canon 7 (*Canon perpetuus super Thema Regium*)   Again, this is a canon at an interval with an accompanying third voice featuring the Royal theme. The title indicates that it may be repeated. Here is its (modern) score:

Canon 8 (*Canon perpetuus*)   Strictly speaking this is not a canon but *two* canons spliced together. The first 13 measures contain one accompanied mirror canon. Then, measures 14 to 18 are a transition leading to the second canon whose canonic voices repeat, in contrary motion, the melody of the first 13 measures. Graphically (i.e. using our simple scheme), Canon 8 looks as follows:

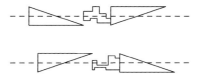

Here we have used, as usual, a dashed line to indicate an axis of reflection. Also, we draw irregular polygons to describe the non-structured bridge between the two (canonical) halves.

Canon 9 (*Canon a 2 Quaerendo invenietis*) In the last two canons Bach returns to the unacommpanied form. In the ninth, a canon *a 2*, there are only the two canonical voices. This is again a canon in contrary motion (we omit the explanation of why this is so). The expression *Quaerendo invenietis* (seek and you shall find) alludes to the fact that Bach did not indicate where the follower should start. This has to be sought by the reader. Curiously, there is more than one solution and a cursory trial shows that a zero delay is not one of them. A commonly played solution makes the follower start 2 and 1/4 measures after the leader.

Canon 10 (*Canon a 4*) As the title makes clear, in this canon there are four voices in canonical relation: the leader and three followers. The time delays are the same. That is, the second voice enters seven measures after the leader, the third seven after the second and the fourth seven after the third. It is also a canon at intervals. The leader starts on G above middle C and so do the second and third voices. The fourth one, in contrast, starts on G but two octaves lower.

One can hardly imagine that the use of geometric transformations in the canons composed by Bach could evolve without the latter being familiar with these transformations in their more common, spatial, sense. Figure 8.5 shows a seal designed by Bach in 1722 and used by him through his years in Leipzig. It contains the letters "JSB" at the left and a reflection of them (with respect to a vertical axis) at the right. More interestingly, perhaps, the emblem below the central crown is obtained

Figure 8.5 **Bach's seal.**

by overlapping these same three letters with their mirror image. It is thus a graphic version of his crab canon.

8.4 Symmetries in music

The canons considered in Sections 8.2 and 8.3 provide examples of geometric transformations in musical composition. In some cases, such as in the crab canon or in a perpetual canon, these transformations are used so that the resulting composition possesses some form of symmetry. These canons, however, by no means exhaust the display of symmetry in music.

Before giving additional examples for this display, a few remarks are due. First, there is the issue of how does one measure distances or, in technical terms, the choice of a metric. In the previous sections we did not address this issue. This is a neglect we may now redress.

For the measure of distances in the horizontal (i.e. temporal) axis there is no choice to do other than the unit of measure. Distances on the vertical axis, that is between pitches, present in contrast different alternatives. The question to be answered is: What is the distance between two given pitches? A possible answer consists of considering all pitches in the chromatic scale as consecutive ones separated by one unit. This is called the *chromatic metric* and is a reasonable choice when all such pitches are being used (as in dodecaphonic music). But when a diatonic key is selected, and the composition features only the pitches of that key,[9] one may be induced to consider only these seven pitches and, again, separate consecutive ones by one unit. This is called the *diatonic metric* (in the appropriate key). In this way, and to give an example, the distance between C and E is 2 in the diatonic metric for C-major and is 4 in the chromatic metric.

A second issue that deserves attention is: Which are the possible symmetries in musical composition? Even restricting ourselves to metric

---

[9] To be precise, one should say "mostly" instead of "only" since most compositions include accidentals.

symmetries (that is, to invariance under isometries) it quickly becomes apparent that not all symmetries make sense in a score. For instance, the only reflections that make sense have either a horizontal or vertical axis (any other position for the axis would not respect the staff). Similarly, the only rotations that make sense are the identity and the half-turn, which can be seen as the composition of two reflections with vertical and horizontal axes. Let us call these constraints *musical restrictions*. They immediately lead to the following result.

Proposition 8.1 *Let S be such that* $\mathsf{Sym}(S)$ *has no translations and satisfies the musical restrictions. Then* $\mathsf{Sym}(S)$ *is of one of the five different kinds:*

$$M1: \quad \mathsf{Sym}(S) = \{\mathsf{Id}\},$$

$$Mh: \quad \mathsf{Sym}(S) = \{\mathsf{Id}, R_\mathsf{h}\},$$

$$Mv: \quad \mathsf{Sym}(S) = \{\mathsf{Id}, R_\mathsf{v}\},$$

$$M2: \quad \mathsf{Sym}(S) = \{\mathsf{Id}, \mathsf{rot}_{180}\} \quad and$$

$$Mhv: \quad \mathsf{Sym}(S) = \{\mathsf{Id}, R_\mathsf{h}, R_\mathsf{v}, \mathsf{rot}_{180}\}.$$

Here, R_h *and* R_v *are reflections with horizontal and vertical axes respectively and* rot_{180} *is a half-turn.*

Proof There can be at most one reflection with horizontal axis in $\mathsf{Sym}(S)$ since the axes of two different such reflections are parallel and its composition would yield a translation (by Proposition 3.10). A similar argument shows that there can be at most one reflection with vertical axis and at most one half-turn. Therefore, we have the following eight possibilities:

	Vertical reflection	Horizontal reflection	Half-turn	
1	Y	Y	Y	*Mhv*
2	Y	Y	N	
3	Y	N	Y	
4	Y	N	N	*Mv*
5	N	Y	Y	
6	N	Y	N	*Mh*
7	N	N	Y	*M2*
8	N	N	N	*M1*

Row 2 is not possible because the composition of a horizontal reflection with a vertical one yields a half-turn. Neither is row 3 because the composition of a vertical reflection with a half-turn yields either a horizontal reflection or a glide (which composed with itself yields a translation). Finally, row 5 is not possible because the composition of a horizontal reflection with a half-turn yields either a vertical reflection or a glide. The remaining five rows correpond to the five groups in the statement. □

8.4.1 The geometry of motifs

Which of the five possible groups in Proposition 8.1 can we find in actual musical compositions? All of them, even though, it goes without saying, some are less common than others. It is possible to find them in small, but recurring, fragments of a composition (appropriately called *motifs* in music) or as a feature of the whole musical piece.

Obviously, the most common kinds of motifs are those having type *M1*; that is, those having no non-trivial symmetries. The Royal theme in the *Musical Offering* is an example.

We pointed out at the end of Section 8.2 that table canons have central symmetry. The score in Figure 8.7 is an example of this kind of composition. Called *Der Spiegel* (*The Mirror*), it is a duet for violins attributed to Wolfgang A. Mozart. Because this duet is a table canon, by construction, it has central symmetry. Since it possesses no other symmetry, its group is of type *M2*. Wilfrid Hodges, who researched the presence of various kinds of symmetries in music (Hodges, 2003), notes that "this is not at all a common pattern, and generally it is not easy to hear".

For the type *Mh* the work of Béla Bartók provides several examples. For instance, the bar in Figure 8.8 is from his fifth string quartet. And the bars shown in Figure 8.9 are a piece for piano where the left hand plays a reflection of what the right hand is playing.

We have already observed that a crab canon (of which the Canon 1 in the *Musical Offering* is an example) has bilateral symmetry with respect to a vertical axis. That is, in the absence of any other symmetry (as is the case with Canon 1) they have a symmetry group of kind *Mv*.

Finally, we are left with the group *Mhv*. Hodges writes that "interesting motifs of this type are extraordinarily rare. [...] It seems that any composer regards this group as too crass to use". Nevertheless, he gives an example of such a motif in a piano exercise by György Kurtág (Figure 8.6).

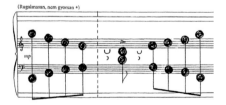

Figure 8.6 *Játékok for piano 1, Hommage à Eötvös péter*, György Kurtág. (ⓒ Copyright by Editio Musica Budapest Music Publishers Ltd. Reproduced by permission of Boosey & Hawkes Music Publishers Ltd.)

Figure 8.7 **Duet for two violins *Der Spiegel*, attributed to W.A. Mozart. (From (Benson, 2007) with permission.)**

Figure 8.8 **A bar from Béla Bartók's *String Quartet No. 5*.** (ⓒ Copyright by Universal Edition A.G., Wien/UE 10737. Reproduced by permission. All rights reserved.)

Figure 8.9 **Béla Bartók's *Mikrokosmos, No. 141, Subject and reflection*. (ⓒ Copyright 1940 by Hawkes & Son (London) Ltd. [Definitive corrected edition ⓒ Copyright 1987 by Hawkes & Son (London) Ltd.] Reproduced by permission of Boosey & Hawkes Music Publishers Ltd.)**

8.4.2 The ubiquitous seven

Proposition 8.1 shows that there are only five possible kinds of finite symmetry groups satisfying the musical restrictions and the above discussion exhibited examples for all of them. A natural next step would be to consider symmetry groups that include translations; to be more precise, frieze groups and wallpaper groups.

It is not impossible to think of an example of wallpaper: a perpetual canon at an interval (say, with the follower singing an octave higher). Using our representative triangles this canon looks like

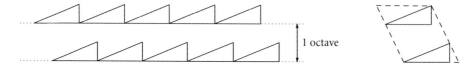

where now the position of the upper row of triangles not only denotes the leader's voice but also shows that this voice is an octave higher than the leader's (the parallelogram on the right shows the unit cell). One remains unconvinced, however, as the repetition in the vertical direction is very much limited. To further explore the realization of wallpaper groups seems somehow artificial.

Repetition in the horizontal direction, in contrast, occurs more frequently. Furthermore, it is the form of repetition most likely to be recognized as such by the listener. Considering translations in the horizontal direction only brings us to frieze groups. One may thus wonder which of the seven frieze groups occur in musical composition and, as a first step towards an answer, observe that all of the seven groups satisfy the musical restrictions (this is the content of Propositions 3.21 and 3.22). It then probably comes as no surprise that all the seven groups occur in music, even though some are quite rarely found. Almost all the examples below are taken from Hodges' article. We present them in the order Hodges originally presented them, putting the three less commonly found at the end.

Canons with frieze group of type *p111* ($\cdots$ p p p p p $\cdots$) are common (*Row, row, row your boat* or *Frère Jacques* are of this type). An example is found at the beginning of Beethoven's *Moonlight Sonata*; see Figure 8.10.

Figure 8.10 **First bars from Beethoven's *Moonlight Sonata*. (From Benson (2007) with permission.)**

Hodges points to birdsongs as characteristic instances of the *p111* pattern and mentions the nightingale portrayed at the end of *Quejas o la Maja y el Ruiseñor* in the piano suite *Goyescas* by Enrique Granados and the *Catalogue d'oiseaux* (*Bird's catalog*) of Olivier Messiaen.

The example for the *pm11* group (··· p q p q p q p q ···) is taken from the third symphony of Jean Sibelius (Figure 8.11). Quoting Hodges:

> that genius of orchestration, gives [...] an arch shape to his violins to play over and over again. What makes this an interesting passage is that [...] he divides the violins into four groups and makes each group start its arches at a different time. The effect is a throbbing sound that repeats at a quarter of the length of the arch; the arch lasts four bars but the combined pattern repeats at each bar.

In Figure 8.11 the upper staff displays the arch-shaped theme of the violins and the lower staff the result of sounding all four groups of violins together. We see that the theme itself is of type *pm11* but the combination of the four groups sounds like a *p111* frieze.

For the *pma2* group (··· p q b d p q b d p q b d ···) our example is taken from Claude Debussy's *Prélude, Feux d'artifice* (Figure 8.12). Hodges notes that "the symmetries are not quite exact, but with music like this, who's counting?"

One may think that since friezes of type *pmm2* (··· $\begin{smallmatrix} p\,q\,p\,q\,p\,q\,p\,q \\ b\,d\,b\,d\,b\,d\,b\,d \end{smallmatrix}$ ···) are the richest in terms of symmetries they are also the most difficult to be found. The endless repetition of a single note, however, constitutes an easily recognizable instance of this pattern. The song *Die liebe Farbe* by Franz Schubert (Figure 8.13) exhibits this repetition.

Figure 8.11 **Jean Sibelius' *Symphony No. 3* (last movement). (Reprinted by permission of Peters Edition Limited, London on behalf of Robert Lienau Musikverlag, Frankfurt.)**

Figure 8.12 **Claude Debussy, *Préludes for piano II, Feux d'artifice*. (From (Hodges, 2003) with permission of W. Hodges.)**

Figure 8.13 **Franz Schubert, *Die Schöne Müllerin, Die liebe Farbe*. (From (Hodges, 2003) with permission of W. Hodges.)**

Figure 8.14 **Igor Stravinsky, *The Rite of Spring* (14 in score). (ⓒ Copyright 1912, 1921 by Hawkes & Son (London) Ltd. Reproduced by permission of Boosey & Hawkes Music Publishers Ltd.)**

Figure 8.15 **Igor Stravinsky, *Petrushka* (53 in score). (ⓒ Copyright 1912 by Hawkes & Son (London) Ltd. Revised version: ⓒ Copyright 1948 by Hawkes & Son (London) Ltd. Reproduced by permission of Boosey & Hawkes Music Publishers Ltd.)**

The three remaining groups are, according to Hodges, more rarely found. As an example of the *p1a1* group (··· p b p b p b p b ···) Hodges shows a part of Igor Stravinsky's *The Rite of Spring* (Figure 8.14) on which "in the lower staff two bassoons alternately play the frieze motif the right way up and inverted". Hodges remarks that the inversion is chromatic, meaning that, in order for this inversion to respect distances, all the 12 pitches in the chromatic scale need to be used and considered as "equally spaced".

The same author provides an example of a canon with *p1m1* group (···$\frac{\text{p p p p p p}}{\text{b b b b b b}}$···) in Igor Stravinsky's *Petrushka* (Figure 8.15). Now,

Figure 8.16 **Claude Debussy, *La mer* (second movement, bar 72). (From (Hodges, 2003) with permission of W. Hodges.)**

however, the metric is diatonic in D minor. That is, only the seven pitches making this key are used (and considered as "equally spaced").

To exemplify a canon of type *p112* (··· p d p d p d p d ···) Hodges returns to Debussy, now to the second movement of *La mer* (Figure 8.16). "The motif is made up of whole tones, so that again we are rotating in a chromatic metric."

It should be mentioned that in some of the examples above there are other musical events happening at the same time. We mentioned this for the case of Sibelius' *Symphony No. 3* (Figure 8.11). We now add that this is also true for the theme in Figure 8.16, which is an accompaniment to a trumpet figure.

8.5 Perception, locality and scale

The mechanisms underlying the perception of symmetries according to Gombrich (which we summarized Section in 5.3) refer to spatial perception and do not immediately transfer to musical phenomena, which occur in time. There are no accents or breaks distinguishing salient parts of the piece and no mechanism akin to peripheral vision which would confirm a symmetry hypothesized from these accents. A measure of how different the situation is now is obtained by comparing the perception of bilateral symmetry in visual arts and in music. We have argued that, in the former, bilateral symmetry is easier to detect (since the fovea guesses it from a small region in the percept and peripheral vision then verifies this guess) than translational symmetry is. This seems unlikely to be so in the perception of music. Take the two figures given by Gombrich:

$$((((O(((O((O(O((((O(((O((O(O$$

and

$$((((O(((O((O(OO)O))O)))O)))).$$

Recall, Gombrich noted that we take in the bilateral symmetry of the second with an immediacy which is missing for the translational correspondence of the first. An aural equivalent of these two figures would consist in replacing the symbols "(" and ")" by a pitch (say C) and the symbol "O" by another pitch (say F). Each of the two lines above would, therefore, correspond to a (surely uninteresting) melody. In contrast with the visual situation, however, the translational correspondence of the first is easier to take in than the bilateral symmetry of the second is. Now both require us to store the melody to be repeated in our memories, but in the second case there is additional effort in retrograding (i.e. reflecting) it. A reflection which is present in the visual context (and which therefore needs only to be confirmed) but needs to be reconstructed in the aural context. Hodges writes that "some [...] musical palindromes seem to have been written for the challenge. If the symmetry is obvious enough, the performers can enjoy it as much as the composer". The listener, however, may be left out of this affair.

Similar remarks may be made for bilateral symmetry with respect to pitch (as in Bartók's scores in Figures 8.8 and 8.9), a form of symmetry whose detection would require identifying the "centres" of a sequence of chords. It is also to be expected that central symmetry, being the invariance under the composition of a vertical and a horizontal reflection, is hard to detect (recall, Hodges writes that this kind of symmetry is in general "not easy to hear").

We have already remarked that we regard translational symmetry in a non-horizontal direction as somehow artificial. This leaves translation in a horizontal direction as the transformation which is most likely to be perceived as such by the listener. And indeed, this form of repetition seems to be the easiest to perceive, at least for unit cells which are of a moderate length, and it is certainly the one that has been most used by composers. Oliver Sacks (2008: 52), a neurologist and writer, refers to them when he states that

There are, of course, inherent tendencies to repetition in music itself. Our poetry, our ballads, our songs are full of repetition. Every piece of classical music has its repeat marks or variations on a theme, and our greatest composers are masters of repetition; nursery rhymes and the little chants and songs we use to teach young children have choruses and refrains. We are attracted to repetition, even as adults; we want the stimulus and the reward again and again, and in music we get it.

The mention of the length of the unit cell brings us to a distinction in the different occurrences of translational invariance (in a horizontal direction) between what we can call *local repeats* and *global repeats*.[10]

---

[10] Peter Kivy, a leading contemporary philosopher of music, some of whose ideas we will discuss in Section 8.6, calls these repetitions *internal* and *external* respectively.

In the first case, the repeats have a short unit and affect a relatively small part of the whole composition like in the first bars of Beethoven's *Moonlight Sonata* shown in Figure 8.10. In the second case, in contrast, repetition affects the whole (or a large part) of the musical piece. An example at hand is the *Cancrizans* in the *Musical Offering*. The difference between local and global repeats is not so clear as to whether a part of a composition is small or not is a matter of scale. It is, nevertheless, sufficiently clear to be useful.

Global repetition permeated the common practice by occurring in many of the musical *forms* of that period. Probably the simplest such form is the *binary form*. This structures the musical piece in two sections, say *A* and *B*, both of which are usually repeated, thus yielding the pattern *AABB*. In its strictest form, these two sections have approximately the same duration and they satisfy some tonal constraints. The first section starts in a given key, the tonic, and modulates to a related key. The second section starts in the latter and eventually modulates back to the original key. If the original key was in major mode then the modulation was to the perfect fifth. A concise way musicians would use to describe this form is the following:

Note the occurrence of the vertically aligned pairs of dots. These act as delimiters indicating that everything between ‖: and :‖ should be repeated once, thus making these repetitions apparent in the notation. Another structure showing global repetition is *ternary form*, which consists of three sections, the first and third being identical and in sharp contrast with the second. They exhibit the pattern *ABA*. A particular case of the ternary form structures its sections as ternary (or binary) forms themselves. That is, *A* has the form *CDC* and *B* the form *EFE* so that the overall pattern is *CDCEFECDC*. This sort of ternary form is said to be *compound* in contrast with the *simple* one showing no particular structure in its sections. As examples of simple ternary forms we can mention the *da capo arias* which were common in Baroque operas. As examples of compounded ternary forms we mention the *minuets*, which usually occurred as the third movement in the symphonies of the classical period. A more complex musical form where global repetition occurs is the *sonata form*, which at its origin in the seventeenth century implied a binary form *AABB* (and of which Domenico Scarlatti composed several hundred[11]).

---

[11] One must be aware, though, that contemporary recordings of these sonatas sometimes omit the repetitions.

8.6 The bare minima (again and again)

The *Vexations* of Satie aside, the amount of repetition in musical composition was limited, until very recently, to doses that would ensure the conveyance of a theme. Beyond the satisfaction of this goal, undue repetition risked boredom. The event at the origins of a change in this pattern was, according to Robert Fink (2005: Chapter 4), a commercial war between Columbia Masterworks and RCA-Victor which took place in 1949–50. At that time, both companies developed technologies to cope with the limit of approximately four and a half minutes of listening which 78 rpm discs could deliver. Columbia came up with the 33 rpm LP and RCA with the 45 rpm disc plus the record changer. The LP allowed 22 minutes of music per side and the record changer to stack up to 10 discs of 45 rpm which would be played consecutively. The eventual outcome of this battle was the fusion of both technologies and, as a result, the possibility to arrange for the consecutive playing of up to 10 LPs.

Such a possibility brought a change in music-listening practices: attention did not have to be put, in an undivided manner, on the piece being played. Music could be deployed as sonic background: aural wallpaper. The wave of "easy listening" music which ensued was a consequence of this change. Another such consequence was a thriving of Baroque music. Collections of concertos (such as the *Brandenburg Concertos* of J.S. Bach or *L'Estro Harmonico* of A. Vivaldi) were sold in multiple-disc albums which could provide, literally, hours of uninterrupted baroque musical background, and such a background became omnipresent in the 1950s. By the end of that decade a considerable reaction against this change had built up, a reaction epitomized in the article *A Pox on Manfredini* by H.C. Robbins Landon (1961) which, under the general goal of criticizing the avalanche of concertos being put on record, made the most to clarify that these concertos ("Albinoni, Geminiani, Corelli, Locatelli, and of course, the father-figure of barococo music: Antonio Vivaldi") lacked intellectual depth; they were shallow music for shallow people. And the reason is that this music was "sufficiently old to be respectable, and sufficiently boring not to need listening to". In the trade-off between boredom and confusion Robbins Landon appears to be favouring solutions that lean towards confusion, since these are the solutions that will require undivided attention from the listener. One needs to say (to put Robbins Landon's reaction in perspective) that undivided attention had been considered inseparable from musical listening. As Philip Alperson (2004: 262) puts it in his description of Hanslick's ideas:

Understanding a piece of music is a matter of attending with extreme vigilance to the composer's designs in the composed work, following the composition as it unfolds in the context of what has been heard and of expectations concerning what might yet be heard.

In spite of the efforts of Robbins Landon (and many others) to demean the new attitude in music listening, this attitude established itself. Fink calls it *repetitive listening* and a central point in his argument is to differentiate it from the non-listening we associate with the so-called "elevator music" (played in dentists' waiting rooms, shopping malls, etc.). In the former, a "musical 'space' [is created] within which our minds can wander [in] a sonic experience that is somewhere between all-consumingly intense and alienatingly dull". These intermediate solutions to the trade-off between boredom and confusion create a variety of moods which allow, and possibly facilitate, the listener to concurrently perform other tasks. As an example, Fink quotes the following passage by the novelist[12] Compton Mackenzie (1955: 106–107):

The privilege of putting one's mind in order by what surely at its best is the most orderly medium of art in existence – a string quartet – [is] of inestimable value. [That is] why I have made a habit of working to the accompaniment of chamber music. [. . .] Every writer, whatever his capacity for concentration, is peculiarly exposed to attacks of the irrelevant and trivial when he is most fain to be free of them. And these petty assaults are somehow warded off much more successfully when the background of my mind is occupied by great music. [. . .] It is the background of the conscious mind for which I crave an occupation. The music is playing the same part as the beads of a rosary. Usually, in spite of working I am able to follow the music intelligently enough to know when the wrong side of a disc has been placed on the turntable, but there are times, rare alas, but all the more wonderful for their rarity, when the determination to hammer some sentence into shape is so tremendous that a long quartet can be played beginning to end without my being consciously aware of it.

The fact that the music "is playing the same part as the beads of a rosary" gives proof of the appropriateness of the expression "repetitive listening".

Fink further elaborates on the quality of the music allowing for repetitive listening: "The movements must be severely abstract and patterned, repetitious, perhaps even a little bit boring; constant figuration so the mind finds something to hold whenever concentration wanes; no lyrics, no programs, no strong emotional eruptions".

Baroque music as it tended to be played and recorded in the 1950s could fit this description (not the way it is played nowadays, mostly for excitement). A music fully doing so emerged in the 1960s in the United States – as a response to the esoterica of post-Webern serialism – which has been variously described as "repetitive music", "acoustical art", "meditative music" and "minimalist music" (Mertens, 1983: 11) and is commonly known today by the latter expression. Musicians such as Terry Riley, Steve Reich and Philip Glass are commonly seen as the most representative of this current. Their compositional styles exhibit a

---

[12] He was also the co-founder and first editor of the *Gramophone Magazine*.

number of differences, yet they share a common feature: a specific use of repetition. Quoting Fink (2005: 20), this is "repetition with a regular pulse, a pulse that underlies the complex evolution of musical patterns to alter listener perceptions of time and *telos* in systematic, culturally influential ways". A detailed explanation of the procedures used in minimalist music to implement this repetition transcending, rather than creating, boredom is outside the scope of this chapter. It suffices for our purposes to note that the notion of "gradual variation" we first met in Section 6.2 (exemplified in some of Warhol's paintings) is here of the essence and we give two representative examples.

Composed in 1964, *In C* (Figure 8.17) is one of Terry Riley's best-known works. It consists of 53 musical phrases (Riley describes them as "melodic patterns") which are sequentially played. These phrases are short and meant to be repeated a (not specified) number of times.

Figure 8.17 *In C* of Terry Riley.

In the Performing Directions accompanying the score,[13] Riley describes the way these repetitions are carried out and combined to make the piece. He first stipulates that "any number of any kind of instruments can play", although "a number of 35 is desired if possible". Then he proceeds:

Patterns are to be played consecutively with each performer having the freedom to determine how many times he or she will repeat each pattern before moving on to the next. There is no fixed rule as to the number of repetitions a pattern may have, however, since performances normally average between 45 minutes and an hour and a half, it can be assumed that one would repeat each pattern from somewhere between 45 seconds and a minute and a half or longer.

The randomness in this structure is attenuated by a number of instructions (such as "the group should aim to merge into a unison at least once or twice during the performance" but "if the players seem to be consistently too much in the same alignment of a pattern, they should try shifting their alignment by an eighth note or quarter note with what's going on in the rest of the ensemble" or yet with some directions for performance after reaching the 53rd phrase). The repetitive character is reinforced by a note (needless to say, a C) being played at constant intervals by "a piano or on a mallet instrument". This is the *pulse* referred to by Fink in the quote above. The gradual variation is apparent in the passage from most phrases to the next one (e.g. see the first five).

The use of gradual variations is also systematic in *Music for Eighteen Musicians* (Figure 8.18), composed in 1976 by Steve Reich. The opening of its Section II clearly displays a way developed by Reich and known as "gradual replacement of rests with notes" together with the simple, unadulterated, repetition of short motifs.

Figure 8.18 *Music for Eighteen Musicians* of Steve Reich, opening of Section II. (From (Fink, 2005) with permission.)

---

[13] The score of *In C* and the Performing Directions provided by Riley are publicly available at http://www.otherminds.org/SCORES/InC.pdf.

The first eight measures in its score show this process. A short motif is built up by beginning with a single harmonic interval and some rests and replacing, on the passage from each measure to the next, some of the rests for notes. Furthermore, each measure is played between three and five times. To convey an idea of the effect of this kind of process we quote Fink (2005: 49) again:

The phenomenological impression of each section is of a painstaking, inexorable flow and ebb of tension ("something that, when heard in the proper spirit can churn up the gut – *very* slowly, but with terrifying control") [...]

8.7 A blurred boundary: II

So designs *à la greque*, foliage for frameworks or on wallpapers, &c., have no intrinsic meaning; they represent nothing – no Object under a definite concept – and are free beauties. We may also rank in the same class [...] all music that is not set to words.   I. Kant (1952: 72)

We argued in Section 6.3 that the tendency towards abstraction brought an emphasis on "decorative" aspects in visual arts. The abstract character of music – at least of music "divorced from its original purpose as an accompaniment to dance, the march, or the religious service" – suggests a similar emphasis, which could not pass unnoticed to the genius of Kant. But in the nineteenth century the notion of music as aural decoration had to meet resistance, since music had already established itself as a fine art (Kristeller, 1965) and it was not until the middle of that century that such a resistance would begin to be overcome in the work of Hanslick. A contemporary defence of music as (at least partially) aural decoration is presented in a brilliant essay by Peter Kivy (1993), which we summarize in what follows.

The early history of European music is dominated by vocal music. In contrast, at the end of the eighteenth century and during the nineteenth century, purely instrumental music experienced an enormous growth and eventually became the dominant form. This "absolute music" (as it came to be called during the Romantic period) or "music alone" (as Kivy calls it) was therefore dissociated from the representational content naturally present in the lyrics of vocal music. And this dissociation naturally raised the question of how music, now reduced to no more than some pleasant but apparently meaningless sound, could share the Pantheon of the "major" arts with forms of expression so heavy on content such as painting, sculpture and literature.

Two possible answers to this question rely on the likening of music to a literary discourse and to a biological organism. In both cases the metaphor is facilitated by the temporal flow of musical works which imposes a linear order. Just as a poem is read or an embryo develops, the musical work appears as an ordered sequence of (musical) events. And once this likening is accepted, the meaning in a literary discourse or the life in the development of an organism would naturally drift from any of the two to the musical piece.

A goal in Kivy's essay is to refute these two answers (which he calls the "literary model" and the "organism model" respectively). The musical feature he will rely on to do so is repetition, a choice that not only fits our development, but also adds evidence to the relationship between symmetry and decoration.

Kivy recalls the features of the *da capo* aria in which "a leading emotion is expressed in the first, main section of the aria, a contrasted or related emotion is expressed in the [. . .] second section, with an instruction to repeat the first section appended to the end, thus making a symmetrical three-part form: ABA". He then describes the reactions the *da capo* aria raised in the eighteenth century, epitomized in the following 1755 comment by Francesco Algarotti (see Strunk (1980: 669)):[14]

> when the sense of an air is finished, the first part of it ought never to be sung again, which is one of our modern innovations and quite repugnant to the natural process of our speech and passions, that are not accustomed to thus turn about and recoil upon themselves.

Kivy notes however that "even Gluck, to whom it [this repetition] was dramatically repugnant, could not help, as a musician, almost wistfully referring to"

> those symmetrical forms, those periodic repetitions which give the arias a piquant and pleasant effect (Gluck, 1962: 117).

Kivy then concludes that the *da capo* aria was considered "perfect as music, absurd as musical speech, because musical repetition makes sense where linguistic[15] repetition does not".

One could guess from this description that the advent of absolute music would have granted repetition (as in the *da capo* aria) a more favourable reception within musical thought. This was not the case, and the reason it was not is the acceptance at the time of the literary model which supported the following implication: if instrumental music is to

---

14 Kivy (1993) does not mention a reference for the literary model as an explicit theory for music, but does so in a more recent work (Kivy, 2009) describing the evolution on the model (in Chapter 1) as well as several contemporary supporters (and referring to Robson (1997), especially its Part II, as a reference for their work).

15 Here, by "linguistic" Kivy likely means "semantic". As we saw in Section 6.5, phonetic repetition as rhyme or metre certainly makes sense in poetry. It simply does not add meaning. The adjective "absurd" applied to the literary discourse, in the rest of the section, is meant to apply to the semantic aspect of this discourse, which is the one the proponents of the literary model transfer to music.

be understood as a literary discourse and repetition is absurd in the latter, then it is also so in the former. But the fact remains that repetition is integral to several musical forms (e.g. the sonata form or the scherzo). Kivy then restates the implication above to dismiss the literary model: if repetition is absurd in the literary discourse but is of the essence in instrumental music, then the latter cannot be likened to the former.

Repetition is also at the heart of Kivy's argument to dismiss the organism model. And more obviously so. After all, while repetition is absurd in the literary discourse, it is nevertheless possible. In contrast, a return to a previous development stage for an embryo is utterly unthinkable.

It is after having refuted both the literary and the organism models that Kivy turns his attention (and ours) to a third model for music, which he calls the "wallpaper model" and whose origin he traces back to no less than Immanuel Kant in the famous passage from the *Critique of Judgement* which opens this section. There is an implicit contempt in Kant's putting of absolute music on a level with decoration. But, it appears, Kant had actually no high regard for music.

A second source for the wallpaper model is the Bohemian–Austrian music critic Eduard Hanslick, one of the most influential writers on music in the nineteenth century. Hanslick (1986: 29) writes

How music is able to produce beautiful forms without specific feelings as its contents is already to some extent illustrated for us by a branch of ornamentation in the visual arts, namely arabesque. [. . .] Now let us think of an arabesque not dead and static, but coming into being in continuous self-formation before our eyes. . . Does this mental impression not come close to that of music?

And later on

As children all of us have enjoyed the play of colour and shape in a kaleidoscope. Music is a kind of kaleidoscope, although it manifests itself on an incomparably higher level of ideality.

Hence, again, we find music described as sonic decoration. Unlike Kant, however, Hanslick was not happy with putting music at the level of decorative arts and attempted to elevate music from its lesser companions by recourse to the literary model: music is a discourse, it has a content. But we have already dismissed this assumption. As Kivy remarks, "Kant had more philosophical nerve. His high estimation of poetry, and low estimation of music alone, were based on the premise that poetry had content, particularly moral content, and music, being sonic wallpaper, had none".

And it is this acceptance of music as "sonic wallpaper" that brings us again to the question we started with: How can music, not more than a sonic decoration, share the Pantheon of the "major" arts? It is at this stage that Kivy proposes "to regain Kant's original insight and try to match his nerve: to follow, as Socrates would urge, where the argument leads". Furthermore – and this brings us back to the heart of the discussion in Section 6.3 – he adds

The original insight [. . .] is that music, indeed, possesses no content. It is pure, empty decoration: arabesque. [. . .] Instead of seeing [this] identification of music with the decorative arts as a trivialization of the former, perhaps, rather, we will come to see it as an ennoblement of the latter: the redeeming of the decorative arts from the Western prejudice for knowing as the only goal worthy of the human intellect.

This passage suggests the blurring of boundaries between fine and decorative arts we first argued in Section 6.3, a thought made more explicit in the latest of Kivy's essays (Kivy, 2009: 245) which deserves quotation in full:

I do not want to make, and argue for the claim that absolute music *is* a decorative art. I have no theory of my own about what fine art *is*. And so I do not have the philosophical means to distinguish *it* from decorative art in the first place. What I *do* intend is to rely on our intuitive notion of what decorative, as opposed to fine art is, and to explore the hypothesis that absolute music belongs to the former category.

To "follow where the argument leads" Kivy begins by noticing that the wallpaper model which, up to now, is no more than the acceptance that absolute music possesses no content, is (unlike the literary and organism models) consistent with the musical repeat. It is by evolving from mere logical consistency to a full-fledged analogy that Kivy then embarks on a description of the wallpaper model. And he does so by analysing a Persian carpet: after all, if music is sonic wallpaper, a viable approach to understand music is by understanding first a visual wallpaper. To be precise, Kivy focuses on the *frieze* of his carpet and on the motif which is repeated along this frieze. He analyses the way we perceive this frieze when we explore the carpet and finds it to correspond with the way we perceive the musical repeat[16] to conclude as follows:

Thus repeats, both the external ones, in which whole sections of works are literally played again, or internal ones, where small musical figures or patterns are reiterated, are the means by which the composer of the sonic carpet makes his design, in the large and in detail [. . .] Musical repeats, then, perform an obvious and vital function in that they are the composer's way of allowing us, indeed compelling us to linger; to retrace our steps so that we can fix the fleeting sonic pattern; they allow us to grope so that we can grasp.

---

[16] We cannot give here the detail of Kivy's argument. For this, the reader is encouraged to turn to Kivy's essay.

One may still be tempted to eliminate the repetitions after we have grasped this "fleeting pattern"(for instance, by eliminating it from repertoire pieces which we know well enough). But this would have the same logic as cutting out a part of a carpet because we have already grasped its design, an action that would have as a consequence the destruction of the design itself. As Kivy puts it:

repetition is the means of grasping pattern; but, by definition, pattern is that very repetition, and to dispense with the remainder after it has been grasped would be to dispense with *it*, whereas *it*, the *pattern*, is the whole point of the exercise.

The wallpaper model having given the content that absolute music otherwise lacks, Kivy returns, once again, to our original question to attempt some answer grounded in this content. We say "attempt" since there is no pretension to give a definitive answer, but, rather, the awareness that such an answer will only come when we will have a "more fully understanding [of] all those things about the sonic patterns we call absolute music that make them intriguing to human beings, and capture their attention; that make music orders of magnitude more compelling than what customarily decorates our walls and floors". And, as a possible commencement of the list of "all those things", Kivy enumerates four.

First, music is multidimensional. To the obvious two dimensions of the musical frieze we must add the dimensions inherent to polyphonic composition. And, on top of them, those derived from orchestration and the variety of sound textures corresponding to different musical instruments.

Second, music has a syntax (Kivy hastens to make clear that, unlike the syntax in human languages, it is independent of any semantics; also, Kivy writes "quasi-syntax" to emphasize the differences with the strictures of linguistic grammar).

Third, absolute music is, at least most of it, highly expressive of a variety of emotions (e.g. it is agreed that a tonic in a major key conveys joy or cheerfulness, in contrast with the melancholic mood found in minor keys, even though there is no agreement on whether this is culturally acquired or innate).

Fourth, absolute music is moving, "it provides a profoundly emotional experience to its devotees". Kivy is quick to point out that these emotions are not those expressed by the music, referred to in the point above, but, I suspect, emotions that may be particular to the listener and possibly of a different nature.

It is this multidimensional, syntactical, expressive and deeply moving wallpaper which claims its place in the Pantheon of the fine arts.

9 The dawn of perspective

Oh, che dolce cosa è questa prospettiva![1]

Paolo Uccello, quoted in (Vasari, 1991: 83)

Perspective is to Painting what the bridle is to a horse, and the rudder to a ship.

Leonardo da Vinci (2002: §113)

Figure 9.1 shows a photograph taken by Edward Weston in 1937 of a tomato field on the Monterey Coast of California. The multitude of tomato shrubs draws a pattern that we quickly identify (disregarding imperfections on the placement of individual shrubs) as a wallpaper and it is the contrast between this pattern and the gently irregular landscape of the hill in the background that, most probably, first attracts the eye.

This impression of regularity that we identify with a wallpaper pattern is, however, put to test by the fact that the image does not evenly spread the shrubs: those in the lower part are separated by bigger distances in the surface of the photograph than those close to the hill. This is certainly in contrast with the wallpapers shown, for instance, in Figures 3.12, 3.13, 4.20 and 6.5.

The reasons for this disagreement come quickly to mind, and one may express them with several wordings: the distances in the plane where the

Figure 9.1 **Edward Weston, *Tomato Field*, 1937, gelatine silver print.**

---

[1] Oh, what a sweet thing this perspective is!

Figure 9.2 **Bernat Martorell,** *Annunciation,* **1427, oil and tempera on panel. (The Montreal Museum of Fine Arts, F. Cleveland Morgan Bequest. Photo: The Montreal Museum of Fine Arts, Denis Farley.)**

tomato shrubs grow are not respected in the plane of the picture, distant objects appear smaller than ones close to the observer, or (encompassing the previous phrases in a common expression) the picture shows the landscape in perspective. Even without a precise definition of what perspective is, we understand its immediate relation with the pursuit of a realistic representation, not only in depictions where realism[2] is achieved, such as Weston's photograph, but in those where it fails to do so. The painting by Bernat Martorell in Figure 9.2, crafted in Catalonia in the early fifteenth century, provides an example.

Although there is an intention of perspective (evidenced, for instance, in the contour of the walls) a number of points of execution mar this effort. One such point is the tiling of the room floor. It shows a beautiful

---

[2] The word "realism" has been employed with different intentions in art theory. The use we will make of it in this, and the next, chapter should be, however, not controversial and associated with the optical effect achieved when taking a photograph with a standard lens.

wallpaper pattern on which the different copies of the motif have the same size and are correctly placed. But this respect for the isometries on the floor collides with the wish for realism elsewhere. For instance, the baseline of the wall at the left of the painting cuts the copies of the wallpaper's motif diagonally, in a clear violation of the fact that the wall runs parallel to the translation direction of the motif towards the background.

The discussion above gives evidence that the rules of symmetry as discussed in Chapter 3 may not be appropriate to depict symmetric objects in the context of a perspective viewpoint. More generally, it appears that the notion of distance underlying metric geometry is not sufficient to provide the rules for such a kind of depiction. A goal of this chapter is to describe which new notions come into play and how they came to be. Remarkably, the makers of these new notions were the Italian artists of the Renaissance. It took a few centuries for the mathematicians to integrate them in the overall system of mathematics. Remarkably as well, this integration encountered the discussion on the fifth postulate of Euclid we mentioned in Chapter 1 and sowed the seed of doubt on its necessity.

9.1 Alberti's window

The establishment of Christianity in Europe, and the consequent proliferation of churches, raised the question of how these churches had to be decorated. The issue at stake was the observance of the second commandment which banishes images. A strict observance of this interdiction, as it occurs in the Islamic world, would confine church decoration to the display of geometric patterns. The possible use of images, instead, would allow for an easily accessible account of the Christian story. This was not a minor issue. To understand why, one needs to remember that the natural access to this story, the reading of the Bible, was off limits for most people in those days (and would stay so for many centuries to come). A narration through images, in contrast, could make the life of Christ widely accessible.

This was understood, at the end of the sixth century, by Pope Gregory the Great who sanctioned in favour of displaying paintings in churches. The paintings to be displayed, however, had to conform to the purpose of Biblical narrative, a requirement that imposed a style dominated by a number of restricting features: objects in the painting should be limited to the main characters of the narrated episode; gestures drawn from a small catalogue (e.g. there was a hand gesture for teaching, one for denoting awe, etc.); characters identified by a distinctive signal (e.g. Moses by the tablets of the law, St Matthew by a key, etc.) or replaced by a symbolic figure (e.g. St Mark by a lion, St Matthew by an angel, St Luke by an ox and St John by an eagle).

Figure 9.3 **Duccio di Buoninsegna,** *The Calling of the Apostles Peter and Andrew* **(from the** *Maestà*)**, Siena 1308–11, tempera.**

Among the several features considered as inessential was the sense of depth, the set of relationships between the sizes of the objects depicted and their positions on the painting which allows the observer to infer the positions of these objects in the space where the painted image takes place. Erwin Panofsky (1971: 247) writes that before the fourteenth century "the picture had been conceived as a material surface covered with lines and colors which could be interpreted as tokens or symbols of three-dimensional objects". The relative size of these tokens was determined by their symbolic weight more than by their position on the physical space they were inhabiting. *The Calling of the Apostles Peter and Andrew* as painted by Duccio bears witness that this was still the case in the early fourteenth century (Figure 9.3). The position of Christ's head and the direction of his sight make certain that the boat where the apostles stand is not behind Him. Yet, the size of the apostles in the painting is considerably smaller than the size of Christ.

The above notwithstanding, there is evidence in Duccio's painting of a search for realism. Testimony to this is the fish, which are painted in different orientations and those that offer a frontal view, or nearly so, are correctly foreshortened. Such a foreshortening would have been considered irrelevant to the episode in the few centuries that followed Gregory the Great and all of the fish would, most likely, have been shown in profile. In his *Story of Art*, Gombrich (1989: Chapters 6–11) describes the sequence of changes that led to this search for realism. We will not attempt to summarize this evolution, but content ourselves by noting that during the fourteenth century "the forms appearing *on* the [material] surface came to be thought of as something existing *behind* that surface", a concept whose evolution would lead Leone Battista Alberti, one century later, to liken the picture with a "transparent window through which we look out into a section of the visible world" (Panofsky, 1971: 247). The triumphal resolution of this search for realism, the set of rules governing the depiction of this "section of the visible world", was reached in the fifteenth century and we know it today under the name of *perspective*.[3]

The origins of perspective, however, can be traced back to the study of the mathematical properties of vision done by Euclid, who wrote a book on the subject which he called *Optics*. A basic postulate since then is that the eye emanates "visual rays" which meet the objects in sight. The ensemble of these rays forms the *visual cone*, its apex the *vantage point* (or *viewpoint*) and the image the beholder perceives arises from the intersection of the visual cone with the retina.

We may leave aside the physical nature of these rays and focus on pictorial representation. The influence of Euclid's *Optics* for the latter stems from the following simple observation. Assume we intersect the cone of light with a plane (in the following called the *picture plane*) and replace each intersection point on this plane by a point coloured as the extreme of the intersecting visual ray. In doing so we colour (a part of) the plane, thus creating an image on it. The key observation is that, for the eye at the apex of the visual cone, the original scene (say the octagon at the right of Figure 9.4[4]) and its corresponding image on the picture plane are indistinguishable.

The consequences of this observation for painting were to be monumental. It reduced (a key aspect of) the search for realism to the understanding of relative sizes in such an intersection. And yet, even though the work of Euclid was followed by scholars in the Middle Ages, under the name of *Prospectiva*, medieval painters seem to have largely ignored this development as much as these scholars were unaware of the painter's concerns: "Classical 'Optica' and medieval 'Prospectiva', then, were no more concerned with problems of artistic representation than the representational methods of Jan van Eyck, Petrus Cristus or

---

[3] The word "perspective" derives from the Latin *perspicere* meaning "to see through". It perfectly fits the context of the "transparent window" of Alberti.

[4] This figure is taken from a book on perspective written by da Vignola, whom we have already encountered, as an architect, in Section 7.5.

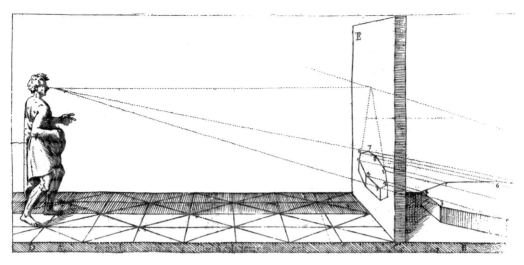

Figure 9.4 **Jiacomo Barozzi da Vignola, drawing in** *Le Due Regole della Prospettiva Practica,* **1633. (ⓒ The British Library Board,** *Le due regole della prospettiva*/**Jiacomo Barozzi da Vignola; ed. I. Danti.)**

Dirk Bouts were based on the doctrines of scholastic writers" (Panofsky, 1971: 249).

The quest for realism among painters took the form of a sequence of trials and errors which were embodied in simple recipes. Recipes that were inaccurate as often as not. Panofsky mentions that, still in 1435, the successive intervals in a sequence of equidistant points in the depicted scene were "mechanically, and of course mistakenly, diminished by one-third each" (Panofsky, 1971: 247).[5]

Nonetheless, in this process of trials and errors, not all trials led to errors. Paramount among those which did not was the finding, already established by the mid-fourteenth century, that lines perpendicular (also called *orthogonal*) to the picture plane are represented in the latter by lines meeting at a common point, which came to be called *central vanishing point*.

It is common, and we will not differ, to illustrate the vanishing point with the *Flagellation* of Piero della Francesca (Figure 9.5). This painting, described by Kenneth Clark as "the greatest small painting in the world" (quoted in (Owen, 2008)) shows, in the background, at the left, Christ being scourged by the Romans during his passion while, in the foreground, at the right, three men appear to be chatting and paying no attention to the scourging.

To give a sense of depth, and allow for the figures to have the right proportions, Piero frames the sacred scene with architectural elements

---

[5] This method had already been disproved by Euclid in Proposition 8 of his *Optics*.

Figure 9.5 **Piero della Francesca, *The Flagellation of Christ*, tempera on panel, 1455–60. (Galleria Nazionale delle Marche, Urbino, Italy/The Bridgeman Art Library.)**

which appear to be aligned orthogonally to the picture plane. Thus, lines separating columns of tiles (we have highlighted two of them), or signalling the edge of a ceiling beam or a roof eaves (both highlighted as well), meet at a vanishing point which is located half-way between the left and right borders of the painting.

The awareness of the vanishing point represented a major step in the quest for realism. It solved the issue of how to draw the lines orthogonal to the picture plane. It did not, however, provide answers to a number of related questions, such as, for instance, how far away from each other should the lines separating consecutive rows of tiles be in Piero's *Flagellation*. This problem, how to draw a tiled floor or *pavimento*, was going to become, as we shall see soon enough, a central one in the fifteenth century.

It was Filippo Brunelleschi whom, around 1425, produced the first paintings (a frontal view of the Florentine Baptistery and an oblique view of the Palazzo Vechio) bringing to the arena of artistic representation the ideas of optics. Those paintings were intended to be compared with the depicted scene,[6] for which the beholder had to stand in a precise position. According to contemporary accounts, both images – that in the scene and that in the painting – were then identical. Brunelleschi did not, however, put in writing the method he used to produce the paintings, and even the paintings themselves have, unfortunately, been lost since at least the middle of the sixteenth century.

The impact of Brunelleschi's work in Florence was widespread and immediate. On the one hand, many of the painters of the time adopted Brunelleschi's techniques and by the middle of the fifteenth century all great Florentine artists painted in perspective. On the other hand, it triggered a number of treatises on perspective. The first such treatise was *De Pictura*, written by Alberti, published in 1435 in Latin and translated to Italian (*Della Pitura*) one year later (and eventually to English (Alberti, 1991)). This book was not easily read, however, since it had no illustrations and its first part is rather technical. A first comprehensive exposition of perspective, lavish with illustrations, generous with worked-out examples, is Piero della Francesca's *De Prospectiva Pingendi* (written in the 1470s). Piero's book explains with every detail the set of rules, known at the time as *costruzione legittima* (the true method), which allowed painters to correctly deduce the proportions of the object to be painted in the plane of the picture.

We may illustrate the *costruzione legittima* by drawing a prisma with rectangular basis; that is, a rectangular box. To do so one needs first to execute two preliminary drawings with the box viewed from its side (the *elevation*) and from above (the *groundplan*[7]). In both cases we do not attempt any effect of depth but simply draw the visible side of the box (lateral in the elevation, top in the groundplan). Furthermore, in each of these drawings one places three objects: the box, the picture plane ℓ and the eye O. It is clear that the latter occurs as a point in the drawings. The box, as we mentioned, is drawn as a rectangle and the picture plane as a line.

Next, one traces the lines, in each of the drawings, from the four vertices of the (visible side of the) box to the eye. The intersection with the picture plane determines four points in each case. If we label the eight vertices of the box with the letters A, B, C, D, E, F, G and H then,

---

[6] Actually, at least for the Baptistery painting, the comparison had to be done between the scene and the painting as reflected in a mirror. The beholder would look through a small hole in the painting and by placing or removing a mirror he would see the painting or the original scene respectively. This allowed one for a quick alternance between both views and hence for an undemanding comparison between them. The details of how this was achieved are preserved in a 1480s text by Antonio Manetti (1970: 42–44) (see also (Edgerton, 2009: Chapter 6)).

[7] The terms "elevation" and "groundplan" come from architectural practice, a domain where the *costruzione legittima* was already in use, although not with this name.

for instance, the line corresponding to the edge *GH* (which is seen as a rectangle vertex in the elevation) cuts the picture plane at a point *G′H′*, and the same can be done for the other three vertices. Similarly, the line corresponding to the edge *DH* (which is seen as a rectangle vertex in the groundplan) cuts the picture plane at a point *D′H′* and, again, the same can be done for the other three vertices.

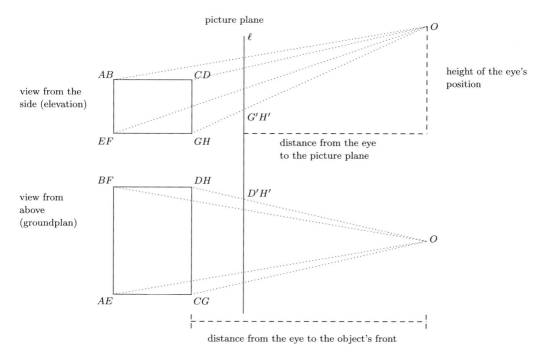

From these preparatory drawings one can now draw the box as follows. Set two coordinate axes and mark in the horizontal axis the four points in ℓ obtained in the groundplan, respecting the distance between them. Similarly, mark in the vertical axes the four points in ℓ obtained in the elevation. Each of the eight vertices of the box is now placed in the point of the plane with coordinates determined by the only pair (horizontal and vertical) having the vertex in both. For instance, the point *H* is placed at the intersection of the vertical line through *D′H′* and the horizontal line through *G′H′*.

Note that the box in this final drawing (see below) appears to be smaller than the rectangles in the elevation and groundplan. This has to be so. The dimensions of these rectangles are the true dimensions of the box, which are independent of the positions of the eye and the picture plane, while the dimensions in the final drawing are meant to depend on these positions. The box front *CDGH* would be seen with its true dimensions if it were "touching" the picture plane. The more distant it is behind it, the smaller it will appear.

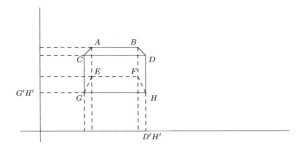

The procedure just described was too complicated and, as Panofsky (1971: 251) points out,

In practice, not even the most conscientious of painters would even construct individual objects, let alone figures, by first developing two diagrams and then projecting these on the picture plane. It was deemed sufficient to build-up a three-dimensional system of coordinates in foreshortening which enabled the artist to determine the relative magnitude, though not the shape, of any object he might wish to render.

Assume, to fix ideas that the object to be painted (e.g. a column, a person) is represented by a black rectangle and that we have already painted a similar object (or, at least, one with similar dimensions). How large should be the one we want to paint? More precisely, how can one approximately compute its width and height depending on the position of the object on the picture? The following figure explains the method. There, the given rectangle (the one which gives sense to the word "relative" in Panofsky's description) is at the bottom right of the figure, and the one whose dimensions are to be computed is above and at the left. The figure also shows the "three-dimensional system of coordinates" in the form of two grids for the floor plane and a lateral vertical plane.

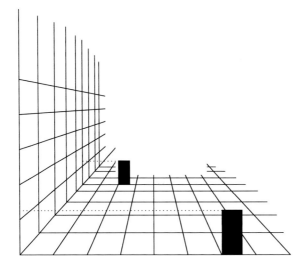

The way one estimates the dimensions of the rectangle at the back is simple. Recall, we want the two of them to be of comparable size. The rectangle at the front has a basis of about 0.6 units (the measure of the tile's side) and a height of about 1.25 units (this is obtained by looking at the intersection of the horizontal dotted line with the vertical rule at the left). Hence, once we have decided the position of the second rectangle, its dimensions will be given by taking a basis of 0.6 units and a height of 1.25 units (both measured at the rectangle's position).

To use this procedure one needs at hand the three-dimensional system of coordinates; that is, the two grids. The accurate drawing of such *pavimenti* became in this way a central task among the painters of the Quattrocento.

In contrast with the *costruzione legittima*, however, these grids are easy to obtain. Once we have fixed a unit (the length of a tile's side), in order to draw the grids the only information we need is the collection of the distances between the successive rows of tiles. This is so because the receding lines in the grid can be drawn by joining points separated by a unit length in the horizontal and vertical boundaries with a vanishing point P (which is located midway between the vertical boundaries of the canvas and at the height of the eye's position):

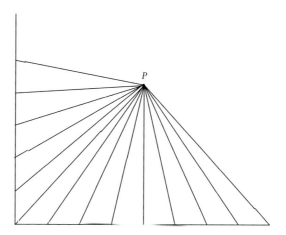

We are thus left with the task of computing the distances between the successive rows of tiles. This can be done with a ruler as follows. First, draw the separation lines between rows of tiles as well as the picture plane ℓ as on an elevation. That is, these separating lines appear as equidistant points on a horizontal line and the picture plane as a vertical line. Mark as well the position O of the eye.

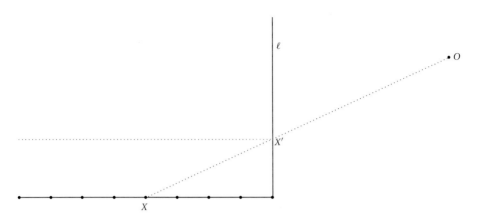

Take the line separating any two rows of tiles (for instance, in the figure above, the point X separating the fourth row from the fifth). The plane joining X with the eye O interesects the picture plane on a line X'. In our lateral (i.e. elevation) view, this plane is seen as a line and the intersection X' as a point. By construction, the height of X' in ℓ is the height that the line separating the fourth row of tiles from the fifth should have on the picture plane. Doing so for all the separation lines gives the distances between them.

The two arguments above – giving the receding lines and the distances between rows of tiles – can furthermore be merged into a single figure which completes the grid:

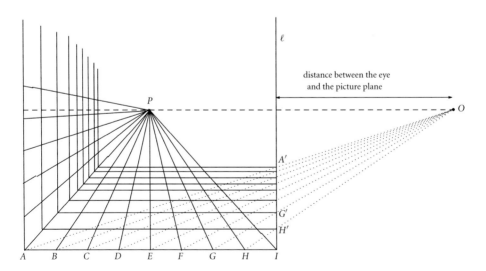

It is this *abbreviated construction* what was used by most Italian painters in the Quattrocento. Both Alberti and Piero della Francesca describe it in their writings.[8] Subsequently, other constructions making use of new ideas (such as, for instance, the use of "lateral vanishing points") further simplified, and added understanding to, the drawing of *pavimenti*. We will not pursue these developments in our account.

The onset of perspective not only allowed for a more realistic depiction, but also became a way of organizing the space of the painting. Paintings such as Martini's *Maestà* (Figure A.1) or Duccio's *Calling* (Figure 9.3) organized the figures within the pictorial space but not the space itself, which was identified with the flat surface of the picture plane. This is in contrast with Piero's *Flagellation* (Figure 9.5), where, as we have already pointed out, the presence of buildings, tiled floors and other elements was meant to render on the picture a sense of measure of the three dimensions of the space depicted.

The search of realism was not a sole preserve of Italian painters. The process of trials and errors that led to the use of perspective in fifteenth-century Florence also took place in other parts of Europe, notably in the Netherlands, where some painters, notwithstanding a lack of proper perspective rules, managed to produce extremely convincing work (see Figure 9.6). This is the case of the painters mentioned by Panofsky in our quote above – Jan van Eyck, Petrus Cristus and Dirk Bouts – and this lack of proper perspective is at the core of Panofsky's meaning in that quote. Also, even though unrelated with our guiding theme, it must be noted that neither was the use of perspective the only path in the quest for realism. The employment of oil as a medium to bound pigments is an example at hand. This technique was certainly known in the first millennium of the common era. But it was not until the fifteenth century that it spread in European painting, pioneered by Netherlandish painters and, fundamentally among them, by Jan van Eyck. The slow drying of the oil allowed for a translucency that was not possible in the, most commonly egg-yolk-based, water-soluble, fast-drying medium previously used (known as *tempera*). This translucency not only conferred the painting a characteristic gloss, but also endowed the artist with the capacity of gently shading colours, a capacity that would be at the heart of the *sfumato* mode characterizing Leonardo's style, a few decades after van Eyck.

Oil painting was quickly adopted in northern Europe and, shortly after, in Italy. Concurrently with this phenomenon, the methods and

---

[8] The method used by Brunelleschi to produce his two 1425 paintings is, however, a matter of discussion. Richard Krautheimer (1970: Chapter XVI) suggested that Brunelleschi used the *costruzione legittima*, and this thesis was recently supported by David Summers (2003: 514). Even more recently, though, Samuel Edgerton (2009: Chapter 7) argued that a different, simpler, method was used.

Figure 9.6 **Jan van Eyck,** *The Arnolfini Portrait*, **oil on panel, 1434.**
(National Gallery, London, UK/The Bridgeman Art Library.)

ideas around perspective developed in the fifteenth century in Italy spread to Northern Europa. Among the most enthusiastic in embracing them was Albrecht Dürer, who in 1506 travelled to Bologna "for the sake of 'art' in secret perspective which some one wants to teach me" (Panofsky, 1971: 248). Nearly two decades later, Dürer published a treatise on geometry[9] where, among other things, he popularized perspective among German painters. Besides this main goal, the book of Dürer provides us with evidence of some of the mechanical ways that were alternatively used in the search for accurate pictorial representations. The woodcut in Figure 9.7 shows one such way. It consisted of a frame on which two strings could glide, one horizontally and one vertically. The painter would consider a point in its subject (a lute in Dürer's woodcut) and stretch a third string between this point and a fixed position, say in a wall, behind the plane of the frame (a pulley at one end

[9] *Underweysung der Messung mit dem Zirckel uñ Richtscheyt*, 1525.

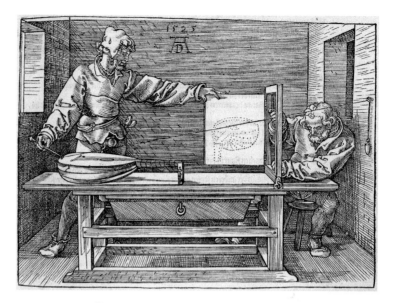

Figure 9.7 **Albrecht Dürer, *Underweysung der Messung*, Book 3, Figure 64, woodcut. (© The British Library Board. C119h7(1).)**

of this string and an assistant at the other ensured that the curve made by the string was a straight line). Then he would move the horizontal and vertical strings so that their crossing touches the third string. Finally, he would remove the third string, turn round the canvas (which was fixed to one of the lateral sides of the frame and can rotate around it) so that it was just behind the frame and mark a point on it where the two gliding strings crossed. In doing so sufficiently many times, one obtains a collection of points in the canvas which provides more than enough information to finish the drawing.

No matter whether due to an adequate control of the laws of perspective or to the use of a mechanical device such as the one described above, the final result would provide the viewer with the "transparent window" mentioned by Alberti.[10]

---

[10] One may object to this statement for a number of reasons. For instance, true vision is stereographic (we look with both our eyes through the window) while perspective relies on a single position for the eye. Also, this position should coincide with that of the observer's eye(s): the accuracy of the representation diminishes when one distances oneself from it. This decrease, however, is mostly compensated by our sensorial apparatus for moderate deviations from the vantage point and, in most of the cases, the effect of a well-crafted painting is certainly convincing. These issues are discussed, for example, by Panowsky (1991) in his best known book and by Gombrich (1982) in his article "The mirror and the map". We will not enter the discussion here.

9.2 The dawn of projective geometry

> In the fifteenth and sixteenth centuries artists took geometry out of its
> elements by using it for concrete pictorial ends. Along the way they laid
> the foundations for the projective geometry of the following centuries, in
> which they played no further part.   A. Flocon and A. Barre (1987: 34)

We ended Section A.4 in the Appetizers maintaining that, occasionally, art has returned in kind the influence received from mathematics.
The ebullience around perspective which set up in Europe during the
fifteenth century was at the core of a number of new ideas in the development of geometry. Some of these ideas will occupy us in this and the
next section. In doing so, neither historical nor mathematical detail will
be the goal; conciseness will. The reader interested in (the first 300 years
of) the relation between perspective and mathematics will find pleasure
in (Andersen, 2007).

Even though some of the painters in the Italian Renaissance had a
considerable knowledge of the mathematics of their time (this is the
case of Piero della Francesca, as shown by Judith Field (2005), of Dürer
and of Leonardo da Vinci), most of them lacked a basic education in
mathematics. This explains why an array of purely mathematical questions naturally arising from the perspective constructions blooming in
the Renaissance were not pursued during that period. In fact – leaving
aside the work of Desargues, Monge and a few other precursors working between the seventeenth and the nineteenth centuries – this pursuit
would have to wait until the nineteenth century to be carried out, and its
circumstances are not without interest.

Jean-Victor Poncelet studied at the École Polytechnique, a school that
allowed him to become a mathematician, an engineer and a militar. It was
in this last capacity that he participated in Napoleon's invasion of Russia
in 1812, where, to his misfortune, he was taken prisoner and confined
in a small village near Saratov, on the Volga. Poncelet would not recover
freedom until 1814. During this time, however, confinement conditions
were apparently gentle and allowed Poncelet to lecture geometry to a
group of fellow officers who, like him, had also learned the subject at
the École Polytechnique. The content of these lectures would eventually
take the form of a book, called *Traité des Propriétés Projectives des Figures*
(*Treatise on the Projective Properties of Figures*), which is today considered
as the first organized exposition of projective geometry (the name itself
being due to Poncelet).

What exactly are these "projective properties"? Recall that the basic
construction used in perspective consisted of producing the visual cone
(associated with a given scene and viewpoint) and then intersecting it
with the picture plane. To put this construction in a mathematically

convenient way we can imagine two planes lying in Euclidean three-dimensional space. Let us denote the planes by π and π' and the three-dimensional space (in all what follows) by $\mathbb{E}^3$. Now consider a point O which is neither on π nor on π'. To a point P in π we can associate a point P' in π' by doing the following:

(i) draw the line ℓ (in $\mathbb{E}^3$) determined by P and O;
(ii) let P' be the point where ℓ intersects π'.

Here, π corresponds to the scene, π' to the picture plane and O to the viewpoint. The drawing in Figure 9.4 illustrates this construction. The plane π is the top of the lying polygonal figure, the plane π' is the vertical one (denoted by E in that drawing), the point O the location of the viewer's eye and the point P any of the vertices numbered 3, 4, 5 or 6. One can see the associated points in π' numbered in the same manner.

This association, mapping points on π to points on π', is called a *projectivity*, and the question above can be restated as: Which properties of plane figures are preserved by projectivities?

We will have to wait until § 9.2.3 to be in a situation to tackle this question. Nonetheless, we may now note that distances, angles and simple ratios (quotients of distances between collinear points) are not preserved. This is immediately verified by looking at any *pavimento*, such as the one on Piero's *Flagellation* (Figure 9.5) or in the simple drawing below (which shows the floor obtained with the abbreviated construction in Section 9.1).

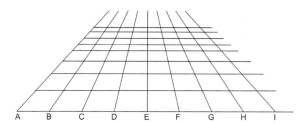

It is apparent in this drawing that angles are not preserved (otherwise all tiles in the drawing would be rectangular), nor are distances (since those tiles do not have the same side-length in the picture), nor are simple ratios (by the same reason).

We now reach a key issue in the development of projective geometry.

The construction described in steps (i) and (ii) above for the projectivity from plane π to plane π' is faulty in two, somehow similar, ways. First, there are points in π for which we cannot associate any point in π'. Indeed, let f be the intersection of π with the plane α (not drawn below) parallel to π' passing through O. We call f the *foot line*.

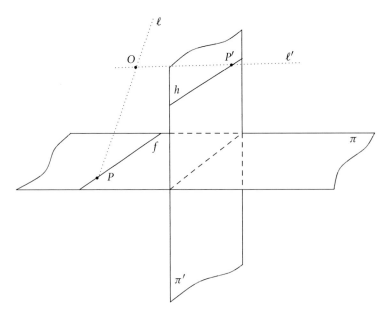

For any point P in f the line ℓ passing through P and O is included in α, which is parallel to π'. Therefore, part (ii) of the construction fails since ℓ cannot intersect π'.

Second, there are points in π' which are not the image of any point in π. Indeed, let h be the intersection of π' with the plane β parallel to π passing through O. We call h the *horizon*. As before, for any point P' in h the line ℓ' passing through P' and O is included in β, which is parallel to π. Therefore, ℓ' cannot have been produced in part (i) of the construction since ℓ' does not intersect π and, consequently, P' cannot be the image of any point in π. In pictorial terms, this translates to the fact that, while we can normally ask for any point on the picture plane which point of the world is represented, we cannot do so with the horizon line in the picture: this line corresponds to no points in the depicted scene.

This feature of projectivities did not inconvenience Renaissance painters. For, on the one hand, they never intended to paint the neighbourhood of the foot line. And, on the other hand, they considered points on the horizon line as entities lacking the ability to reflect any part of the painted scene (just as the perceived horizon in this scene does not correspond to any physical location). Besides, some points in the horizon line were useful to the painter in their role of vanishing points, and this service largely compensated the fact that those points did not correspond to any point in the scene. Renaissance painters' pragmatism on this issue is exemplified by da Vignola, whom we met in Sections 7.5 (as an architect) and 9.1 (as the author of a book on perspective). His fifth definition in this book (da Vignola, 1987) characterizes "perspective parallel lines" as those meeting at the horizon line. Immediately after this, da Vignola states (translation is mine): "This definition may seem at a first glance false and contrary to Definition 3.5 in the first [book] of Euclid: but those who will consider it sound [...] will find it very accommodating and appropriate for this art".

The same feature, however, could not fail to disturb mathematicians. To understand why, we need to describe a concept that helps us to understand the nature of transformations.

9.2.1 Bijections and invertible functions

In Section 2.1 we defined the notion of a function. Recall, we wrote $f : A \to B$ to denote a function f associating an element of the set B with every element of the set A. The functions we have dealt with thus far – translations, reflections, rotations, glides, homothecies, strains – are not, however, "simply" functions; they also happen to satisfy some additional properties.

We say that f is *injective* when, for all $x, y \in A$, if $x \neq y$ then $f(x) \neq f(y)$. We say that f is *surjective* when, for all $z \in B$, there exists $x \in A$ such that $f(x) = z$. Finally, we say that f is *bijective* (or that it is a *bijection*) when it is both injective and surjective. Bijections are sometimes referred to as *one-to-one correspondences*.

An intuition for the meaning of these notions might be obtained by considering a hypothetical theatre and the happy bunch of people who will attend a given performance there. Denote by A the set of these attendees and by B the set of benches, or seats, at the theatre. A way of allocating the seats among the attendees consists of associating, to each attendee $x \in A$, a well-defined seat $f(x) \in B$. This defines a function $f : A \to B$.

For the allocation given by f to be acceptable it is a must that different attendees will be given different seats. Hence, we want the function f to be injective. Also, the surjectivity of f amounts to the fact that every seat in the theatre is assigned to one attendee, a feature we may again consider as desirable (it certainly is for the theatre's managers). Let us therefore assume for a while that, for this performance, the function f above is both injective and surjective; that is, f is bijective.

A first consequence we draw from f being a bijection is the certainty that the number of attendees coincides with the number of seats in the theatre. Remarkably, we are certain of this equality even in the absolute ignorance of the actual theatre capacity. It is this notion that bijectable sets must have the same cardinality that allowed for a formal development of a theory of cardinality for infinite sets.

A second consequence, more relevant to this section, is the fact that the function f can be inverted in a sense to be made clear next. Since f is surjective, any seat in the theatre is allocated to some spectator. Since f is injective, no seat is allocated to more than one spectator. It follows that every seat is allocated to one and only one spectator. Consider the function $g : B \to A$ associating with every seat $z \in B$ the only spectator $x \in A$ to whom seat z is allocated; that is, the only $x \in A$ such that $f(x) = z$. One can consider the compositions (recall Definition 3.5)

$$g \circ f$$

$$A \xrightarrow{\ f\ } B \xrightarrow{\ g\ } A$$
$$x \qquad f(x) \qquad g(f(x))$$

and

$$f \circ g$$

$$B \xrightarrow{\ g\ } A \xrightarrow{\ f\ } B$$
$$z \qquad g(z) \qquad f(g(z))$$

and easily get convinced that $(g \circ f)(x) = x$, for all $x \in A$, and $(f \circ g)(z) = z$, for all $z \in B$. That is, if we denote by $\mathrm{Id}_A : A \to A$ and $\mathrm{Id}_B : B \to B$ the identity functions on A and B respectively, we have $g \circ f = \mathrm{Id}_A$ and $f \circ g = \mathrm{Id}_B$. In this situation, consistently with our notation in Section 3.2, we say that g is the *inverse* of f, that f is *invertible*, and we write f^{-1} instead of g. Invertibility has a close relationship with bijectivity.

Proposition 9.1 *A function $f : A \to B$ is invertible if and only if it is bijective.*

Proof Assume first that f is invertible. We will show that it is bijective, proving first that it is injective. To do so, consider $x, y \in A$ with $x \neq y$. If $f(x) = f(y)$ then

$$x = f^{-1}(f(x)) = f^{-1}(f(y)) = y,$$

which is a contradiction. This shows that f is injective. Also, for any $z \in B$ we have $z = f(f^{-1}(z))$, which shows f is surjective.

Assume now that f is bijective. For every $z \in B$ there exists $x \in A$ such that $f(x) = z$ (since f is surjective). Furthermore, if $f(x_1) = f(x_2) = z$ then $x_1 = x_2$ (since f is injective). Hence, for every $z \in B$ there exists a unique $x \in A$ such that $f(x) = z$. Define $g : B \to A$ by $g(z) = x$. Then, for all $z \in B, f(g(z)) = f(x) = z$. Also, for all $x \in A, g(f(x))$ is the only element y in A such that $f(y) = f(x)$. But then y needs to be x and we have $g(f(x)) = x$. This shows that g is the inverse of f. $\qquad\square$

We have seen in previous chapters that translations, reflections, rotations, glides, homothecies, shears and strains have an inverse (a translation with opposed vector, the same reflection, a rotation with opposed angle, etc.). It follows from Proposition 9.1 that all these transformations, when considered simply as functions, are bijective.

9.2.2 The projective plane

The reason that projectivities, as defined above, had to produce discomfort among mathematicians is that these functions, unlike isometries or affinities, were not necessarily bijections between planes. This complicated the issue of composing projectivities.

It soon became apparent that the problem did not lie with the notion of projectivity but with the nature of the plane the projectivities are acting on. A new conception of plane was needed making room for points "in the horizon". What follows is a way of constructing this extended conception of plane.

The general idea is briefly stated: if the Euclidean plane has no points at the horizon, we simply add them to it. The question is, what is, mathematically speaking, a point at the horizon? A glance at the following drawing suggests an answer:

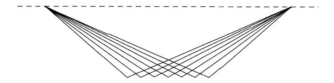

We see on it a *pavimento* – say on a plane π – viewed from a position that makes its tiles "point" toward the observer. In particular, the lines where the tiles join divide naturally into two sets, and for both of them the lines in the set appear to meet at the horizon (shown as a dashed line in the drawing). Meetings at the horizon aside, the only feature shared by all lines in each of these sets is a common direction: they are all parallel in π. Hence, parallel lines meet at the horizon. And conversely, if two lines meet at the horizon then they must be parallel in π. Sets of parallel lines with different directions (such as those corresponding to perpendicular sides of the tiles) meet at different points at the horizon line and to any point in the horizon line we can associate a set of parallel lines that meet on that point. A first emerging conclusion is that points at the horizon correspond with directions on the plane π.

A question, however, arises from this conclusion: How many directions has a line? In other words, does a line have one or two points in the horizon?

A naive approach to this question would take the latter for an answer. When I look at a (straight) road I see its sides meeting at a point on the horizon. But if I turn and look at the same road "in the other direction" I will see its sides meeting at a different point on the horizon, opposite, as it were, to the first. Yet, a more considered reflection about our original context shows that these two points actually coincide in the picture plane π'.

Recall the drawing in page 241 where the foot and horizon lines were introduced. Consider a projectivity from plane π to plane π' with projection point O. Consider as well a line p lying on π and its image p' lying on π'. The following drawing shows p and p' (as dotted lines) and the image P' of a point P in p:

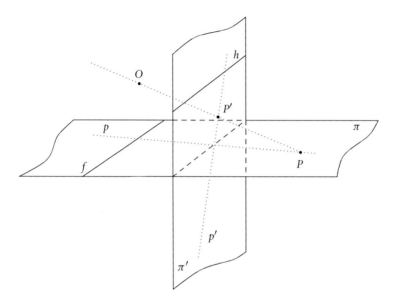

Consider what happens to the point P' when P moves towards the "horizon of π," that is, when P moves away from the scene, either to the right or to the left.

To see that the two apparently opposite directions are, within this projective context, the same direction, look at the plane γ determined by p and O. The line p' is in γ as well, since it is the intersection of γ and π'. We denote by F the intersection of p with the foot line and by H that of p' with the horizon line. Now think about several points – say P_1, P_2, Q_1 and Q_2 – in p and their images in p':

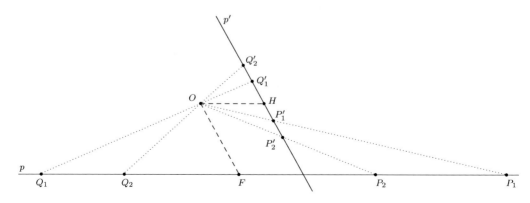

A glance at the drawing (maybe accompanied by a moment's reflection) shows that, at the right of F, when P moves to the right its image P' approaches H from below (compare the images of P_1 and P_2). Likewise, at the left of F, when P moves to the left its image P' approaches H from above (compare now the images of Q_1 and Q_2). That is (no matter whether to the right or to the left), when P approaches the horizon of π its image P' approaches the point H. We only need to add one point (and not two) to the line p to fill the gap created by H in the correspondence between p and p'.

Having now at hand an inkling of what the missing points look like, we can specify how to add them to the Euclidean plane.

Fix a point O on the Euclidean plane $\mathbb{E}$. Each line passing through O determines a direction (which, unlike the direction of a vector, is not oriented) and we can therefore look at this set of lines as a set of directions (the figure on the left shows a few of them; on it, for instance, the vertical line corresponds neither with the direction towards north nor with that towards south, but rather with the idea of "vertical direction").

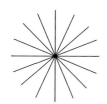

Furthermore, because of the axiom P5$'$, for each line ℓ on the plane there is a unique line r through O parallel to ℓ. We can thus say that every line has a unique direction. In addition, two lines on $\mathbb{E}$ are parallel if and only if they have the same direction. We denote by ℓ_∞ this set of directions.

Definition 9.2 We define the *projective plane* $\mathbb{P}$ to be the union of the Euclidean plane $\mathbb{E}$ with the set ℓ_∞. We say that ℓ_∞ is the *line at infinity* or the *improper line* of $\mathbb{P}$.

A reader meeting the definition above for the first time may perfectly well feel disconcerted: we tend to associate the word "point" with an idea of "dot" or "position", not with a direction. The linking of a direction with the position ultimately reached by following that direction could ease this feeling of disconcertment. The circumstance that this position does not actually exist is a rewording of the fact that $\mathbb{P}$ actually extends $\mathbb{E}$. Similar considerations apply to the use of the word "line" for ℓ_∞.

In Section 12.5 we will have a new look at the projective plane which, though bearing little relation with Renaissance perspective, will bring all points in this plane closer to the common idea of point. For the time being, we proceed with our exposition of the features of $\mathbb{P}$.

Definition 9.2 constructs the projective plane $\mathbb{P}$ and makes the Euclidean plane $\mathbb{E}$ a subset of $\mathbb{P}$. In doing so, it raises a number of natural questions, such as: What are the lines in $\mathbb{P}$? What is the relationship of these lines with lines in $\mathbb{E}$? Which properties do lines and points in $\mathbb{P}$ possess?

The next definition tackles the first two questions. The proposition following it is a first attempt at an answer for the third one.

Definition 9.3 For any line r on $\mathbb{E}$ we define its *projective closure* $\bar{r}$ to be the union of r and the point R_∞ in ℓ_∞ given by the direction of r. A *projective line* is either the improper line ℓ_∞ or the projective closure of a line r on $\mathbb{E}$. We also say that r is the *Euclidean part* of $\bar{r}$, which we write as $\text{Euc}(\bar{r})$. By convention, $\text{Euc}(\ell_\infty) = \varnothing$, the empty set. Sometimes, to avoid confusion we will talk about "Euclidean lines" to mean a line in $\mathbb{E}$ (as opposed to a projective line on $\mathbb{P}$).

We thus look at $\mathbb{P}$ as a Euclidean plane $\mathbb{E}$ to which we have added a line, ℓ_∞, corresponding to the horizon; hence the expression "at infinity" for this line. Every Euclidean line has a point in ℓ_∞ and it is important to keep in mind that this point is reached no matter which direction the line is traversed: the projective closure of a Euclidean line has only one point at infinity, not two.

The following result summarizes the basic incidence relationships of lines and points on $\mathbb{P}$.

Proposition 9.4

(i) *Two different points determine a unique projective line in $\mathbb{P}$.*
(ii) *Two different projective lines determine a unique point in $\mathbb{P}$.*

Proof To prove (i) we consider three possible cases: the two points are in $\mathbb{E}$; one of them is in $\mathbb{E}$ and the other in ℓ_∞; and both are in ℓ_∞.

If both points are in $\mathbb{E}$ then they determine a unique line by axiom P1. If one of them, say P, is in $\mathbb{E}$ and the other, say v, is in ℓ_∞ then there is a unique line passing through P with the direction v (by axiom P5'). Finally, by construction, if the two points are in ℓ_∞ then the only line passing through both is ℓ_∞ itself.

To prove (ii) we also consider the following two possible cases: both lines are proper or one of the lines is ℓ_∞. For the former, there are two possibilities: either the two lines are not parallel – in which case they intersect at a unique point – or they are – in which case their only intersection is their common direction (which belongs to ℓ_∞). For the latter, the only intersection of a proper line with ℓ_∞ is the direction of the proper line. $\qquad\square$

Remark 9.5 While part (i) of Proposition 9.4 is a projective version of axiom P1, part (ii) is in contradiction with P5': it states that there are no parallel lines in $\mathbb{P}$. A first, rushed, conclusion of this contradiction is that a geometry satisfying axioms P1–P4 but not P5 is possible. A moment of reflection dispels this idea. Axioms P3 and P4 involve distances and angles and we do not have an obvious projective version of these measures. Indeed, what should be the distance between a proper point and an improper point? And between two improper points? Which angle would a proper line and ℓ_∞ make? And two lines meeting at ℓ_∞? These questions do not have an easy answer and we will leave them unanswered for a time to come. We will return to them in Section 12.5.

At this point we claim that the projective plane solves the problem mentioned above in the sense that it makes projectivities bijective. A first step to prove this claim consists of extending projectivities from functions defined on Euclidean planes to functions defined on projective planes. Recall that the definition of projectivity we gave above involved two different planes in three-dimensional Euclidean space $\mathbb{E}^3$. Extending projectivities to the projective closure of these planes will, therefore, require the consideration of a three-dimensional projective space $\mathbb{P}^3$ where projective planes could live (in the same manner that Euclidean planes live in Euclidean three-dimensional space $\mathbb{E}^3$). To proceed with this roadmap we need at hand a few basic incidence relations in three-dimensional Euclidean geometry. We list them below (and consider them as axioms).[11]

A1 Two different points determine a unique line in $\mathbb{E}^3$.

A2 A line and a point not in that line determine a unique plane in $\mathbb{E}^3$.

A3 Given a plane π and a line ℓ not cutting π there exists a unique plane π' containing ℓ and not cutting π.

If two planes in $\mathbb{E}^3$ do not intersect we say that they are *parallel*. We say that two lines in $\mathbb{E}^3$ are *disjoint* if they do not intersect. We say that they are *parallel* if they are disjoint and there exists a plane containing them both. For instance, in the box with vertices A, B, C, D, E, F, G and H which we drew in perspective in Section 9.1 the lines passing through AB and EG, respectively, are disjoint, whereas those passing through AB and GH are parallel. Disjoint lines may have different directions in $\mathbb{E}^3$; parallel lines have the same. Similarly, parallel planes have the same set of directions (because of A4 below).

A4 Given a plane π, a line ℓ on π, a plane π' parallel to π and a point P on π', there exists a unique line ℓ' on π' passing through P and parallel to ℓ.

The set of all possible directions in $\mathbb{E}^3$ is denoted by π_∞. We define the three-dimensional *projective space* $\mathbb{P}^3$ to be the union of Euclidean space $\mathbb{E}^3$ with the set π_∞. We say that π_∞ is the *plane at infinity* of $\mathbb{P}^3$. Euclidean parts and projective closures of lines and planes are defined as in Definition 9.3.

---

[11] The attentive reader will notice that we have already used these axioms at the beginning of this section.

Proposition 9.6

(i) *Two different points determine a unique line in $\mathbb{P}^3$.*

(ii) *A plane and a line not included in the plane determine a unique point in $\mathbb{P}^3$.*

(iii) *Two different planes in $\mathbb{P}^3$ intersect on a line.*

(iv) *A line and a point not in that line determine a unique plane in $\mathbb{P}^3$.*

Proof Part (i) is shown as in Proposition 9.6 (using A4 instead of P5′). To show part (ii) let π be a projective plane and ℓ a projective line. Denote by $\mathsf{Euc}(\pi)$ and $\mathsf{Euc}(\ell)$ their Euclidean parts. If $\mathsf{Euc}(\pi)$ cuts $\mathsf{Euc}(\ell)$ then we have proved part (ii). This intersection is the point we are looking for.

Assume now that $\mathsf{Euc}(\pi)$ does not cut $\mathsf{Euc}(\ell)$. Then, by A3, there exists a unique Euclidean plane γ passing through ℓ and parallel to $\mathsf{Euc}(\pi)$. Let $\bar{\gamma}$ be the projective closure of γ and P_∞ the point at infinity of ℓ. Then P_∞ belongs to the improper line of $\bar{\gamma}$ and, since this line is the intersection of $\bar{\gamma}$ with π, it also belongs to π. Again, P_∞ is the point we are looking for.

In both cases, the point we found in the intersection of ℓ and π is the only one possible, since otherwise ℓ would have two points in π and this would imply, by part (i), that ℓ is included in π.

For part (iii) let π and π' be two different planes in $\mathbb{P}^3$. Take any line p on π. If p is included in π' we have found a line in $\pi \cap \pi'$. Assume p is not included on π'. By part (ii), p and π' intersect at a single point P. Consider any point Q on p, $Q \neq P$, and any line r on π with $Q \in r$ and $r \neq p$. Again, if r is included on π' we are done. So assume it is not. Then, by part (ii), r and π' intersect at a single point R. Furthermore, $R \neq P$ since otherwise p and r would be the same line by part (i) (both lines would pass through Q and P). Since P and R lie on both π and π' it follows that so does the line determined by P and R. We have therefore proved that π and π' determine a line. They cannot intersect at two different lines since otherwise they would have at least three points in common, contradicting part (i).

Part (iv) is a projective version of A2 above. We will not give the details of the proof. □

We now can describe the extension of projectivities to projective planes. Consider π and π', two different projective planes in $\mathbb{P}^3$, and a point O not belonging to any of them. Then, to any point P in π we can associate a point P' in π' as follows:

(i) draw the projective line ℓ determined by P and O;

(ii) let P' be the point where ℓ intersects π'.

The association above is well defined (parts (i) and (ii) are ensured, respectively, by parts (i) and (ii) in Proposition 9.6). In what follows we will abuse notation and use the word *projectivity* as well for the extension of these functions to the projective plane described above. This convention should not introduce ambiguities and is more convenient than the cumbersome "projective projectivities".

We can now prove the main result in this subsection.

Proposition 9.7 *Projectivities are bijections on* $\mathbb{P}$.

Proof Let $\varphi : \pi \to \pi'$ be a projectivity between projective planes and let O be its projection point (which is neither on π nor on π'). Because of Proposition 9.1, to show that φ is bijective, it is enough to show that it has an inverse. But the inverse φ^{-1} of φ is obvious. To any point P' in π' we associate the point P in π obtained as follows (we use again Proposition 9.6):

(i) draw the projective line ℓ passing through P' and O
(ii) let P be the point where ℓ intersects π.

We leave to the reader to check that $\varphi \circ \varphi^{-1} = \mathsf{Id}_{\pi'}$ as well as $\varphi^{-1} \circ \varphi = \mathsf{Id}_{\pi}$. $\qquad\square$

9.2.3 A Kleinian view of projective geometry

That projectivities on projective space are bijective may suggest the idea that projectivities are a group of transformations. There is an obstacle to this notion, but, as we shall see, it can be easily overcome. The problem lies in the fact that transformations are (bijective) functions from a space into itself. Projectivities, in contrast, are functions between two different projective planes. A natural way to fix this problem consists of moving (i.e. composing with an isometry) the image plane to make it coincide with the domain plane (the one the projectivity is acting on). That is, we take a projectivity $\varphi : \pi \to \pi'$ between projective planes π and π' and an isometry $\psi : \pi' \to \pi$ and consider the composition

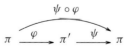

which gives a transformation $\psi \circ \varphi : \pi \to \pi$ of the plane π. A minor point still requires attention: we have defined isometries only as transformations of the Euclidean plane $\mathbb{E}$ and not as functions between possibly different projective planes (in three-dimensional projective space $\mathbb{P}^3$).

To formally develop a theory for the latter would require an amount of work which is disproportionate for the point we want to make here. We therefore limit ourselves to state the following known facts:

(1) Isometries in Euclidean 3-space $\mathbb{E}^3$ are defined as transformations of $\mathbb{E}^3$-preserving distances. Appropriate versions of Propositions 2.2, 2.3 and 2.4 hold.[12]

(2) Isometries in $\mathbb{E}^3$ extend in a natural way to transformations of $\mathbb{P}^3$ (which, abusing language, we will also call isometries[13]). Indeed, if $\psi : \mathbb{E}^3 \to \mathbb{E}^3$ is an isometry we only need to define what is $\psi(P)$ for a point P at infinity. This is simple. Take any line ℓ with direction P. We consider its image $\psi(\ell)$, which is a line as well since isometries preserve lines, and define $\psi(P)$ to be its point at infinity. The image $\psi(P)$ does not depend on the choice of ℓ. If r is a different line having P as point at infinity as well then ℓ and r are parallel. But then $\psi(\ell)$ and $\psi(r)$ are parallel as well (isometries preserve parallelism) and, hence, have the same point at infinity. In particular, we have that $\psi(\pi_\infty) = \pi_\infty$.

(3) If $\psi : \mathbb{P}^3 \to \mathbb{P}^3$ is an isometry and $\pi \subset \mathbb{P}^3$ is a projective plane then $\pi' = \psi(\pi)$ is also a projective plane in $\mathbb{P}^3$ and the restriction $\psi : \pi \to \pi'$ is bijective. If ℓ_∞ and ℓ'_∞ are the lines at infinity of π and π' respectively, then $\psi(\ell_\infty) = \ell'_\infty$. We will again abuse language and refer to such a restriction as an *isometry between projective planes*. The inverse ψ^{-1} of ψ is also an isometry between projective planes.

We can now make our point.

Definition 9.8 Let π be a projective plane. We say that a function $f : \pi \to \pi$ is a *projective transformation* when there exist a natural number $k \geq 1$, projective planes $\pi_0, \pi_1, \ldots, \pi_k$ and functions $\varphi_1, \ldots, \varphi_k$ such that:

(i) $\pi_0 = \pi_k = \pi$;

(ii) for all $i = 1, \ldots, k$, $\varphi_i : \pi_{i-1} \to \pi_i$ is either a projectivity or an isometry between projective planes; and

(iii) $f = \varphi_k \circ \varphi_{k-1} \circ \cdots \circ \varphi_2 \circ \varphi_1$.

We denote by $\mathrm{Proj}(\mathbb{P})$ the set of projective transformations (on a projective plane $\mathbb{P}$).

---

[12] Isometries in $\mathbb{E}^3$ can be classified just as those in $\mathbb{E}$ were classified in Theorem 2.10. A few more basic isometries occur. We will not use this classification.

[13] The abuse of language is because we are not meaning transformations preserving distances in $\mathbb{P}^3$. We are not considering any such distance.

The following result is proved as Theorem 7.6.

Theorem 9.9 *The set* $\mathrm{Proj}(\mathbb{P})$ *is a group (of transformations on* $\mathbb{P}$*).*

The endowment of projective space $\mathbb{P}$ with a group of transformations defines a geometry in the sense of Klein, which, it goes without saying, we will refer to as *projective geometry*. Objects which can be transformed into one another by a projective transformation have the same shape in this geometry. This is the case of a well-painted *pavimento*, but also, approximately, of the two fish in Figure 9.8 (taken from (Thompson, 1961: Chapter IX)).

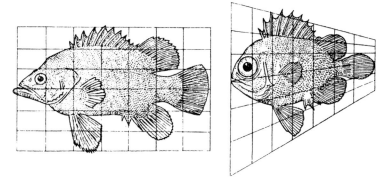

Figure 9.8 *Polyprion* **(left) and** *Pseudopriacanthus altus* **(right). (From (Thompson, 1961) with permission.)**

It is worth remarking that, unlike the situation in Section 7.6, where all three of the geometries considered shared the underlying (two-dimensional) space and differed only in their groups of transformations, projective geometry differs from them, in addition, because of a distinct underlying space. In this geometry, not only does the shape of things change, but so also does the nature of space.

The reference to the shape of things brings us to an obvious question, namely: What are the essential features of projective geometry? Or, in other words, which properties are preserved by projective transformations?

9.2.4 Essential features of projective geometry

The question above brings us back to the "projective properties" studied by Poncelet. A first such property is alignment.

Proposition 9.10 *Projective transformations preserve straight lines.*

Proof We know that isometries preserve straight lines. Therefore, we only need to show that so do projectivities.

Consider, thus, a line ℓ in a projective plane π, another projective plane π' and a point $O \in \mathbb{P}^3$ neither on π nor on π'. Then, the lines passing through O and ℓ form a projective plane in $\mathbb{P}^3$ (Proposition 9.6(iv)) and the intersection of this plane with π' yields a projective line (Proposition 9.6(iii)). $\qquad\qquad\qquad\qquad\qquad\qquad\square$

We mentioned in Remark 9.5 the lack of notions of distance and angle extending the Euclidean. We will therefore ignore features related to these notions[14] and proceed with a second property preserved by projective transformations.

A cursory inspection of many paintings or photographs shows that (Euclidean) circles are not preserved by projectivities. Circular objects in these images appear, most of the times, as ellipses (the sound hole of the lutes in Figures 9.7 and 11.3 give examples). Since a circle can be seen as a special case of an ellipse, we may wonder whether projectivities preserve ellipses (as affinities do; recall Theorem 7.25). It is, however, easy to verify that this is not the case. Indeed, consider the circle in the lower middle part of Figure 7.6. That same middle part shows the cone obtained by drawing the lines passing through points in the circle and a different, fixed point (the apex of the cone). Its upper part shows an ellipse produced by cutting the cone with a plane. But the left and right parts of Figure 7.6 show instances of planes intersecting the cone in a parabola and hyperbola respectively. Hence, projectivities may transform circles into parabolas or hyperbolas as well.

This is not a chance fact. Projective transformations preserve conics. To grasp the meaning of this statement, however, we need to understand what are the projective closures of ellipses, parabolas and hyperbolas. We will see that in § 9.3.2.

9.3 A projective view of affine geometry

We have thus far used, maybe inadvertently, two different conceptions of the projective plane. On the one hand, deriving from the definition of $\mathbb{P}$, one that sees the projective plane as an inhomogeneous space with points of two different natures: proper (Euclidean) and improper (at infinity). On the other hand, one implicit in statements such as those in Proposition 9.4 or Theorem 9.9 which do not make any reference to this inhomogeneity and offer instead an unvarying view of the projective plane.

We will have an opportunity to return to the second of these two views of projective geometry in Section 12.5. In the meantime we will take heed

---

[14] We note, nevertheless, that a quantity defined, for four aligned points, as a quotient of simple ratios and known as the *cross-ratio* is preserved by projective transformations.

of the consequence of the first, the fact that, by looking at the Euclidean plane as a subset of the projective plane, one can revisit a number of Euclidean notions as particular cases of projective situations.

9.3.1 A distant vantage point

Our first definition of projectivity considered the Euclidean planes π and π' and a point O in $\mathbb{E}^3$ neither on π nor on π'. We subsequently extended this notion by taking π and π' to be projective planes and allowing O to be in $\mathbb{P}^3$. If O is proper, then the consequence of doing so is that the projectivity maps points in the foot line to improper points and improper points to points in the horizon line. But what happens if O is a point at infinity in $\mathbb{P}^3$? Also, in this early discussion, and even though we never made it explicit, we assumed the Euclidean planes π and π' to be non-parallel. What happens if this is not the case?

A simple property of a projectivity ψ as in any of the two cases above is that it will preserve points at infinity. That is, if ℓ_∞ and ℓ'_∞ denote the lines at infinity of π and π' respectively, then $\psi(\ell_\infty) = \ell'_\infty$ (we already noticed this property for isometries between projective planes). Indeed, assume that O is at infinity. In this case, if P is any point at infinity in π then the line determined by O and P is contained in π_∞ and its intersection with π' will be at the infinity line of π'. Now assume instead that $\mathrm{Euc}(\pi)$ and $\mathrm{Euc}(\pi')$ are parallel. This means that their lines at infinity are the same. Hence, if P is a point at infinity on π then P is also a point at infinity on π' and the line joining O and P will, obviously, intersect π' at P itself.

Let us have a closer look at these special situations beginning with the case of parallel projection planes.

Consider the line ℓ perpendicular to π (and therefore also to π') and passing through O. Let Q and Q' be the intersection points of ℓ with π and π' respectively. Also, let P be any point in π and P' its image in π'.

Let r be the line passing throug O, P and P'. Since r and ℓ meet at O they determine a plane, say γ. Drawing (at the right in the picture below) the points and lines on this plane only, and using Thales' theorem (Theorem 1.2), we deduce that

$$\mathrm{dist}(P', Q') = \frac{\mathrm{dist}(O, Q')}{\mathrm{dist}(O, Q)} \mathrm{dist}(P, Q).$$

If we translate π with vector $\overrightarrow{QQ'}$, so that π overlaps π' and Q becomes Q', we can see ψ as a homothecy of centre Q and coefficient $\lambda = \frac{\mathrm{dist}(O,Q')}{\mathrm{dist}(O,Q)} = \frac{\mathrm{dist}(O,\pi')}{\mathrm{dist}(O,\pi)}$.

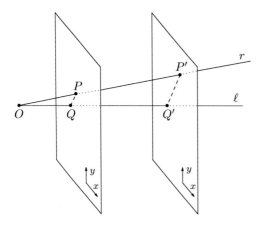
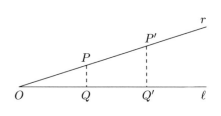

So, projectivities with parallel projection planes are, essentially, homothecies and the coefficient of such a homothecy is given by the ratio of the distances between the vantage point and the projection planes.

What about projectivities with projection point O at infinity? From a Euclidean point of view, the set of lines passing through O is the set of all lines having a given direction. In particular, they are all parallel. If the plane π is perpendicular to this direction then the function $\psi : \pi \to \pi'$ we obtain is familiar to us. The upper drawing on the right shows the image P' in π' of a point P in π.

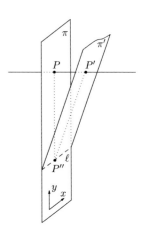

We have marked two perpendicular directions in π and denoted them by x and y, with x parallel to the intersection line ℓ of π and π'. One may take coordinate systems in π and π' with ℓ as a common x-axis. If we denote by P_x and P'_x the x-coordinates of P and P' respectively we then have $P_x = P'_x$ (since both P and P' project onto the same point P'' in ℓ). This shows that ψ preserves distances between points with the same y-coordinate.

What happens with distances along the y-coordinate? To get an answer to this question we consider two points in π (say P and Q) with the same x-coordinate (see right).

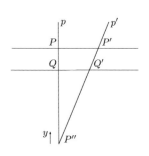

Let P' and Q' be their images in π'. Then the points P, Q, P' and Q' are all contained in a plane, perpendicular to ℓ, and cutting π and π' in lines p and p' respectively. Let P'' be the intersection of p and p'. The resulting image is familiar to us. Thales' theorem (Theorem 1.2) tells us that distances in p are increased in p' by a factor $\lambda = \frac{\text{dist}(P,P'')}{\text{dist}(P',P'')}$. Note that λ depends only on the angle θ between π and π' (or, what is the same, between p and p'). Readers familiar with trigonometry will immediately notice that $\lambda = 1/\cos\theta$.

In summary, π is an expansion in the y direction and the identity in the x direction. If we discard the fact that π and π' are different planes then the emerging result is that ψ is a strain. Furthermore, by appropriately choosing the angle between π and π' we conclude that any strain of coefficient $\lambda > 1$ can be described as a projectivity with projection point at infinity perpendicular to π. Similarly, those with coefficient $\lambda < 1$ can be described by such a projectivity with now π' perpendicular to the direction given by O. We state this result as a proposition for further use.

Proposition 9.11 *Let π and π' be non-parallel Euclidean planes, ℓ their intersection line and θ the angle between them. Let $\varphi : \pi \to \pi'$ be the projectivity with projection point at infinity, in the direction perpendicular to π. Then φ (or, to be precise, φ composed with the rotation in $\mathbb{E}^3$ of angle θ around ℓ which maps π' to π) is an expansive strain with axis ℓ and factor $\lambda = 1/\cos\theta$. Its inverse $\varphi^{-1} : \pi' \to \pi$ is, therefore, a contractive strain with the same axis and factor $\lambda = \cos\theta$.*

Interestingly, the geometric transformations captured by the Euclidean part of projectivities with projection point at infinity are essentially strains. To make this claim precise we need the following definition.

Definition 9.12 Let π be a projective plane. We say that a function $f : \pi \to \pi$ is an *affine transformation of π* when there exist a natural number $k \geq 1$, projective planes $\pi_0, \pi_1, \ldots, \pi_k$ and functions $\varphi_1, \ldots, \varphi_k$ such that:

(i) $\pi_0 = \pi_k = \pi$;

(ii) for all $i = 1, \ldots, k$, $\varphi_i : \pi_{i-1} \to \pi_i$ is either a projectivity with projection point at infinity, or a projectivity with parallel projection planes, or an isometry between projective planes; and

(iii) $f = \varphi_k \circ \varphi_{k-1} \circ \cdots \circ \varphi_2 \circ \varphi_1$.

We denote by $\mathsf{Aff}(\mathbb{P})$ the set of affine transformations of the projective plane.

We can use the notion of affine transformation to make precise our claim above. The following result states a bijection between the sets $\mathsf{Aff}(\mathbb{E})$ and $\mathsf{Aff}(\mathbb{P})$.

Theorem 9.13 *Let π be a projective plane.*

(i) *For every affine transformation $\psi : \pi \to \pi$ the restriction of ψ to the Euclidean part $\mathsf{Euc}(\pi)$ of π is an affinity. In particular, ψ associates proper points to proper points and points at infinity to points at infinity.*

(ii) *Conversely, for every affinity $\varphi : \mathsf{Euc}(\pi) \to \mathsf{Euc}(\pi)$ there exists an affine transformation $\bar{\varphi} : \pi \to \pi$ of π such that the restriction of $\bar{\varphi}$ to $\mathsf{Euc}(\pi)$ coincides with φ.*

One of the consequences of Theorem 9.13 is that the set $\mathsf{Aff}(\mathbb{P})$ of affine transformations of the projective plane is a subgroup of the group

Proj($\mathbb{P}$) of projective transformations on that plane. Therefore, we can regard projective geometry as more general (and poorer in theorems) than affine geometry. This statement is meaningful, for instance, in the framework for comparison of species set out by D'Arcy Thompson. Within this framework the two pairs of species in Figures 7.1 and 9.8 are related but we may now say that those in the first figure are more closely so than those in the second.

Remark 9.14 A very special case, blending the two we considered above, is given by placing the vantage point at infinity and the planes π and π' parallel. By modifying any of the two arguments above (or by taking $\theta = 0$) one can show that this situation corresponds with a homothecy (or a strain) of coefficient 1; that is, with an isometry.

9.3.2 Conics revisited

In § 9.3.1 we described how lines in $\mathbb{E}$ are added to a point to make them projective lines. A similar closure construction can be done with conics in $\mathbb{E}$, but some thought is needed about the number of points at infinity we add to a conic. This number depends now on the kind of conic we are considering: two for an hyperbola, one for a parabola and none for an ellipse. The following figure gives an idea. In its upper part we see the "projective version" of the three conics. In each of them the conic is represented by a circle and the improper line by a short line segment. Below each of the three cases we see the corresponding "Euclidean version".

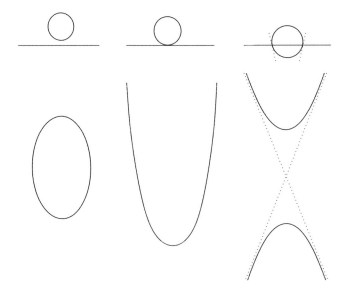

The dotted lines in the drawing of the hyperbola (which already appear in the picture with examples of conics in Section 7.4) are its *asymptotes*.

One has the feeling that when the branches of the hyperbola diverge from its centre they approach the asymptotes, the distance between branch and asymptote becoming arbitrarily small. This feeling is confirmed in the upper part of the drawing, where one observes that the lines corresponding to the asymptotes on $\mathbb{P}$ are tangential to the circle corresponding to the hyperbola precisely at the points where the circle intersects ℓ_∞.

Similarly, observation of a parabola shows that its two branches tend to align with its axis when diverging from its focus. In other words, tangent lines tend to be parallel to the axis when the point of tangency moves away from the focus. Hence, even though in Euclidean space these branches appear to separate from each other, they actually do not on the projective closure of this plane: they meet at a point in the line at infinity. Furthermore, this is the only point at infinity of the parabola, and the improper line is tangential to the parabola at this point.

We can now summarize the observations above and define the *projective closure* of a conic in $\mathbb{E}$. If the conic is an ellipse, it coincides with its closure. If it is a parabola, its closure is obtained by adding the point at infinity of the parabola's axis. If, finally, it is a hyperbola, the closure is obtained by adding the points at infinity of its asymptotes.

A *conic* in the projective plane is the projective closure of a conic in the Euclidean plane. With this definition in our hands we can shed light on a statement we made in § 9.2.4. We will still refer to ellipse, parabola or hyperbola as the type of a conic in $\mathbb{P}$ according to whether it has zero, one or two points at infinity respectively.

Theorem 9.15 *Projective transformations preserve conics.*

Furthermore, isometries between projective planes preserve conic types (i.e. they map ellipses to ellipses, parabolas to parabolas and hyperbolas to hyperbolas) and so do affinities.

Finally, let π and π' be projective planes, $O \in \mathbb{E}^3$ a point lying outside both planes and $\psi : \pi \to \pi'$ the projectivity with O as projection point. Let f be the foot line on π, C be any conic on π and C' be the image of C on π'. We have:

(i) *If C does not cut f then C' is an ellipse.*
(ii) *If C cuts f in one point only then C' is a parabola.*
(iii) *If C cuts f in two points then C' is a hyperbola.*

In all three cases, C' has a point in the horizon h on π' if C is a parabola, has two points in h if C is a hyperbola and has none if C is an ellipse.

It is worth mentioning, even though we will not enter into the details, that the unified view of conics afforded by projective geometry has provided both a better understanding of these figures and some simplified proofs for its properties.

A repertoire of drawing systems

10

On those remote pages it is written that animals are divided into (a) those
that belong to the Emperor, (b) embalmed ones, (c) those that are trained,
(d) suckling pigs, (e) mermaids, (f) fabulous ones, (g) stray dogs, (h) those that
are included in this classification, (i) those that tremble as if they were mad,
(j) innumerable ones, (k) those drawn with a very fine camel's hair brush,
(l) others, (m) those that have just broken a flower vase, (n) those that resemble
flies from a distance.

J.L. Borges (1964: The analytical language of John Wilkins)

The artist drawing a scene faces the choice of a number of possibilities regarding vantage point and position of the picture plane. These
choices naturally determine the way the artist sees the scene; for instance,
whether a subject is portrayed frontwards, or side face, or somehow in
between. It is apparent that the number of substantially different choices
for our artist depends on, and increases with the complexity of, the scene.
Since the nature of this scene is not necessarily amenable to mathematical terms, the idea of a catalogue of such choices akin to the catalogues we
described in Section 3.8 is out of place. But there are a number of choices
having a mathematical character which can be considered independently
of the depicted scene.

10.1 Projections and drawing systems

Among the choices that the artist has at hand there is, first, the position
of the vantage point. This can be in the Euclidean space or at infinity. A second fundamental choice, in the case the vantage point is at
infinity – which, we recall, amounts to having parallel projection rays –
is to place the picture plane perpendicular to these rays or oblique to
them. There are three resulting possibilities. First, the vantage point is
in $\mathbb{E}^3$. This is referred to as *perspective projection*. Second, the vantage
point is in π_∞ and the picture plane is perpendicular to the projection rays. This is referred to as *orthogonal projection* (or *orthographic
projection*). Third, the vantage point is in π_∞ and the picture plane is
not perpendicular to the projection rays. This is referred to as *oblique
projection*. The ensemble of orthogonal and oblique projections is sometimes called *parallel projections* in allusion to the fact that projection rays
are parallel.

We can summarize our first (and most fundamental) division between drawing systems in the following flowchart:

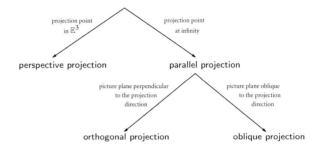

The reason to describe oblique projections as "parallel" is expressed in the following proposition.

Proposition 10.1 *Parallel projections preserve parallelism. That is, if p and q are parallel lines in $\mathbb{E}^3$ then their representations in the picture plane are parallel as well.*

Proof Since p and q are parallel there exists a plane π containing both p and q. Denote by π' the picture plane. We saw in Proposition 9.11 that the projectivity from π to π' with vantage point at infinity is a strain. In particular, it is an affinity and by Theorem 7.18 it preserves parallelism. □

An immediate consequence of Proposition 10.1 is the following result.

Proposition 10.2 *If an object is translated in $\mathbb{E}^3$ then its image (under a parallel projection) on the picture plane is also translated. In particular, the size of this image does not depend on the distance from the object to π'.*

Proof Denote by $\vec{v}$ the vector of the translation (which is now a vector in $\mathbb{E}^3$) and by F the object to be projected onto π'. Let P be any point in F, $\bar{P} = P + \vec{v}$ its image under the translation, and P' and $\bar{P}'$ the projection of these two points on the picture plane π'. Let $\vec{w}$ be the vector with origin at P' and end at $\bar{P}'$. We want to show that the projection of the translated object $\text{trans}_{\vec{v}}(F)$ is $\text{trans}_{\vec{w}}(F')$, where F' denotes the projection of F onto π'.

To do so, it is enough to see that, for all other points Q in F, if $\bar{Q}, Q'$ and $\bar{Q}'$ are as above then $\bar{Q}' = Q' + \vec{w}$. To show this, consider the plane Γ determined by the line passing through P and $\bar{P}$ and the direction of projection. Then all four points $P, \bar{P}, P'$ and $\bar{P}'$ are in Γ. Furthermore, since the lines passing through P, P' and through $\bar{P}, \bar{P}'$ are parallel (they are projection rays) the quadrilateral defined by these four points is a trapezium.[1] Let p be the intersection of π' and Γ, so that the points P' and $\bar{P}'$ are in p and let α, β, γ and δ be the angles of the trapezium at $P, \bar{P}, \bar{P}'$ and P' respectively (note, $\beta = 180 - \alpha$ and $\gamma = 180 - \delta$). The left part of the following picture shows all these elements on the plane Γ.

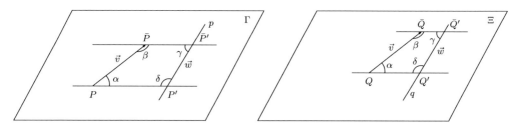

Its right part shows a similar situation constructed from the point Q. The plane Ξ containing the corresponding elements for Q is parallel to Γ since its direction is also determined by the projection direction and the vector $\vec{v}$, just that instead of passing through P it passes through Q.

Because of this, the line q (where Ξ intersects π') is parallel to p. Since the other three sides of the trapezium in-are also parallel to their corresponding sides of the trapezium in Γ. Therefore, the angles at $Q, \bar{Q}, \bar{Q}'$ and Q' are also α, β, γ and δ respectively. But the length of the segment $P\bar{P}$ is the same as the length of the segment $Q\bar{Q}$ (namely, the length of $\vec{v}$). We finally use that if two trapezia have the same four angles and also the same lengths for one of their non-parallel sides, then the other such sides have the same length as well. That is, $\mathrm{dist}(P', \bar{P}') = \mathrm{dist}(Q', \bar{Q}')$ as we wanted. □

Remark 10.3 Proposition 10.2 sets a remarkable difference between perspective and parallel projections. In the former, objects receding from the picture plane are projected onto this plane in increasingly smaller sizes. In contrast, in parallel projections, objects are represented with the same size independently of their position (assuming, of course, that they are only translated, not twisted or tilted in any manner).

In addition to the choices producing these three types of projection, a number of options have been considered arising from assuming that the object to be represented is a cube. These options produce a refined repertoire of projection systems by partitioning the types above into subclasses. We next provide a brief account of this repertoire.[2] In doing so, it will be helpful to denote by C the cube in $\mathbb{E}^3$ that is to be drawn and by C' its representation; that is, its projection onto the picture plane. Also, in all that follows, we will keep denoting the latter by π'. We will also

---

[1] To have a pair of parallel sides is precisely the definition of a trapezium.

[2] We have not found in the literature a unified, consistent, use of names for the different drawing systems. The names and logic in our exposition attempt to maximize clarity.

refer to the lines passing through the centre of the cube and parallel to its edges as *principal axes* (or simply *axes*) of C and to the length of any of these edges by L.

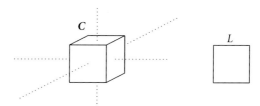

10.1.1 Orthogonal projections

Recall that in an orthogonal projection the vantage point is at infinity and the picture plane is perpendicular to the direction defining this point. Needless to say, the position of the vantage point does not require the artist to abandon the universe and infinitely distance themselves from the scene. It merely requires him to represent the scene as if he was at such an infinite distance and endowed with an infinitely powerful prospect-glass. In practice (but not in theory), this is indistinguishable from looking at the scene from very far away with a very powerful telephoto lens.

The intersection with the picture plane, together with the selection of a rectangular area (the *picture*) within this plane, is translated in this analogy into the assumption of a rectangular frame for the view through the prospect-glass. A further assumption, the possibility to adjust its power, leads to the notion of *scale*. Proposition 10.2 ensures that the size of an object's representation in the picture is independent of the position of the object in the scene. But depending on the size of the painting and the nature of the scene to be painted one may want to apply to the projected image a homothecy of a conveniently chosen factor to make the scene fit into the picture.

We are interested in discriminating different drawing systems within the class of orthogonal projections. We do so by noting that C can be positioned in three different ways with respect to the picture plane according to whether the resulting number of visible faces in C' is one, two or three. The resulting types of drawing systems are plans, orthogonal two-sided projections and orthogonal axonometric projections respectively.

Plans

A *plan* is an orthogonal projection in which a side of C is parallel to the picture plane. The four faces adjacent to this one are then parallel to the projection direction and represented as line segments in the picture plane. In a plan C' is a square.

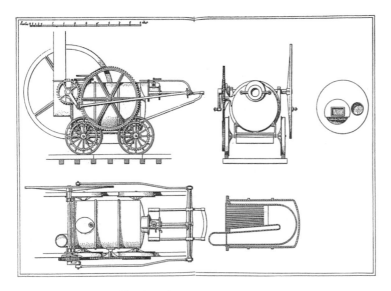

Figure 10.1 **Locomotive by Richard Trevithick, c. 1804.**

We have already found plans in the elevations and groundplans in Figure 3.3 as well as in the preparatory drawings for the *costruzione legittima*. They are actually the bread and butter of all forms of technical drawing due to the fact that, in case one represents only those lines and curves contained on planes parallel to the picture plane, the resulting drawing respects proportions. Indeed, the situation corresponds to the special case described in Remark 9.14 and the representation is therefore an isometry. As it happens, large subjects are commonly represented at a smaller scale and small subjects at a larger one (and hence indications such as `scale 1:25` which tell the observer that lengths in the drawing have to be multiplied by 25 to yield lengths in the depicted object). But except for this change of scale, the isometric character allows one to determine true lengths from measurings on the drawing. In addition, the representation not only preserves parallelism but also angles.

The names *groundplan* and *elevation* correspond to the projections obtained by placing the bottom or a lateral face respectively of C parallel (and closest) to the picture plane. Often in technical drawings several plans of the same object – corresponding to different but, in general, perpendicular projection directions – are grouped together. Such groupings are sometimes called *multiviews*. For instance, the locomotive in Figure 10.1 is shown in five different views.

Plans serve a different purpose than perspective projections, and even masters of the latter used them when such a purpose was to be served. During the last years of his life Dürer devoted considerable time to establishing average proportions of different parts of the human body. To this goal, he measured a large number of instances (from "two to three

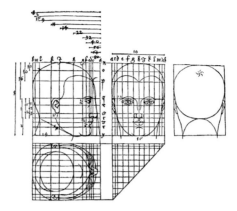

Figure 10.2 **Drawing by Dürer, from** *Hierinn sind*
begriffen vier Bücher von menschlicher Proportion, **1528.**
(From (Booker, 1963).)

hundred of living persons" (Panofsky, 1971)) of these proportions and
recorded them in four books, *Hierinn sind begriffen vier Bücher von menschlicher Proportion*, which were published in 1528, the year of his death.
Disregarding the numerical values of these proportions, which are of no
relevance to our present discussion, we note that in the head drawings in
Figure 10.2, Dürer used, as in Figure 10.1, four orthogonal projections
with mutually perpendicular directions.

Figure 10.3 **Joaquín Torres García,** *Puerto de Montevideo al Amanecer*, **oil on cardboard, 1940.**

A less technical use of plans is found in the work of Joaquín
Torres García (and in many of the painters following his school), of which
Figure 10.3 offers an example, where, we observe, accuracy is not the
purpose.

Orthogonal two-sided projections

Assume now that the projection direction is not perpendicular to a face of C (so that we are not talking about a plan). If one of these faces is parallel to the projection direction then C' shows exactly two of them. Typically, these two faces are either side by side (*horizontal projection*) or one on top of the other (*vertical projection*).

They are drawn as (non-square) rectangles with longer side measuring L. The lengths of their shorter sides depend on the inclination of C. The section on the left shows the picture plane (as a line), the projection rays (as dashed lines) and the shorter sides of the two rectangles (as the two thick segments measuring e_1 and e_2).

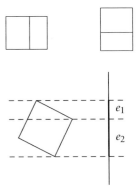

These two rectangles are acually obtained from their corresponding squares by two different strains (with coefficients e_1/L and e_2/L respectively). We are therefore familiar with the sort of visual distortion they produce.

An example of orthogonal two-sided projection is provided by Dürer, who used the plans in Figure 10.2 to produce the drawing in Figure 10.4. Note that the inclination of the (ideal cube containing the) head is 45° with respect to the projection direction. This yields similar distortions for the face and neck parts of the drawing.

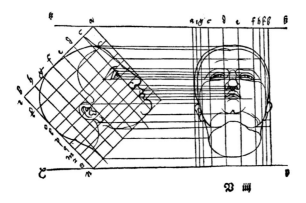

Figure 10.4 **Drawing by Dürer, from *Hierinn sind begriffen vier Bücher von menschlicher Proportion*, 1528. (From (Booker, 1963).)**

Orthogonal axonometric projections

A projection showing three sides of C is said to be *axonometric*. This is the case for orthogonal projections if none of the faces of C is parallel to the projection direction. Each of these faces is represented in the picture plane by a parallelogram, since parallel projections preserve parallelism (by Proposition 10.1). But because the projection of each of these faces over the picture plane is a strain (by Proposition 9.11) these parallelograms are not squares. Actually, they are not even rectangles since, for this to be the case, the axis of the strain should be perpendicular to a face of C. Typically, C' looks as the drawing on the left.

Within the class of orthogonal axonometric projections a natural division can be made. If all three visible faces in C' are congruent we say that the projection is *isometric*. If two are congruent but they are not to the third one we say it is *dimetric*. Finally, if no two faces are congruent we say it is *trimetric*.

Isometric projections. In a cube drawn in isometric projection the apex of the cube which is shown in the interior of C' is such that the three angles made by the edges at this apex (α, β and γ in the middle drawing on the left) are equal (and therefore measure 120° each).

Isometric projections were formalized in 1822 by William Farish (but had been in use in Asian art for centuries). During the nineteenth century they became of common use for technical (engineering and architectural) drawing due to the fact that distances in the three axis-directions of C are measured with the same ruler (as one can see in the drawing, where the three edges of C' incident at the central apex have the same length). Figure 10.5 of an optimal-grinding engine model is taken from Farish's foundational article (Farish, 1822).

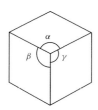

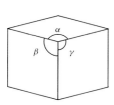

Dimetric projections If C is drawn in dimetric projection then two of the angles made by the edges at the central apex of C' (β and γ in the bottom drawing on the left) are equal (and different to the third one).

Figure 10.5 **W. Farish, Optical-grinding engine model, 1822.**

Parallel projections, both orthogonal and oblique, are common in classical Asian art. A genre where they often occur is the *Ukiyo-e*. These were woodblock prints derived from a master design painted by an artist. Because of the affordability allowed by its mass production, ukiyo-e acquired great popularity. They first appeared in the seventeenth century in Edo (nowadays Tokyo) and were meant to portray the pleasures of life. Even though ukiyo-e depicted a variety of subjects, two of them became characteristic of the genre: *yakusha-e* (pictures of actors performing in *kabuki* roles) and *bijin-ga* (which literally means "pictures of beautiful women").

It was common for bijin-ga to portray the beautiful women within well-decorated rooms which, just as the architectural elements in the Italian Renaissance paintings, were used to organize the pictorial space. Almost invariably (but there were exceptions) these elements betray a parallel projection. Figure 10.6 is an example by Kunichika Toyohara of the use of dimetric projection.

A carefully selected, beautifully printed, selection of ukiyo-e appears in Fahr-Becker (1994).

Figure 10.6 **Kunichika Toyohara,** *Scene in a Villa*, **Ukiyo-e, nineteen century. (Prints & Photographs Division, Library of Congress, LC-USZC4-8450, LC-USZC4-8451.)**

Trimetric projections   If no two visible faces of C' are congruent then the projection is trimetric. The drawing of C given on page 267 to illustrate orthogonal axonometric projections is in this kind of projection. The chest shown in Figure 10.7 is also represented this way.

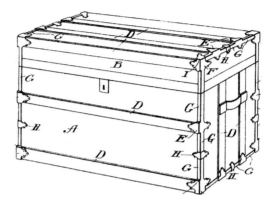

Figure 10.7 **Drawing of a chest, 1895. (*The Official Gazette of the United States Patent Office* (Washington, DC: Government Printing Office, 1895). "Rectangular Trunk Trimming". Retrieved 8 February, 2012, from http://etc.usf.edu/clipart/76700/ 76792_trimming.htm.)**

10.1.2 Oblique projections

If the projection rays are oblique (neither perpendicular nor, needless to say, parallel to the picture plane) we are dealing with an *oblique projection*.

Strictly speaking, oblique projections do not naturally correspond to a view of the scene, not even allowing for infinitely powerful prospect-glasses. Instead, they can be understood as a skewed view of an image in orthogonal projection. Indeed, let π'' be any plane perpendicular to the direction of projection:

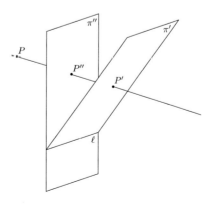

A point P in the scene projects onto the picture plane at a point P' and the line defined by P and P' intersects π'' at a point P''. The function mapping P into P'' is an orthogonal projection. Therefore, the projection of the scene onto π' can be understood as first projecting the scene onto π'' (an orthogonal projection) and then straining this projected image by a factor $\lambda = 1/\cos\theta$, where θ is the angle between π'' and π' (because of Proposition 9.11).

Despite the fact that images drawn in oblique projection do not correspond with well-defined views of the scene, these images may have a pleasant effect and convey a sense of realism. Because of this, oblique projections have been prevalent in Asian painting and commonly used in the West for a number of purposes, from architecture to advertisements to military drawings.

The class of oblique projections appears to be too general for our exposition, for which we may consider only the situation having one face of C parallel to the picture plane. This restriction has an important consequence.

Proposition 10.4 *Let π be the plane containing the side of C parallel to π'. Then the projection of π onto π' is an isometry.*

Proof Take any two points P and Q on π and let γ be the Euclidean plane determined by P, Q and the direction of projection. Let p be the intersection of π with γ and similarly p' that of π' and γ. Then p and p' are lines, P and Q are both in p and their images P' and Q' on π' by the projection are in p'. The situation (drawn in the plane γ) is as follows:

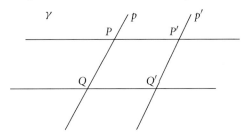

Since the projection lines (joining P with P' and Q with Q') are parallel and so are p and p', the quadrilateral $PP'QQ'$ is a parallelogram. Therefore, $\text{dist}(P, Q) = \text{dist}(P', Q')$, as we claimed. $\square$

Because π projects onto π' isometrically, the face of C included in π is drawn on π' as a square. If the projection is axonometric (i.e. the represented cube shows three faces) the drawing looks as that on the right.

As we have mentioned, oblique projections are ubiquitous in Asian art.

One of the masters of bijin-ga is Torii Kiyonaga. Space in his *Women in a Bathhouse* (Figure 10.8) is structured around the wooden elements in

Figure 10.8 **Torii Kiyonaga,** *Women in a Bathhouse***, Ukiyo-e, 1780s.**

the room (floors, walls, benches, etc.), which follow three perpendicular directions and are obliquely projected onto the picture plane.

Chinese art also offers a wealth of works using different forms of parallel projection, and in particular of oblique projection. Dubery and Willats (1983: 33) write that

> Oblique projection seems to have arrived in China from Rome by way of India round about the first or second century AD. During the journey the perspective element, perhaps never very firmly established, was lost. Throughout its whole history up until the eighteenth century Chinese painting is remarkable for the consistent way in which oblique projection was used, at any rate so far as rectilinear objects were concerned.

Figure 10.9 shows an archetype of the classical use of oblique perspective in Chinese painting. Houses, pavilions and enclosures are aligned following two perpendicular directions and lines along these two directions are scrupulously represented on the picture plane following two (now oblique) directions. Furthermore, consistently with Proposition 10.2, similar objects – and notably all depicted persons – are represented with similar measures independently of their position on the scene or the painting.

Figure 10.10 shows an example of what began to occur in the eighteenth century. The houses in this woodcut are oriented in the same manner, and each of them is drawn in oblique projection with one side parallel to the picture plane. Interestingly, the representation of the whole

Figure 10.9 *Wen-chi Returns Home*, copy after an anonymous twelfth-century painting, China, late fourteenth century. (From (Fong and Fu, 1973).)

scene is not, strictly speaking, an oblique projection, as these kinds of projections, being parallel, show as equally heighted objects which are so in the depicted scene (Proposition 10.2). But a cursory glance at Figure 10.10 reveals that persons and houses decrease in size as they recede from the picture plane. The artist combined oblique projection for each individual house (as well as boats, etc.) with elements of perspective for the depiction of the whole scene. The image thus obtained does not result from any choice of positions for the vantage point and the picture plane, but nonetheless it is strangely convincing.

Drawings in oblique projection allow for a variety of viewpoints – with correspondingly varied visual experiences – depending on the direction of projection. This direction can be easily specified in terms of two angles as follows. The cube C has four faces perpendicular to π'. Fix one of them and let Φ be the plane containing this face, which we will call the *floor plane*. A drawing (in oblique projection, of course) of the situation is the following:

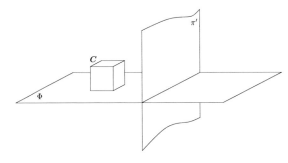

Figure 10.10 **Colour print from woodblocks, China, *c.* 1734. (Courtesy of Umi-Mori Art Museum, Hiroshima, Japan.)**

A projection direction is determined by the angles θ and v that any line ℓ with this direction makes with π' and Φ respectively. A multiview (with groundplan at the left and elevation at the right) of this idea is the following:

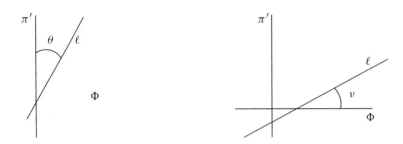

Orthogonal projections correspond with $v = 0°$ and $\theta = 90°$. Increasing v "raises" the viewpoint whereas decreasing θ "skews" it. Changes in v and θ are reflected in changes in two other parameters (on the picture plane and, therefore, with more direct relevance to the drawer): the length s of the drawn lateral edge and the angle ς this edge makes with the horizontal edge of the drawing.

We already established that the front face of C' is a square congruent with the faces of C. In contrast with this necessity, we now note that the value of s can be any non-negative number. It is small when θ is close to 90° and large when θ is close to 0°.

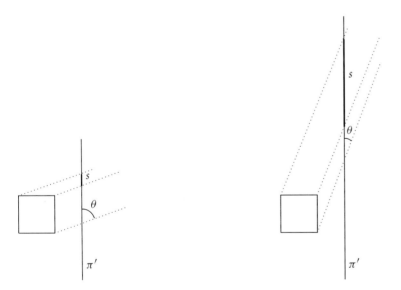

Tradition, through the service of well-defined purposes, has established two particular families of projection directions: Cabinet and Cavalier.

Cabinet projections   The class owes its name to its extended use for drawing this kind of furniture (Figure 10.11). The value of s is one half of L and the angle ς may vary but it is usually taken to be 30° or, less frequently, 45°.

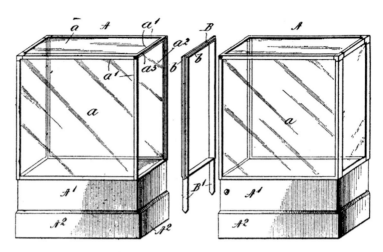

Figure 10.11 **Drawing of an expandable cabinet, 1912. (*Official Gazette* (Washington, DC: Government Printing Office, 1912). "Expandable Cabinet". Retrieved 8 February, 2012, from http://etc.usf.edu/clipart/77400/77480/77480_expandable.htm.)**

Cavalier projections   Used within the military, this kind of projection attempts to provide a useful picture of fortifications, ramparts, walls and other military constructions, a purpose which is best served with a raised viewpoint, close to an aerial view (Figure 10.12). This raising is obtained by setting the value of ς to be at least 45°. In addition, to ensure true measures in all three edge directions, θ is taken to be 45°, a choice which implies that $s = L$.

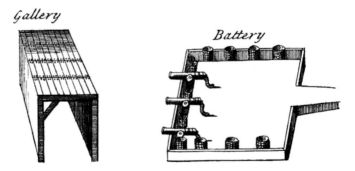

Figure 10.12 **Pieces of fortification drawn in cavalier perspective. From the *Cyclopaedia*, 1728.**

There is no agreement on the origin of the name "cavalier" for this drawing system. A possible explanation for it is suggested by an acception of this word, and now we quote the *Oxford English Dictionary*, as "A work generally raised within the body of the place, ten or twelve feet higher than the rest of the works [. . .] to command all the adjacent works and country round". Another suggested origin is the meaning of cavalier in French (horseman) and the consequently raised (in comparison with infantry troops) viewpoint.

Oblique two-sided projections

When $v = 0°$ there are only two visible faces in C', one on top of the other. We say that C is drawn in *oblique vertical projection*. Similarly, when $\theta = 90°$ we see two faces in C', one beside the other, and we say that C is drawn in *oblique horizontal projection*.

It is worth noting the differences with orthogonal two-sided projections. In the latter the two rectangles in C' are both non-square and have longer side measuring L (see the drawing on the left on page 266). In oblique two-sided projections, in contrast, one of the faces of C' is a square (isometric to the faces of C and, therefore, with sides of measure L) and the other is a rectangle with one of the sides of measure L and the other having an arbitrary length. For instance, in an oblique vertical projection, this length is less than L when $v < 45°$, larger than L when $v > 45°$ and equal to L (hence making C' the union of two squares) when $v = 45°$.

A beautiful example of a drawing in oblique horizontal projection is the ink-painting *The Orchard Pavilion* by the fourteenth-century artist Huang Kung-wang (Figure 10.13). The houses in this piece are similarly oriented and show two of their walls, both drawn as rectangles. The shape of the visible windows suggests that both of them are drawn isometrically (the frontal face would be so because of Proposition 10.4 and the lateral one by choosing the projection direction so that $\theta = 45°$).

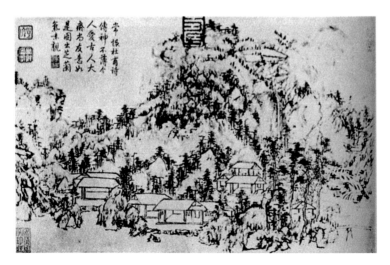

Figure 10.13 **Huang Kung-wang, *The Orchard Pavilion*, ink-painting, China, *c.* 1342. (From (Willats, 1997). Reprinted by permission of Princeton University Press.)**

Figure 10.14 **Kamāl ud-Dīn Behzād,** *Iskander and Syrens,* **miniature in the Nizami manuscript, Persia, 1495–6.**

Drawings in vertical oblique projection may look odder than draw-ings in their horizontal counterpart. Yet, they are common in art schools such as Persian miniatures. This form of art is in the background of the novel *My Name is Red* by Orhan Pamuk (2001: 20), a novel mentioning "Bihzad, the master of masters, patron saint of all miniaturists". Among Bihzad's works are the illustration of *Hamse*, five poems by Nizami, a copy of which is in the British Library. The painting in Figure 10.14 is from this manuscript. It is in vertical oblique projection, with $\nu = 45°$. We thus see the floor plane – with the pool where the sirens play – isometrically projected on the drawing and similarly with the front face of the tower. The fact that we see a lateral side of the tower should mean that the tower has not a rectangular basis, but other details, such as the position of the protruding balcony at this lateral side, suggest that, instead, the execution of the oblique vertical projection is far from accurate.

10.1.3 On tilt and distance

Before proceeding with the description of the subclasses of perspective as a drawing system it will be useful to briefly discuss the nature of some visual distortions and their characteristics in both parallel and perspective projections.

We begin with the latter; that is, assuming a viewpoint in Euclidean space. In this case, the simplest visual distortion arises from the mechanisms for size perception, and the simplest of these mechanisms is the one decreasing the size of the viewed object as it recedes from the viewer. The following diagram conveys the idea:

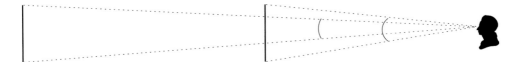

The viewer is looking at a (to simplify the issue) line segment which lies vertically, some distance away. When the segment is moved away from the viewer the angle of the visual cone determined by the segment's extremities decreases. It is this angle, called *angle of vision* that determines the absolute size of the perceived segment. We use the word "absolute" because its relative size – the one resulting from the comparison of sizes for the segment and other objects in the scene – will result from adjusting the perceived absolute sizes of all these objects in terms of possible depth cues in the scene.

The angle of vision is not only determined by the distance from the object to the viewer, but also by a possible tilt of the object. In the following diagram the segment closest to the viewer is bigger than the more distant one and in spite of these facts both segments determine the same visual angle. The reason, it is apparent in the picture, is the *tilt* of the longest segment.

It is not difficult to evaluate the extent of this tilt. Consider the line ℓ passsing through the segment, the line r passing through O and perpendicular to ℓ, and the intersection point C of r and ℓ (we call C the *central point* for the segment with respect to the viewpoint O). The more distant the segment is from C (assuming it lies always in ℓ) the more tilted it is. The least tilted segment is the one having C as middle point. Departure from this situation yields increasingly tilted segments.

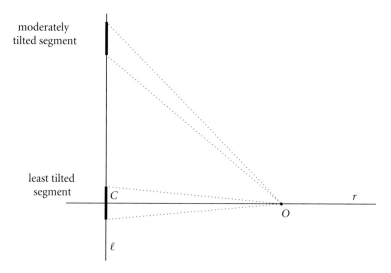

At this point we observe that tilt varies within a given segment. This is patent when looking at a long segment, as we may consider a number of smaller segments, contained in the longer one and at different locations within it, and each of them will have its own tilt.

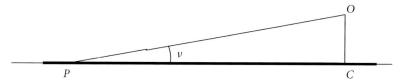

We can actually talk of the *tilt at a point P* in the segment, and this notion can be precisely quantified by the angle ν formed by the line passing through P and O and the line where the segment lies. Thus, the central point C for the segment is the one with $\nu = 90°$. As we have seen, segments having C as the middle point minimize their tilt.

To understand the way distance and tilt combine in the visual experience we may now add thickness to our segment and make it into a rectangle, such as ▬▬▬▬. Then we can consider tilt with respect to both main axes of the rectangle. Assume that the rectangle and viewer are such that the rectangle is least tilted. That is, that the line perpendicular to the rectangle's plane and passing through the viewpoint O intersects the rectangle at its centre C. Assume as well that the rectangle is elongated so that its long side is substantially longer than its short side and, for convenience, that it is divided into square sections which will take turns to be the focus of our regard.

When we look (from O) at the square centred around C the two segments parallel to the sides and passing through C have both minimal tilt, since they both have C as middle point. In contrast, when we look (always from O) at the top square in the rectangle the corresponding vertical segment is substantially more tilted than the horizontal one. This is because when we displace focus upwards along the vertical axis of the rectangle (shown as a dashed line in the picture) the central point for the horizontal segments move upwards as well, whereas the central point for the vertical segments remain at C. This difference in tilt between the horizontal and the vertical axes in the top square produces a difference in the relative size of the horizontal and vertical sides of the square. We say that the top square is *foreshortened* (when seen from O). The effect is akin to that produced by a strain. It goes without saying, foreshortening is a quantifiable phenomenon and, in our example at hand, it gradually increases when we displace focus upwards. We have already mentioned foreshortening (in a casual way, without being precise about its meaning) when describing the fish in Duccio's painting in Figure 9.3.

The displacement of the eyesight upwards along the vertical axis also changes the distance from the viewpoint to the viewed square and, hence, the absolute size of the perceived square. But this change affects both horizontal and vertical sides of the square in approximately the same manner. This diminution in absolute size is a cue for depth (or distance to the viewpoint) but does not produce foreshortening.

It is the combination of size variations produced by distance to the eye and tilt that conjures the visual effect of a scene under a determined viewpoint. And it is to be noted that the effects of this combination are on the visual cone and precede any intersection of it with a picture plane.

We find two specific choices of viewpoint among the bread and butter of the photographer and the filmmaker. They correspond to the situation where the observer looks at the object of attention from above (referred to as *high-angle*) and from below (*low-angle*).

The use of low and high angles to create specific visual impressions is common in cinematography. A book by Jennifer van Sijll (2005) lists "the 100 most powerful film conventions every filmmaker must know" with, for each of the hundred, a short description followed by a couple of examples and a statement about its dramatic value. Conventions 64 and 65 are high angle and low angle respectively. The examples for low angle selected by van Sijll are taken from *E.T.*, the 1982 film by Steven Spielberg.

Most of the scenes in this film are shown to the spectator as seen by the small alien (eventually named E.T.) and thus shot in low angle. Van Sijll offers two examples of low angle, both from *E.T.* She notes that "In showing both the redwoods and trucks from this vantage point, power and mass is transferred to them. This contributes to our sense of E.T.'s vulnerability". A film maker who associated his name with both low and high angles is Orson Welles. For many of his movies (e.g. *Citizen Kane, The Trial*) Welles had, for certain scenes, tailor-made stages which allowed the camera to be below floor level, thus creating extreme low angles. This made possible, for instance, a larger-than-life image of Charles Foster Kane.

How do distance and tilt affect the visual cone (which, properly speaking, we should call visual cylinder) when the viewpoint is at infinity?

We proved in Proposition 10.2 (and followed in Remark 10.3) that the size of an object rendered with parallel projection is independent of the position of the object. We may revisit this result under the light of the discussion in § 10.1.4. Indeed, any translation of the object moves it only a finite distance and this displacement is negligible when compared with the distance between the object and the eye. In this sense, it is as if the object was not moved at all. The same constancy occurs if instead of displacing the object we displace the focus of our eyesight: there is no change of absolute size.

A related constancy occurs with respect to tilt. Consider a segment, or a line, s in $\mathbb{E}^3$. The lines joining points in s with the viewpoint are all parallel:

Therefore, the tilt at a point P in s is the same for all points P in s (in particular, a point is central for s if and only if all points in s are central).

Putting together these facts, we conclude that if we displace the eyesight along a segment with the viewpoint at infinity, the visual experience is not affected by either distance or tilt. In particular, low angle and high angle are not defined in terms of a particular object in the scene – as in the case of perspective projection – but are a feature common to the entire scene (and it is determined, as we discussed above, by the value of θ).

But foreshortening is still possible. The reason is that when looking at a two- (or three-) dimensional object, say a rectangle to fix ideas, the tilts for the two axes may be different. Again, the difference with a perspective view is that the extent of this foreshortening is constant along the rectangle, whereas in the perspective view it is not. But we note that, in a perspective view, the change in this foreshortening decreases when the viewpoint distances itself from the rectangle. In the limit, when this distance becomes infinity, the projection becomes parallel and the change non-existent (i.e. constant foreshortening).

10.1.4 Perspective projection

We finish our repertory of drawing systems by returning to perspective. As with parallel projection, we will retain the assumption that the object to be drawn is a cube to further divide perspective into subclasses of projections. The discriminating parameter is now the number of axes of C which are parallel to the picture plane π' (two, one or none).

One-point perspective

If two axes of C are parallel to π' then the face of C given by their directions and closest to π' is drawn in the picture plane as a square. The size of this square depends on the distances of this face to π' and the vantage point (recall § 9.3.1). Lines with the remaining edge direction are perpendicular to π' and, therefore, its images in the picture plane meet at a vanishing point in the horizon line, which, since these lines are perpendicular to the picture plane, is what we called the central vanishing point.

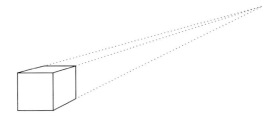

The cube above is drawn in one-point perspective. Note that the viewpoint is at the right and above (i.e. high angle) of the cube since the central vanishing point is so.

One-point perspective was common in the Renaissance. Piero's *Flagellation* (Figure 9.5) is a beautiful example – the edge directions being given by the architectural elements in the painting: columns, ceiling beams, roof eaves, tiles, stairs and so on. It also occurs in East Asia, although infrequently until recent times. In the first half of the eighteenth century, influenced by European art, Okumura Masanobu introduced elements of perspective in some of his ukiyo-e. This eventually resulted in a subgenre called *Uki-e* of perspective-rendered images which reached its apex with the work of Utagawa Toyoharu (in the eighteenth century), Utagawa Kunisaga and Utagawa Hiroshige (both active in the first half of the nineteenth century). Uki-e were almost invariably drawn in one-point perspective.

Two-point perspective

This corresponds to the case when one (and only one) axis of *C* is parallel to the picture plane. It follows that the other two directions are oblique to this plane. Lines with these directions are represented in π' by lines meeting at two different vanishing points in the horizon line. In the following cube these vanishing points are located at left and right of the central vanishing point and are equidistant to it:

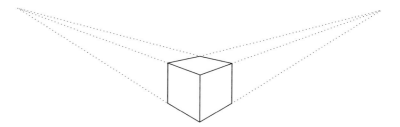

But there is no reason for this to be so (for instance, one may, and it is actually easy, draw a cube in two-point perspective with one of the axes of the cube perpendicular to the picture plane).

The painter Giorgio de Chirico produced a number of works during the 1910s where desolate urban landscapes are inhabited by mannequin-like figures and unnatural-looking objects. The mood of these scenes is oppressive and vaguely ominous. Figure 10.15 shows a work from this period. The building in the background, with its absence of windows and its dark openings, creates a dismal atmosphere around it. The foreshortening created by the two-point perspective rendering, together with the low angle with respect to the building, adds to this feeling.

Three-point perspective

If no axis of *C* is parallel to the picture plane, then, for each of these axes, all lines parallel to it meet at a vanishing point in the picture plane. In the following cube we can produce these three vanishing points by

Figure 10.15 **G. de Chirico, *The Evil Genius of a King*, oil on canvas, 1914–15. (ⓒ DACS 2012.)**

extending the cube's edges and noting the intersections corresponding to parallel edges:

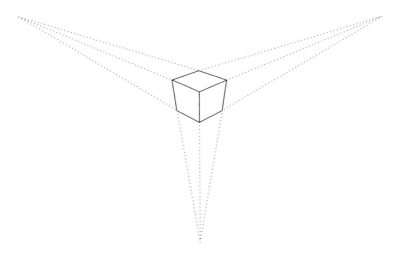

In this particular drawing, two of these intersections lie on a horizontal line above the cube (we are looking at the cube from above) and the third one lies below the cube.

Three-point perspective results from an off-centred situation of the painted object with respect to the picture plane. This eccentricity can be

Figure 10.16 **A. Feininger, Panorama of RCA Building from Rockefeller Center, 1957. (LIFE magazine.)**

very small (this is so when a slight change in the position of the object, or of the picture plane, may yield a two-point perspective rendering). The visual characteristics of three-point perspective are, nonetheless, brought to the fore when the eccentricity is substantial. They are also enhanced when the vantage point is off-centred as well; that is, when it is removed from the axes of C, a situation we know amounts to an observer who looks at C either from above (high angle) or from below (low angle), and either from the left or from the right. The photograph in Figure 10.16, which combines three-point perspective with a high-angle viewpoint, gives an example.

Remark 10.5 The terms "One-point perspective", "Two-point perspective" and "Three-point perspective" should not induce the misconception that any drawing in perspective has either one, two or three vanishing points. A picture can have as many vanishing points as sets of parallel lines are present in the depicted scene. If we simplify the latter to our cube C, then this number is at most three, but it is not hard to conceive scenes more complex than C. Furthermore, many scenes (natural landscapes, irregular objects, etc.) offer no collection of parallel lines. This seems to be the case, for instance, with the string instrument which gets the attention of the artist in Figure 9.7. It is not uncommon to refer to paintings of such scenes as done in *zero-point perspective*.

We finally note that a scene may contain objects placed in a way for which some are in one-point perspective, some in two-point perspective and so on. A well-known example is the painting *The School of Athens* by Raphael (gracing our cover), where almost all the architectural elements are depicted in one-point perspective but the marble cube in the foreground is rotated so that it appears in two-point perspective.

10.2 Voyeurs and demiurges

Perspective techniques became a trademark of European art in the fifteenth century, and remained so until the end of the nineteenth century. In the Far East, in contrast, perspective was rarely used. One could hypothetize that this is due to a lack of knowledge of these techniques, but historical facts show that by the eighteenth century this was certainly not the case. It is well documented that Matteo Ricci – a Jesuit who arrived in China in 1582 and remained there until his demise in 1610 – brought with him several paintings in perspective. Whereas it is just possible that he could have brought as well the knowledge required to paint them, it is without doubt that Giuseppe Castiglione did so in the early eighteenth century. A Jesuit priest as well, but also an accomplished artist, Castiglione arrived in China in 1715 and lived there until his death in 1766. He worked as a painter for three Qing emperors (Kang-xi, Yong-zheng and Qian-long) developing a style which would fuse Chinese and Western stylistic elements. His use of perspective raised the interest of an artist in charge of the manufacture of ceramics for the Imperial palace, named Nian Xiyao, who eventually wrote a book on "Western Perspective" (with the Chinese title *Shi xue*, meaning "theory of vision").

We have also mentioned the introduction of perspective in ukiyo-e by Okumura Masanobu in the first half of the eighteenth century. Thus, both China and Japan possessed by then the theoretical elements of perpective. And yet, they did not apply this knowledge to actual painting except for a few remarkable exceptions. Why was this so? A complete answer to this question surely involves a variety of reasons.[3] A plausible one is based on a recurrent fit between the material framework and the underlying geometry of an art practice.[4] To argue for it we need to describe a specific genre of painting. It is a format of Chinese and Japanese painting known as *handscroll* (a literal translation from the Chinese *shou-juan*, "shou" meaning hand and "juan" meaning scroll[5]).

A handscroll consists of a painting in either silk or paper shaped as a horizontal rectangle whose base is several times longer than its height. The painting is mounted on a slightly larger, stiffer, surface with a round wooden roller attached at its left extremity and a semicircular wooden stave at its right extremity. Handscrolls were not displayed; most of the time they were wound (around the roller) and stocked. But every so often they would be subject to contemplation. The nature of this viewing, being essentially different from that of other formats of painting, calls for an accurate description. We borrow from the experts (Silbergeld, 1982: 13):

> The painting was viewed from right to left, as one reads in Chinese, unrolling a bit at a time from the roller and transferring the excess to a loose roll temporarily maintained around the stretcher on the right. About one arm's length was exposed at a time for viewing. [...] The great distinction between the handscroll and all other formats is its considerable length and its sequential exposure of the painting, allowing the artist to control the pace of visual events and to manipulate the viewer's response with shifts in subject matter and treatment.

Wu Hung claims that the origin of handscrolls can be traced back to lacquer decorations crafted before the Han Dynasty (206 BCE–220 CE) and notes that probably the oldest extant handscroll is the *Admonition of the Instructress to Palace Ladies*, executed by Gu Kai-zhi in the fourth century (Wu, 1996: 63).[6] This painting contains nine sections which are separated by vertically written text so that individual scenes,

---

[3] Fritz van Briessen (1998: 129) suggests cultural reasons, writing, "Oddly enough, the pictures shown in the penny-in-the-slot machines that attract such crowds of children at any Chinese fair [...] make an almost desperate use of central perspective. This observation [...] provides reason for speculating that the Chinese artist has always tended to regard Western perspective as little more than a *trompe l'œil* suitable for the amusement of children but not worth the notice of an inheritor of the glorious tradition of Chinese painting".

[4] An example of such a fit is given by the commonly occurring wallpapers in textile patterns, an occurrence which derives from the use of small-sized wood blocks to print the fabric.

[5] In Japanese they are called *emaki*, which translates as "picture roll". Japanese handscrolls follow essentially the same format of their Chinese counterpart but they have distinctive characteristics (see (Hironobu, 1991) for more on this issue).

[6] This handscroll is currently at the British Museum. A comprehensive book about it is (McCausland, 2003).

Figure 10.17 **Gu Hong-zhong, *The Night Entertainment of Han Xizai*, twelfth-century remake.**

even though merging into a unified narration, are clearly distinguished. A scroll found in Dunhuang, dated around the eighth or ninth century, shows an evolution in visual structure. According to Wu (1996: 66),

the Dunhuang painting rejects texts (which are inscribed on the back of the scroll as the storyteller's script). The structural function of texts in the *Admonition* scroll is assumed by images: a number of trees, rather schematically executed, divide the Dunhuang painting into six sections [...].

Further evolution can be appreciated in the *Night Entertainment of Han Xizai* (Figure 10.17), a scroll painted in the tenth century by Gu Hong-zhong,[7] whose analysis occupies Chapter 1 in (Wu, 1996). The scroll describes a revel at Han Xizai's house which, tradition claims, Gu Hong-zhong was attending on the orders of Emperor Li Yu. The character of these revels had reached the ear of the Emperor, who sent Gu to investigate the truth about them.

Scenes are now marked by standing screens which, Wu notes, "help structure *both* individual scenes and the whole handscroll: while punctuating the motion of unrolling the painting, they also define places – the living-room, the bedroom, and the back chamber – because they are part of these places". The last words are important to our argument. These screens are integrated in the painting. They are part of a unified space (a fact emphasized by details such as the man and woman talking while at different sides of a screen, see the lower part of Figure 10.17). Now the question is posed: Where should the viewpoint for this unified space be?

A viewpoint in Euclidean space would require the viewer to place themselves at or near this location and look at the entirety of the scroll, in opposition to the stop-and-go viewing process described by Silbergeld

---

[7] A twelfth-century remake of this scroll is nowadays housed at the Palace Museum in Beijing.

in the quotation above. A viewpoint at infinity, in contrast, is perfectly suited to this process, since, as we remarked in § 10.1.3, when we displace the eyesight along a segment with the viewpoint at infinity – a displacement equivalent to the unrolling of the scroll under a fixed viewpoint – the visual experience is not affected by either distance or tilt. We thus reckon that parallel projection is the natural drawing system for the handscroll format.

Castiglione was well aware of this fact. In his handscroll *One Hundred Horses*[8] he showed this number of horses along the painting. To give a sense of depth he used elements of perspective – diminution in size for distant objects – together with light variations (in general absent in Chinese painting) but he carefully avoided the organization of space given by architectural elements or, more generally, prism-shaped objects.

The handscroll structure is also subjacent in Escher's *Metamorphosis III* (Figures 6.8 and 6.9). Even though we cannot properly talk about either a unified space or a narration, this woodcut is best appreciated when the viewer displaces his gaze along the image to see bees, birds, fish, houses and many other forms continuously morph into each other.

The handscroll format reveals some characteristics for both author and viewer which relate to the use of parallel projection. We next touch on them, beginning with the latter.

Unlike a framed painting, which can be simultaneously observed by several viewers, handscroll contemplation is a private matter. The viewer sits facing a table purposely designed for holding the scroll during the viewing process. Over-the-shoulder onlookers are not meant to be there. Wu Hung goes beyond this privacy to claim that the viewing of a handscroll is a voyeuristic experience. He cites Ellis (1992: 45) to explain this idea:

Voyeurism implies the power of the spectator over what is seen. Not the power of changing it, but the knowledge that the actions being undertaken are played out for the spectator. [...] The spectacle is separated from the spectator [...] This sense of separation permits the spectator to maintain a particular relation of power over what he or she sees, and constructs the need for a continuous change and development in what is seen. The characteristic voyeuristic attitude in cinema is that of wanting to see what happens, to see things unrolling.

The nature of the handscroll viewing process, Wu Hung observes, conforms to the passage above. Both the temporal dimension of the narration and the separation of the spectator from the scene's characters (the events therein are played by and for these characters, who remain absolutely unaware of the spectator's gaze) are intrinsic to this process.

Jianping Gao maintains that a natural counterpart for this power in the role of the painter gives the latter a likeness with a Creator (a word which he quickly associates with its Taoist meaning as opposed to its

[8] Housed at the National Palace Museum in Taiwan.

Christian acception). He devotes a section of his thesis to elaborate on the fact that a number of "painters and critics announced that the painter was able to create a picture in the same way that the Creator created the world" (Gao, 1996: 171). It is not our intention to go over Gao's arguments (which involve notions developed in China over millennia and proper to Chinese schools of thought). We will content ourselves, instead, by making two observations supporting his claim. The first one merely points that the location of the viewpoint at infinity, together with the choice of an appropriate scale without recourse to any optical device, is fully consistent with the liking of the painter with a demiurge.

The second observation is a ratification coming from what are probably unexpected quarters: the design and playing of video games. This is an arena with a continuously changing taxonomy but on which it is nevertheless immediate to separate between those games with a fixed scene (the game takes place in a space with definite boundaries such as a chessboard, a pinball machine, the green surface hosting the cards of a solitaire) and those with a changing scene (navigated with the help of a mouse or keyboard). Our focus is on this latter kind.

Probably the vast majority of games with a changing scene involve a character who pursues some ultimate goal and faces a sequence of challenges in this pursuit. The player controls this character (their *avatar* (Rollins and Adams, 2003: 37)) and either sees the world through the eyes of the avatar (*first-person* view) or from behind it (*third-person* view). A paradigm of a first-person, role-playing, action game is *Fallout 3*. Action takes place in the year 2277, about 200 years after a nuclear disaster. Due to the disappearance of his father, the player is forced to quit the shelter he lives in and wander in the ruins of Washington, DC. Figure 10.18 shows a screenshot of *Fallout 3*. Disregarding the green rulers (which indicate availability of some resources), what the

Figure 10.18 **A screenshot of the videogame *Fallout 3*.**

player sees (through the eyes of his avatar) is the scene surrounding him rendered in perspective projection. Changes in this view (obeying to purely geometric causes) may be due to either the displacement of the avatar – and hence of the viewpoint – or a movement of his head – amounting to a change in the direction of sight. The use of perspective rendering is appropriate for a realistic display of the player's surroundings and the feelings these surroundings induce. For instance, in Figure 10.18, the low-angle, three-point perspective of the buildings facing the player conveys an inauspicious mood for trespassing these premises.

The game *SimCity* is at the other end of the spectrum of variable-scene games. Probably the most celebrated city-building simulation game, it has not a set of victory conditions whose satisfaction marks the end of the game. Instead, the player builds a city with either some self-imposed objectives to fulfil or no objectives at all. In building the city, one can decide on a vast range of issues, from land use to tax rates or transportation systems. Julie Lew (1998) reported that the creator of *SimCity*, Will Wright, was "intrigued by a story of Stanisław Lem called 'The Seventh Sally' [in which] a banished tyrant returns to his despotic ways after being given control over a simulated city" and that this feeling is at the origin of *SimCity*.

Figure 10.19 **A screenshot of the video game *SimCity*. (SimCity™ image used with permission of Electronic Arts Inc.)**

There is no avatar in *SimCity*. Instead, the player enjoys a view of the city as in the screenshot shown in Figure 10.19. We immediately notice that, in contrast with *Fallout 3*, scenes in *SimCity* are rendered using parallel (to be precise, trimetric) projection. Changes in the viewed scene may now be due either to a variation of scale – allowing the player to zoom in and out of his view – or to one in the "direction of sight" – which amounts to a translation of the picture frame. The latter is comparable to the rolling and unrolling of a handscroll, but with an additional degree of freedom: the player can displace his view not only horizontally but also vertically (actually, in any direction) as well.

The use of parallel projection is not exclusive to *SimCity*. Many strategy games feature this visual presentation. Rollins and Adams (2003: 339) note that, in general in these games "the player is given a godlike view of the game world". A statement that brings us back to the role of a handscroll painter – a distant predecessor of a *SimCity* player – as a Creator. Unlike *Fallout 3*, whose players experience the hardships of this world, a player of *SimCity* may be compared with the tyrant of *The Seventh Sally* additionally endowed with the capacity of scaling (zooming in and out) his view of the city: an all-powerful creator. I had this judgement confirmed when I approached my teenager daughter in search of a description of the playing of *SimCity*. She offered me a detailed description of the game whose increasing complexity was only matched by my equally increasing sense of loss. Eventually, realizing the latter, she summarized the gist of the game: "You play God".

The vicissitudes of perspective

From its dawn in the fifteenth century to well into the second half of the nineteenth, perspective was a trademark of European art. A walk through any modern art museum reveals that, in contrast, during the twentieth century this protagonism was by and large abandoned. A plethora of styles – loaning diverse prefixes to the ending *ism* – placed emphasis on goals independent, when not at odds, with realistic representation and brought with them a shift in drawing techniques.

This falling from grace did not occur overnight. Nor was it the only fate befalling perspective. On the contrary, the techniques of perspective were used for purposes other than the ones which led to their creation and, somehow conversely, a number of alternative techniques were proposed to serve these purposes. The goal of this chapter is to journey through these vicissitudes.

11.1 Deceptions

"Yes," he said, "any deception is a form of magic […]"

Plato (1994: 117) *Republic*, 413c

Probably the first perspective painting that was widely available to Renaissance men was the *Trinity* (Figure 11.1) painted by Masaccio in the Church of Santa Maria Novella, in Florence, around 1427. The rendering of the figures was realistic, their sizes uniform across the painting, their importance highlighted by their position. What apparently most impressed the viewers, however, was the vault shown in the background. According to Vasari (1991: 104), "what is most beautiful, besides the figures, is the barrel vault drawn in perspective, and divided in squares full of rosettes which are so well diminished and foreshortened that the wall appears to have holes in it". The capacity to deceive the observer was therefore present at the very birth of perspective painting.[1] Martin Banham (1995: entry "theatre design") observes that this capacity was put to good use in the design of stages for theatrical performances: "The road to naturalistic design – the attempt to create an illusion of reality

---

[1] A feature going a long way in making this deception possible is the absence of frame. The fresco is painted on a wall and crafted in a way that places no emphasis on the boundaries with its surroundings. The viewer thus includes the painting, without discontinuity, in his reading of the wall and becomes prone to integrate depicted architectural elements into it. Alberti's window may be framed, Masaccio's vault cannot.

Figure 11.1 **Masaccio, *Holy Trinity with the Virgin, St. John the Evangelist, and Donors*, fresco, *c.* 1427. (Santa Maria Novella, Florence, Italy/The Bridgeman Art Library.)**

or verisimilitude – began in the Renaissance; and at the same time theory, design and architecture combined radically to alter the relationship between the stage and the auditorium. The three most significant developments in this period were the relocation of the theatre indoors, the discovery of techniques for creating perspective sets, and the evolution of changeable scenery."

Time would consecrate the French expression *trompe l'œil* (literally, "fool the eye") to denote paintings where an optical illusion of three dimensions is created. Time would also consecrate the ceiling of the nave of the Sant'Ignazio Church, painted by Andrea Pozzo[2] at the end of the seventeenth century, as one of the greatest instances ever of trompe l'œil.

Figure 11.2 **Andrea Pozzo, nave ceiling of the Sant'Ignazio Church, Rome, 1685–94. (Church of St. Ignatius, Rome, Italy/Alinari /The Bridgeman Art Library.)**

The original plans of the church included a large dome. But Dominican monks occupying a neighbouring building raised the complaint that such a dome would cut off the light from their library. The solution found by the Jesuits – behind whom was the initiative of building Sant'Ignazio – was to commission Pozzo to create a false dome (and

---

[2] Incidentally, we note that Giuseppe Castiglione, whom we met in Section 10.2, had been influenced by the teaching of Pozzo before leaving for China.

even an extra floor) by painting it on the flat surface of the ceiling (see (Seckel, 2004: 10)). Figure 11.2 shows a fragment of this ceiling photographed from the correct vantage point (roughly at the centre of the nave's floor). Departure from this location shows, and this is unavoidable, some distortion on the painted figures. Remarkably, however, one needs to significantly distance oneself from this privileged location for the illusion to break down.

11.2 Concealments

'Twere a concealement worse than a Theft.

W. Shakespeare (1951: 835), *Coriolanus*, I.ix.21

In 1532 Hans Holbein returned to England. He had already lived there between 1526 and 1528 and this time he was going to stay longer, until 1540, working as a court painter for Henry VIII. It is during this period that Holbein painted his most celebrated work, a portrait of Jean de Dinteville, ambassador of Francis I of France, and George de Selve, Bishop of Lavaur, known as *The Ambassadors* (Figure 11.3). The portrayed characters stand surrounded by a number of objects with a scholarly character, such as scientific instruments (a terrestrial globe and a celestial one, a quadrant, a torquetum, a sundial, etc.), musical instruments and books. The two figures are doubtless symbolic of secular and religious spheres of scholarship, but there is controversy as to whether Holbein intended to depict their unity or the conflicts which pervaded England's religious life at that time.

Among the collection of objects making the still life in Holbein's *The Ambassadors*, the one that has been most talked about is the least discernible. At the foreground of the painting, in its lower part, lies an elongated form, colourless, with no definite shape. A view of the painting from below and nearly from its left side reveals a skull.

Figure 11.4 shows, at the left, a straining of the painting which approximates the view from the vantage point just mentioned, and at the right, an enlargement of the previously shapeless form. What Holbein was doing, remarkably for the early sixteenth century, was to depict in a single picture plane two scenes observed from different vantage points and rendered with different drawing systems. On the one hand, the ambassadors and their scholar environment, seen from an observer facing them, depicted in faultless perspective. On the other hand, the skull which appears to be rendered in some form of oblique projection (recall § 10.1.2). Indeed, no choice of vantage point would result in such an elongation. The skull was most likely painted (elsewhere) using standard perspective projection and then an oblique view of this painted image was added to *The Ambassadors*.

Figure 11.3 **Hans Holbein the Younger, *The Ambassadors*, 1533, oil and tempera on oak panel. (National Gallery, London, UK/The Bridgeman Art Library.)**

Historians have not agreed on a single explanation for the concealment of this skull on Holbein's painting. The most commonly proposed accounts for it as a memento mori, a reminder of the inevitability of death. The oddest, mentioned by Seckel (2004: 11), is the possibility of Holbein putting it as a signature – as his name means "hollow bone" in German.

The procedure of distorting the image of an object by representing it in a way that requires special conditions for proper viewing (an unnatural viewpoint, reflection in an unusual mirror, etc.) is called *anamorphosis* (from the Greek ἀναμόρφωσις meaning transformation).

Both trompe-l'œil and anamorphosis are effects of foreshortening, but in trompe-l'œil the artist wants to create an illusion of reality (that departing from the vantage point destroys) and in anamorphosis the artist wants to hide some view, which can only be recovered from a correct vantage point. In the first case the vantage point is "natural" while in the second it is "bizarre".

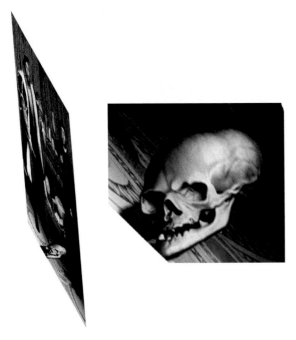

Figure 11.4 **A strained view of** *The Ambassadors* **(left) and an enlargement of the strained skull (right).**

11.3 Bends

In Chapter 10, empowered by the basic ideas of projective geometry, we saw parallel projection as a limiting perspective when the vantage point is at an infinite distance from the painted scene. We noted in § 10.1.3 that this corresponds to an increasingly small viewing angle. We may add, borrowing terms from photography, that this recession towards infinity yields images obtainable with increasingly powerful telephoto lenses.

This remark raises, to anyone familiar with photographic equipment, a natural question. Which sort of image arises at the limit when using increasingly powerful wide-angle lenses? In opposition to the use of tele-photos, this corresponds to an increasingly large viewing angle. In the limit, this angle becomes 180°. The picture on the right shows a canvas C and four different vantage points (with their corresponding viewing angle). A cursory look at this picture points to a problem: the limiting situation (viewing angle equals 180°) is reached when the vantage point is on the canvas. And at this juncture we can no longer represent the scene since all points in this scene project into the vantage point.

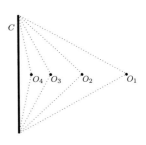

A possible solution for this problem is provided by assuming that the canvas is a half-sphere $\mathscr{S}$: in this case, the 180° view is obtained with the viewpoint O situated at the centre of $\mathscr{S}$, a point which is not in $\mathscr{S}$.

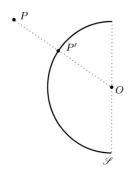

We can picture this setting (with points P in the scene projected to points P' on the canvas) as on the left.

But now we are facing a new problem, namely the fact that canvases are not half-spheric. Nor are photographs taken with wide-angle lenses. And we cannot conceive a curved canvas as a limit of flat ones. A possible solution for this new problem consists of projecting (say, with a parallel projection) the image on $\mathscr{S}$ onto a circular, but flat, canvas C.

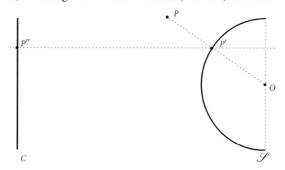

We can postulate this procedure for viewing angles smaller than $180°$ as well. But we observe that, for small angles, the curvature of the relevant portion of $\mathscr{S}$ is in like manner small and, consequently, the resulting image is close to one obtained with a perspective view.

A further parallel projection onto a flat canvas introduces some distortion close to the boundary of the picture area but this distortion is negligible for small viewing angles.

For large viewing angles, however, and more specifically for the $180°$ case which characterizes our limiting situation, the distortion near the boundary will be marked. Parmigianino, whose long-necked Madonna originated the Mannerist style in painting (see Section 7.4) was, perhaps not surprisingly, interested in these distortions. In 1524 he crafted a self-portrait depicting his image as reflected in a spherically curved shaving mirror (Figure 11.5). At the centre of the image we see the face of the then 21-year-old artist, essentially undistorted; at its boundaries, window frames, wall edges and even the artist's fingers, bent themselves to accommodate to the mirror's shape.

Nearly four centuries later, in 1935, Escher drew a lithograph with a similar motive (Figure 11.6). He is holding a sphere which reflects on its surface both the artist and the surrounding room. In a manner

Figure 11.5 **Parmigianino, *Self-portrait in a Convex Mirror*, oil on wood, *c.* 1524. (Courtesy of Kunsthistorisches Museum, Vienna.)**

characteristic of Escher's playfulness, the hand holding the sphere is painted twice: both the undistorted – the actual hand as seen by the artist – and its distorted reflection on the sphere. In doing so, Escher produced an image where two different drawing systems naturally coexist.

In both Parmigianino's and Escher's works the procedure to obtain a 180° viewing angle passes through a material version of our line of thought above. A scene is first reflected on a spherical mirror and then projected onto a (flat) picture plane. Lens manufacturers nowadays produce some extreme wide-angle lenses, known as *fisheye lenses*, which allow the photographer to shortcut this procedure and directly view and, if and when desired, photograph the scene with a 180° viewing angle. The portrait in Figure 11.7 was obtained using one such lens.

From a geometric viewpoint, one of the most immediately noticeable effects of considering a spheric canvas (and a further projection onto a flat one) is the fact that straight lines are no longer necessarily so in the picture. This is true for all viewing angles, but is accentuated at large ones. An example of this phenomenon in painting is the *Bedroom in Arles* by Vincent van Gogh (Figure 11.8). The Dutchman produced three versions of this picture between October 1888 and September 1889.

Figure 11.6 **M.C. Escher's *Hand with Reflecting Sphere*, lithograph, 1935**

The treatment of colour is inspired by ukiyo-e, of which van Gogh was
fond. In a letter to his brother (dated Tuesday, 16 October, 1888 (van
Gogh, 2009: 330)) he says that "The shadows and cast shadows are
removed; it's coloured in flat, plain tints like Japanese prints". The drawing system, instead, could not be farther away. Rather than a viewpoint
at infinity as is the standard in ukiyo-e, van Gogh placed it very near the
scene. So near, indeed, that he found it appropriate to bend some of the
straight lines in the room (such as those joining wooden tiles on the floor,
near to the lower boundary of the picture area). Figure 11.8 gives an idea
(it reproduces the third version of the painting).

   As far as I am aware, van Gogh did not attempt to justify his bent
lines. He simply drew them. Neither did he elaborate a system to make
these kinds of drawings. These endeavours would have to wait until the
1960s, when they were carried out by Albert Flocon and André Barre. In
a book first published in French in 1968 (Flocon and Barre, 1987) they

Figure 11.7 **Portrait of the author at Les Tuileries d'Affiac.**

Figure 11.8 **Vincent van Gogh,** *Bedroom in Arles*, **oil on canvas, 1889. (Van Gogh Museum, Amsterdam, The Netherlands/The Bridgeman Art Library.)**

described a drawing system called *curvilinear perspective* and argued that it approximates retinal vision better than (rectilinear) perspective. The core of their argument relies on the distinction between foveal and peripheral vision. While the fovea has a small viewing angle, the region covered by peripheral vision is close to hemispherical; yet, accuracy close to the boundary of this hemisphere is poor. Retinal vision would then be unlike the everywhere-sharp images produced by Parmigianino or Escher, but it would be equally unlike the small-viewing-angle images produced by rectilinear perspective. Some physiological studies add some light to Flocon and Barre's persuasion. In his treatise on optics, Pirenne (1970: 71) described a number of experiments on vision and states:

These experiments [...] show that if any given set of parallel lines is extended indefinitely in both directions away from a motionless eye in any fixed position, the retinal images of the parallel lines do in general tend to converge in *both* directions. But convergence may be difficult or impossible to observe in more than one direction, simply because in the other direction images of the lines may not be formed in the retina.

The first half of this paragraph seems to confirm the curvilinear nature of the retinal images of lines. Indeed, only such a nature would allow for two different lines to intersect at two different points (the convergence points for the two directions which Pirenne emphasizes by italicizing the word "both"). The second half, however, adds a note of caution.

We should not deduce from the preceding considerations that curvilinear perspective is a more adequate tool than linear perspective in the search for an accurate rendering. Because, even accepting the suitability of curvilinear perspective to describe retinal vision, it does not follow that images obtained using this drawing system are more faithful representations of the object represented. As it happens, whichever transformation is performed to produce the retinal image, this transformation will be at work both when observing the object and a representation of the object. We do not need to do it in the representation itself (indeed, were we to do it in the rendering of the scene this transformation would be performed twice!). What is more, the experiments mentioned by Pirenne confirm the adequacy of linear perspective to representation. In the page quoted above, he further writes

The main principle of linear perspective is illustrated in Experiments 4 and 5 in which it was found that straight lines passing through the centre of the entrance pupil of the eye were foreshortened, each as a single point image on the retina. Consequently, if straight lines are drawn from the centre of the entrance pupil to the various points of the object, the centre of the pupil remaining stationary, and if this pyramid of lines is sectioned outside the eye by a surface, then it follows that the picture defined by the intersection of the pyramid by the surface must form a retinal image coinciding, "point" by "point", with that of the object. The facts that the retina is curved, and that the retinal image is not an exact central projection of external objects, are irrelevant.

We conclude that, for the sake of an accurate representation, linear perspective is appropriate. If the scene is viewed under a wide angle and this would produce a curvilinear image in the retina, then the same will occur when looking at the linear perspective representation from the correct vantage point.

Having said that, the association of art with accurate representation is by no means a necessity – if we want accurate representations then linear perspective is not an arbitrary choice, but the goal of accuracy is itself arbitrary – and curvilinear perspective has no lesser a place in the artist's toolbox than any other drawing system. We may, therefore, have a closer look at it, beginning with a spatial version of a diagram we saw earlier in this section describing the system as a composition of a projection onto a half-sphere $\mathscr{S}$ followed by a projection onto a plane π' (parallel to the plane passing through the boundary of $\mathscr{S}$).

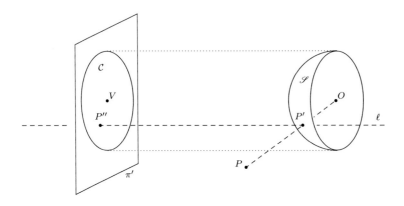

Let O denote the centre of the half-sphere $\mathscr{S}$. A point P in three-dimensional space (or in a plane π living in such space if we want to restrict the whole transformation to between two-dimensional spaces) is mapped first into a point P' in $\mathscr{S}$ via a projection with projection point O, and then the point P' is mapped into a point P'' in π' with parallel projection in the direction perpendicular to π'. This defines a mapping from three-dimensional space (or from a plane π therein) into the interior C of a circle in π'. Lines in three-dimensional space (not passing through O) project into half-great-circles in $\mathscr{S}$ (together with O they define a plane whose intersection with $\mathscr{S}$ yields such a half-great-circle). Further projection onto π' yields either diameters of C (if the half-great-circle passes through the pole of $\mathscr{S}$, as r in the drawing on the right) or half-ellipses having those diameters as major axes (as r' and r'').

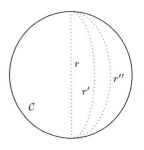

We observe, in passing, that we have already encountered a composition of two different projections in the representation of the skull in Holbein's *The Ambassadors*.

Endowed with this knowledge we can return to our old theme and represent a cube C which, as in the case of one-point perspective, we assume has two axes parallel to π' (and the third one parallel to ℓ). The following picture shows a representation of C seen from the left and from below.

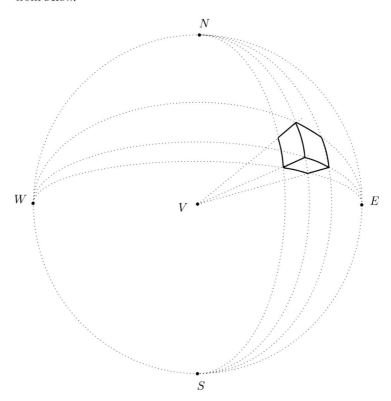

We see that four of the edges of C (the ones parallel to ℓ) are represented as straight segments (lying on diameters of the circle), four more lie on half-ellipses with endpoints N and S, and the remaining four are in half-ellipses with endpoints W and E. Following the naming system used in § 10.1.4, the system we just used is often referred as *five-points perspective*. Yet we note that of the five vanishing points playing a role here, four of them are "paired" into two couples (all lines passing through E must pass through W as well, and the same is true for N and S) and the fifth one has a different nature.

At this juncture one may naturally wonder about the existence of a four-point perspective system. One occasionally finds references to a system with this name. The position of C is as in two-point perspective. Lines parallel to the two non-vertical axes are drawn in the picture plane as straight lines with a single vanishing point for each direction (W and E in the following picture). The vertical direction, instead, is curved so that lines with this direction, meet at two vanishing points (S and N in the picture). Thus, here, only four edges of C are bent in C' (of which one is hidden in the following picture).

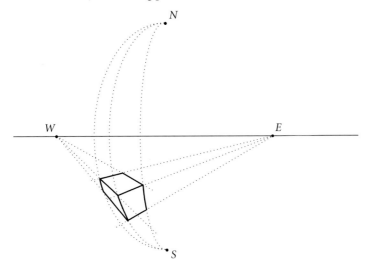

The instances of curvilinear perspective mentioned above are by no means isolated. In his illustrations to the *Grandes Chroniques de France*, done between 1455 and 1460, Jean Fouquet paints in this manner the entry of Charles IV to Saint-Denis. Also, a closer look (see Figure 11.9) at the mirror on the wall of *The Arnolfini Portrait* (Figure 9.6) shows that van Eyck was no stranger to these bends.

11.4 Absurdities

In 1754, William Hogarth, whom we have met in Section 5.3, drew a frontispiece (Figure 11.10) for Joshua Kirby's book *The Perspective of Architecture* to which he added the subtitle "Whoever makes a DESIGN without the Knowledge of PERSPECTIVE will be liable to such Absurdities as are shewn in this Frontispiece". The absurdities mentioned by Hogarth are many. To name a few: the sizes of the sheep in the herd appear to increase with the sheep receding in the drawing (as occurs with the tiles where the gentleman stands), the sign with the crescent moon is affixed to two buildings in an incongruous manner and it is partially covered by two distant trees, the man on top of the hill is lighting his pipe with the candle of the woman in the window, the church in the background appears to have been drawn from different viewpoints, and so on.

Most of the individual components of the scene are plausible. It is their combination that produces a sense of absurdity, as if the whole scene was impossible. This tendency towards illogical combination developed into the string of "impossible objects" pioneered in the work of Oscar

Figure 11.9 **Jan van Eyck, *The Arnolfini Portrait* (detail), 1434. (National Gallery, London, UK / The Bridgeman Art Library.)**

Reutersvärd and followed by artists such as Jos de Mey, Shigeo Fukuda, Sandro Del-Prete and, for sure, Maurits Escher.

A call to a phenomenon well known among psychologists should help to set the framework. Recall the "break-spotter" mechanism and the "assumption of continuity" referred to by Gombrich, which we mentioned in Section 5.3. A simple "impossible figure" created by the absence of these break-spotters is the "Devil's tuning fork" shown at the left in the following picture:

The three elliptical shapes at the left ending of the fork suggest the endings of three rods. This is actually the only possible reading of this left ending. The right ending of the fork, in contrast, shows a ⊔-shaped solid with square sections. Both readings are obvious in the left and right sides of the middle portion in the picture. As it happens, however, these readings are not compatible: no three-dimensional object projects onto the plane as the Devil's tuning fork. Such an incompatibility is manifest in the right portion of the picture.

Figure 11.10 **William Hogarth, *Satire on False Perspective*, 1754. (Private Collection/ Giraudon/The Bridgeman Art Library.)**

What makes this right portion patently impossible whereas the left portion is only so after careful inspection? It appears to be an issue of locality (as discussed in Section 1.1) combined with the features of our perceptual apparatus. Remember that the eye focuses on a small area, the fovea centralis. For the Devil's tuning fork, this means that the eye may focus on either end and it will have to displace the fovea towards the other to reconstruct the object. Our perceptual mechanism will guess a possible reconstruction (the only one possible in the case of an extremity of the tuning fork) and, in the absence of break-spotters, extrapolate this reconstruction along the way. When the displacement does not reach the other end of the fork this extrapolated reconstruction is kept as good. If, in contrast, it does reach this end then the initial guess will show its inconsistency. For a short enough displacement, as is the case of the right portion in the picture, this clash is unavoidable.

The ingredients in the explanation of the Devil's tuning fork play the same role in understanding the left-hand portion of the drawing below:

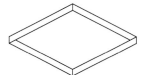
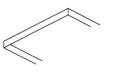
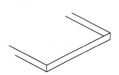

It appears at first sight as a thin, squarely shaped slab. Careful inspection, however, reveals that it is difficult to establish whether the visible side of the slab is facing up or down. The remainder of the drawing splits the slab into two parts. The one at the left – with the upper and left corners – is consistent with the reading of the visible side as facing down (and only with this reading). The one at the right, instead – with the remaining lower and right corners – makes this visible side face up. The illusion is possible because of the use of parallel projection. Sandro Del-Prete made a delightful use of it in his drawing *The Folded Chess Set* (Figure 11.11). Not only do the positions of the pieces highlight the two conflicting orientations of the chessboard, but the ladders also appear to twist to adjust to the reading around their extremities.

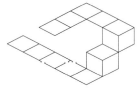

In 1937 Oscar Reutersvard created his first impossible staircase. The idea behind it is captured in the picture on the left.

Because of the characteristics of parallel projection, the end of the path in the drawing appears to be at the same level as its origin, in contradiction with the upward steps found when traversing the right part of the drawing. This coincidence of levels is again reinforced by locality – as well as by the fact that an extra tile would close the circuit – a fact evidenced by erasing the steps in the drawing.

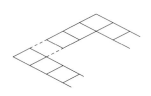

Figure 11.11 **Sandro Del-Prete,** *The Folded Chess Set*, **coloured pencil drawing, 1975. (© Sandro Del-Prete, www.sandrodelprete.com.)**

This idea was later taken by Lionel and Roger Penrose (1958) in their article about impossible objects and shortly after by Escher in his lithograph *Ascending and Descending* (Figure 11.12). A remarkable fact of this artpiece is its (three-point) perspective rendering, in contrast with the other drawings in this section which are produced using parallel projection. It goes without saying, there are small inaccuracies in the perspective rendering purposely made to give rise to the overall illusion. A crucial one is the fact that the staircase itself is drawn with parallel projection: one may check that its sides are parallel and the lines they determine do not pass through the vanishing points of the corresponding building axes.

The whole scene in *Ascending and Descending* is represented in threepoint perspective, but the stairs appear to be a mix of perspective elements with parallel projection (as is the woodblock in Figure 10.9) skilfully construed to produce the never-ending ascent.

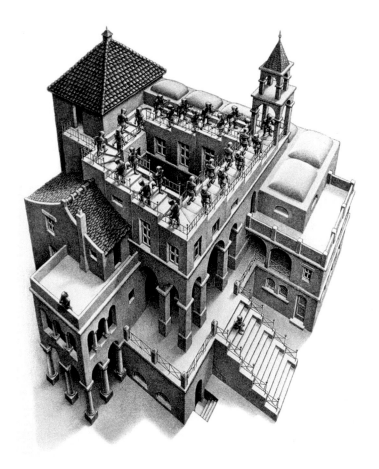

Figure 11.12 **M.C. Escher's** *Ascending and Descending*, **lithograph, 1960** © **2012 The M.C. Escher Company-Holland. All rights reserved. www.mcescher.com.**

11.5 Divergences

Prior to the dawn of linear perspective, image rendering was approached from various standpoints. Our exposition in Section 9.1 emphasized the search for realism evidenced in the works of artists such as Giotto, Duccio or Martini. This endeavour, however, was not universal: its pursuit was mostly western European. In the Far East, India, Persia and a number of other cultures, different forms of parallel projection gave satisfaction to the needs of their art forms. The creation of icons within eastern Orthodox Christianity provides a third approach with distinctive characteristics. A masterpiece within this tradition is the *Trinity* of Andrei Rublev (Figure 11.13), widely considered the highest exponent of Russian medieval painting (and whose life inspired the homonymal film by Andrei Tarkovsky).

Figure 11.13 **Andrei Rublev, *Trinity*, tempera, early fifteenth century. (Scala/Art Resource, NY.)**

The painting shows the three angels who visited Abraham at the oak of Mamre. They are sitting around a table on which a bowl stands, containing roasted lamb. The position of the angels' heads draws attention to the bowl, and probably to its connotations of Communion. But it is to the footstool where the angels' feet lie where our attention should lead us now. The immediate perception of it suggests a trapezoidal shape, but second thoughts point to no plausible reason for such a shape but, instead, to a purposefully weird depiction of a rectangle. These thoughts are ratified by the frequency of such depictions in Orthodox art. The two mosaics at Hagia Sophia (in present-day Istanbul) shown in Figure 11.14, for instance, show a footstool which, again, has been drawn with parallel lines diverging as they recede (at the left) and a similar remark applies to the Book held by Christ and the chair He is sitting on (at the right).

The presence of this phenomenon has been called *reverse* (or *inversed* or yet *divergent*) perspective, the term originating from a work by Oskar Wulff in the early twentieth century (Wulff, 1907). Explanations of this presence are even more varied than the names proposed for it; in a recent

Figure 11.14 **Mosaics at Hagia Sophia, eleventh century. (Vanni/Art Resource, NY (left), Alfredo Dagli Orti/The Art Archive at Art Resource, NY (right).)**

study, Clemena Antonova (2010) identifies six different, often conflicting, such explanations. An obstruction to the agreement on a single stand for reverse perspective is its lack of any physical interpretation (such as, for instance, the assumption of an infinitely distant viewpoint for parallel projections). It is, nonetheless, possible to make geometric sense of reverse perspective. To do so, consider an object – say a square with vertices A, B, C and D – and an observer. Linear perspective relies on the coinciding of two capacities upon the position O of the observer. First, the projection P' of a point P on the picture plane is determined by O (since it is the intersection of this plane with the line through P and O). Second, O determines as well the rules of *occlusion*; that is, the choice of the point to be represented on the picture plane in the case that many points in the object lie on the same projection ray. The standard rule – following the logic that opaque bodies block the view of whatever lies behind them – specifies that, of all the points of the object lying in a projection ray, the one to be represented on the picture plane is the first one met when traversing the ray from O toward the plane.

Reverse perspective separates these two capacities by setting, as it were, individual viewpoints for them, which are placed at opposed sides of the object. But since only one of them, say $O_{projection}$, determines the projection rays, the role of the other, call it $O_{occlusion}$, is minimal. Indeed, this role is limited to the setting of the picture plane π', which is located between $O_{occlusion}$ and the object (the exact location is not relevant) and perpendicular to the line joining $O_{projection}$ and $O_{occlusion}$. The following figure shows how reverse perspective projection works. Projection rays with origin at $O_{projection}$ are determined by the object's points, and the *last* such point, when the ray is traversed from $O_{projection}$ to π', is the one represented on π'.

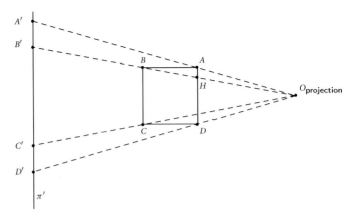

Thus, in the figure, the ray passing through B also hits the point H of the square. But the point represented on π' is B. In this way, we obtain a representation of the square which shows three of its sides ($\overline{AB}$, $\overline{BC}$ and $\overline{CD}$).

The reasoning above is for a two-dimensional situation, but the idea readily extends to three dimensions. The picture below shows divergent perspective drawings of a cube. They correspond to the three relative positions of cube, canvas and viewpoint (or, equivalently, of the cube and the points $O_{\text{projection}}$ and $O_{\text{occlusion}}$) giving place to the examples for one-point, two-point and three-point perspective in § 10.1.4:

A basic requirement in linear perspective is that the observer of a drawing thus made will look at it from the vantage point; that is, the point from which the artist looked at the scene that was drawn. Departure from this requirement allowed for the phenomena which made the subjects of Sections 11.1 and 11.2. For drawings made with reverse perspective, the existence of two points with different roles – or, equivalently, the separation of projection and occlusion – removes the framework on which this requirement stands. Viewpoints are normally chosen around $O_{\text{occlusion}}$.

John Willats mentions the avoidance of idolatry as a reason for the use of reverse perspective. In order "to prevent the worshiper from mistaking the images of God for God himself" medieval artists devised "a spatial system that was clearly different from the everyday corporeal world" (Willats, 1997: 239). It is doubtless due to more secular motives that many cubist paintings show objects drawn in reverse perspective. The intent, however, appears to be similar and could be described, paraphrasing Willats, as "to prevent the observer from mistaking the depicted scene by the scene itself".

Closer to our times, the media artist Héctor Rodríguez has experimented – in his animations *Fruztum: Nomadic Vision* and *Res Extensa* – with the disorientation produced in the observer by the use of divergent perspective and, in the second work, a continuous change of projection systems. An author's statement for these works, along with a concise introduction to divergent perspective, can be found in (Rodríguez, 2006).

11.6 Multiplicities

> Monsieur Braque [...] despises form, reduces everything, places and figures and houses, to geometrical schemas, to cubes. Let us no make fun of him, since he is honest. And let us wait.   Louis Vauxcelles (Fry, 1966: 50)

The decades around the turn of the twentieth century witnessed a sequence of major changes in painting. Led by Édouard Manet, a number of artists who came to be known as *impressionists* proposed a style which gave prominence to colour as opposed to line, dismissing details in the pursuit of capturing instantaneous impressions which, they argued, are closer to the way we actually see things. Other movements – such as the *expressionism* or the *fauvism* – further developed along these tenors. Because of the postulated agreement between their paintings and the casual perception of things, none of these styles raised doubts about perspective as the underlying drawing system. Such a call into question would have to wait for the works of, arguably, the most influential painter of the twentieth century.

In 1907 Pablo Picasso painted *Les Demoiselles d'Avignon* (*The Young Ladies of Avignon*) (Figure 11.15). The canvas shows a brothel scene in the Barcelona of Picasso's youth.[3] While its subject might have been shocking to the Parisian middle classes that were going to see its first public appearance, it was not revolutionary. Its style, instead, assuredly was. In the words of Edward Fry (1966: 13).

what makes *Les Demoiselles* a truly revolutionary work of art is that in it Picasso broke away from the two central characteristics of European painting since the Renaissance: the classical norm for the human figure, and the spacial illusionism of one-[view]point perspective.

---

[3] Brothels at that time were located in and around the *carrer d'Avignó* (street Avignon), in the old part of the city.

Figure 11.15 **Pablo Picasso, *Les Demoiselles d'Avignon*, oil on canvas, 1907.**
(© Succession Picasso/DACS, London 2012. Museum of Modern Art, New York,
USA/Giraudon/The Bridgeman Art Library.)

The breaking of the "norm for the human figure" – disproportionate sizes of body parts, angular shapes, use of African masks – is too obvious to elaborate on. The departure from perspective deserves more attention.

The general impression produced by the painting, concerning space, is that of flatness. No cues for depth are in sight. No indications of volume. A look at the details, in addition, reveals a perplexing rendering of two of the women's faces: a profile view of the nose is imposed upon the frontal view of the face. That is, the vantage point from which the nose is observed does not coincide with the vantage point from which the rest of the face is. Picasso creates an image of an object on which some parts of the object are seen from a vantage point and some others from a different one. He is thus giving up the most basic assumption of perspective (and, by the way, of parallel projection as well): the existence of a unique, well-defined, viewpoint.[4]

---

[4] This is the intended meaning in Fry's text above. He writes "the spacial illusionism of one-point perspective" meaning one *viewpoint*. I have added the word [view] in his text to avoid the possible confusion derived from the meaning of the expression "one-point perspective" we described in § 10.1.4.

Fry (1966: 14) does not hesitate to give precedence to this feature:

> The treatment of space is [...] by far the most significant aspect of *Les Demoiselles* [...]. The challenge facing Picasso was the creation of a new system of indicating three-dimensional relationships that would no longer be dependent on the convention of illusionistic, one [view]point perspective.

This significance was noted shortly after the creation of *Les Demoiselles* by George Braque, who "followed Picasso's lead in combining several points of view into a single image" (Fry, 1966: 17).

The works of Picasso and Braque between 1908 and 1912 form the core of what is known as *analytic cubism*. Multiple viewpoints bloomed in these paintings, often structured through the concomitance of several intersecting planes (each of them with its own vantage point). From 1912 till the end of the decade, joined by Juan Gris, their paintings evolved to have less shading, more colour and more objects in the painted scene. They also introduced the use of *papier collé* (pieces of paper glued on the canvas), textured the painting by using different kinds of brushes and radically decreased the creation of volumes by means of intersecting planes. Paintings of this period, defining a style referred to as *synthetic cubism*, were thus endowed with a heightened sense of flatness.

We are not foreign to the use of several viewpoints. We have actually met a form of them when we discussed multiviews in § 10.1.1. Yet, the component images of a multiview are not merged into a single one but presented as individual entities. And their coexistence has an intent of clarity: the five views of the locomotive in Figure 10.1 attempt to maximize the viewer's understanding of the locomotive's three-dimensional shape. In contrast, the component snapshots of a cubist painting are merged into a single, unified, image without the purpose of an accurate conveyance of a scene or object. The paintings done at the peak of analytic cubism are actually commonly regarded as difficult to read.

Multiple viewpoints would, however, live beyond cubism and occur in paintings markedly less cryptic than a cubist's. Many years after *Les Demoiselles* Picasso would produce a painting of a female figure on which the multiple viewpoints bring to the fore partial views of the observed body. The lying woman in Figure 11.16 is depicted to show a frontal view of her breasts as well as of her Mount of Venus, together with a partially frontal view of her buttocks. Her head is shown in profile in a way that adds yet one more vantage point.

11.7 Abandonment

In 1826 Nicéphore Niépce focused a camera obscura onto a sheet coated with a solution of bitumen dissolved in lavender oil. He allowed the sheet to be thus exposed for eight hours, after which time he washed it with

Figure 11.16 **Pablo Picasso, *Lying Woman*, oil on canvas, 1932. (© Succession Picasso / DACS, London 2012.)**

lavender oil. The bitumen that had been exposed to light darkened and that which had not was washed out. The result of this process, a print of about 20 cm × 25 cm, known as *View from the Window at Le Gras* and today housed at the University of Texas at Austin, is credited to be the first photographic print.

Niépce's technique, further developed by him, Louis Daguerre and William Henry Fox Talbot, among others, took little time to gain popular favour, and by the middle of the nineteenth century photography was an extended form of image creation. The cameras used at that time had a moderate viewing angle which made photographs' visual features correspond to those of (rectilinear) perspective drawings. The view of the Boulevard du Temple taken by Daguerre in 1838 (Figure 11.17) is an early example.

Painters were thus facing an awkward situation: the drawing system perfected along centuries to accurately portray the world was an integral part of a technique which produced images at very mild costs. André Bazin, one of the leading film critics of the twentieth century, was not hesitant in writing about this issue (Bazin, 1960: 9):

So, photography is clearly the most important event in the history of the plastic arts. Simultaneously a liberation and an accomplishment, it has freed Western

Figure 11.17 **Louis Daguerre, *Boulevard du Temple*, Paris, 1838.**

painting, once and for all, from its obsesssion with realism and allowed it to recover its aesthetic autonomy. Impressionist realism, offering science as an alibi, is at the opposite extreme from eye-deceiving trickery. Only when form ceases to have any imitative value can it be swallowed up in color. So, when form, in the person of Cézanne, once more regains possession of the canvas there is no longer any question of the illusions of the geometry of perspective. The painting, being confronted in the mechanically produced image with a competitor able to reach beyond baroque resemblance to the very identity of the model, was compelled into the category of object.

Bazin raises several points. First, that photography, this "competitor able to reach beyond baroque resemblance", liberated Western painting "from its obsession with realism"; that is, that it encouraged the abandonment of the pursuit of realism. Second, he stated the emergence of colour as a main component of painting after its liberation from the tyranny of form. Bazin does not mention it, but – and this cannot be chance – photography is for most of its history a black-and-white affair. Colour may have thus been a haven for painting. Finally, he claimed that painting "was compelled into the category of object". Indeed – and this may be related to the advent of colour in photography – with the progress of the twentieth century, a concern for the materials attached to the canvas – the time-honoured oils were being accompanied by sand, plaster, metals, strings and other elements of bricolage – along with an attitude of asserting the surface of the painting, took relevance in the work of a number of artists. The use of *papier collé* pioneered by Braque in 1912, which we mentioned in Section 11.6, is doubtless a point of reference in this process.

Realism survives as a style within contemporary painting, but it stopped being the dominant style long ago. Contemporary painting, by and large, has closed Alberti's window and is directing our attention to the shutter used to do so.

12   The vicissitudes of geometry

> We come now to the question: what is a priori certain or necessary, respectively in geometry (doctrine of space) or its foundations? Formerly we thought everything; nowadays we think nothing.
>
> A. Einstein (1926)

The period between the fifteenth and nineteenth centuries saw a mutual influence between painting and geometry, with the latter allowing for the dawn of perspective and the former for the dawn of projective geometry. An imperfect narration of this process occupied us in Chapter 9. We then proceeded in Chapter 11 to survey the many vicissitudes that accompanied perspective in its journey from prominence to abandonement.

The words of Einstein opening this chapter suggest that during this period geometry was not free of its own vicissitudes and that the prominence of the Euclidean order was gradually eroded as well. The goal of this chapter, obeying a sense of symmetry, is to describe these vicissitudes. Along the way, we will gain a better understanding of projective geometry and face some perplexities seldom associated with mathematics.

This chapters bears little on art. One may argue that the erosion of the prominence of Euclidean geometry began with the seed of projective geometry planted by Renaissance painters. But this is a weak justification for this chapter, which is otherwise denser than previous ones. Probably a stronger reason for its inclusion is the fact that it will be a basis for the additional material on symmetries in Chapter 13, as well as for the discussion on the shape of the universe which will serve us to return to our early speculations about the nature of space in Section 1.1.

12.1 Euclid revisited

The writing of Euclid's axioms we gave in Section 1.3 is the one occurring in Heath's translation of the *Elements* (Euclid, 1956). For a contemporary reader it is far from immediately understandable. First because, as we mentioned in that section, Euclid's axioms are not without ambiguity. But also because Euclid's formulation simultaneously dealt with purely geometrical questions (properties of plane figures) along with issues of a more practical consequence (whether these figures could be constructed with a ruler, a compass and a straight edge).

There has always been a general agreement, however, on the intended meaning of Euclid's axioms[1] and, consequently, there was no difficulty in rewriting contemporary versions of these axioms. As an example, we quote the version given by Marvin Greenberg (1993) in his book on geometry. Greenberg postulates as primitive notions those of *point, line, to lie on, to be between,* and *to be congruent.* Based on these notions, he states the five axioms as follows.

P1* For every point P and for every point Q not equal to P there exists a unique line ℓ that passes through P and Q.

Here, the expression "ℓ passes through P" is synonymous to "P lies on ℓ". Otherwise, ambiguities are avoided by explicitly stating that $P \neq Q$ and that ℓ is unique.

To state the second postulate, Greenberg defines the notions of segment and that of its endpoints. The *segment* with *endpoints* A and B is the set of points X such that X is between A and B.

P2* For every segment AB and for every segment CD there exists a unique point E such that B is between A and E and segment CD is congruent to segment BE.

We can draw this situation as follows:

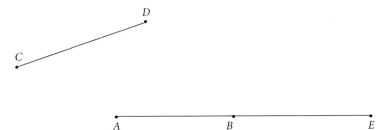

The intuitive meaning of P2* is that every segment can be extended on any direction by any amount. This amount is given by the segment CD to which the extension BE has to be congruent. Note that had we started with a primitive notion of *distance* we could have replaced the requirement "segment CD is congruent to segment BE" by "the distance between C and D equals the distance between B and E".

To state the third postulate, one needs to define the notion of circle with given centre and radius. This is easy: for two different points O and A, the *circle* with *centre* O and *radius* OA is the set of points P such that OP is congruent with OA.

P3* For every point O and every point A not equal to O there exists a circle with centre O and radius OA.

---

[1] We implicitly shared this agreement in Section 1.3.

From a modern viewpoint, P3* is not needed. The existence of the circle with centre O and radius OA does not need to be postulated: its existence is implicit in the definition of circle. In addition, we know that the circle is not empty since A belongs to it. The intended meaning of P3 in the *Elements* is that one can actually draw this circle (with a compass).

To state the fourth postulate we need to define *right angle*. This task requires, in turn, to previously define the notions of *ray* (or *half-line*) with a given *origin*, that of *opposite rays* – those having the same origin and opposite directions – and those of *angle* and *supplementary angles* – in modern words, those adding up to 180°. Once done, this, the fourth postulate, is concise.

P4* All right angles are congruent with each other.

The version of the fifth postulate in Greenberg's book is close to the statement P5$'$ we saw in Section 1.3.

P5* For every line ℓ and for every point P that does not lie on ℓ there exists a unique line m through P that is parallel to ℓ.

The difference between P5$'$ and P5* is that the former asserts the existence of at most one parallel to ℓ through P while the latter requires the existence of exactly one. To see why, or rather in which conditions, these two statements are equivalent we need to further look at the formal contexts introduced above and in Section 1.3.

In doing so, we note a few differences in their corresponding lists of primitive notions. First, Greenberg adds the notions of *to lie on* and *to be between*. Second, Euclid's notions of *distance*, *angle* and *equality* are replaced in Greenberg's exposition by *to be congruent*.

There is little to elaborate around the notion of *to lie on*. All geometric developments since Euclid have agreed on the obvious meaning of this notion. The notion of *being in between*, in contrast, appears to be more subtle and we will return to it, and to its relations with P2, P5$'$ and P5*, in Section 12.5. What we want to emphasize now is the fact that Euclid did not set betweenness as a primitive notion nor did he postulate any axiom enforcing the properties of such a notion. Yet, in some of the proofs in the *Elements*, Euclid inadvertently used betweenness properties by relying on drawings (on which these properties did hold). These proofs would therefore be, strictly speaking, incorrect.

In 1899 the German mathematician David Hilbert proposed a list of axioms replacing Euclid's and such that Euclidean geometry – that is, the contents of the *Elements* – could be formally derived from them. Featuring on this list, were a set of four betweenness axioms filling the gaps left by Euclid. In addition, Hilbert postulated three *axioms of incidence* (for instance, P1*), six *axioms of congruence*, two *continuity axioms*[2] and one *parallelism axiom* (namely, P5'). The original list contained 21 axioms and covered three-dimensional geometry as well. It can be found in Hilbert's original book (Hilbert, 1987). A shorter list, focusing on plane geometry only and providing some explanations, appears in (Greenberg, 1993: Chapter 3). There is no need for us to write down here the complete list of axioms.

Still, it will be useful to list some of them. For instance, the following axioms of betweenness (see (Greenberg, 1993: Chapter 3)):

B1 If B is between A and C then B is between C and A.
B2 For all distinct points B and D there exist points A, C and E lying on the line through B and D and such that B is between A and D, C is between B and D, and D is between B and E.
B3 If A, B and C are three distinct points lying on the same line, then one and only one of the points is between the other two.

We can also state axioms of congruence. Before doing so, let us summarize what we know about this notion. This is not a new notion for us; we stated in Definition 2.1 that two sets are congruent when there is an isometry mapping one into the other. Now, we know that isometries preserve distances (by definition) and angles (this is a theorem in Euclidean geometry). It turns out that if we want instead to have congruence as a primitive notion it is enough to impose a few axioms about congruence of segments (that is, equality of distances) and congruence of angles. Axioms P2* and P4* are examples of them, respectively. Another example is the third congruence axiom in (Greenberg, 1993: Chapter 3).

C3 If B is between A and C, B' is between A' and C', $\overline{AB}$ is conguent with $\overline{A'B'}$, and $\overline{BC}$ is congruent with $\overline{B'C'}$ then $\overline{AC}$ is congruent with $\overline{A'C'}$.

This axiom is quite obvious if we think in terms of distances, since, because B is between A and C, we have

$$\text{dist}(A, C) = \text{dist}(A, B) + \text{dist}(B, C)$$

and, similarly,

$$\text{dist}(A', C') = \text{dist}(A', B') + \text{dist}(B', C').$$

[2] The continuity axioms are meant to impose that lines have sufficiently many points. We visualize lines as continuous, but, failing to force this property through axioms, lines could have "holes" and statements such as the fact that if a point O has distance d to a line ℓ then, for all $R > d$, the circle with centre O and radius R cuts ℓ in two points (recall Proposition 1.4) could not be proved.

The equality $\text{dist}(A, C) = \text{dist}(A', C')$ now follows from the equalities $\text{dist}(A, B) = \text{dist}(A', B')$ and $\text{dist}(B, C) = \text{dist}(B', C')$. But this argument uses the notion of distance (which is not a primitive notion in Hilbert's system) and its properties (which we cannot assume but need to set axioms for them).

Other examples of axioms stated in terms of the notion of congruence are the following:

C2 If $\overline{AB}$ is congruent with $\overline{CD}$ and $\overline{AB}$ is congruent with $\overline{EF}$ then $\overline{CD}$ is congruent with $\overline{EF}$.

C5 If angle $\angle\alpha$ is congruent with angle $\angle\beta$ and $\angle\alpha$ is congruent with $\angle\gamma$ then $\angle\beta$ is congruent with $\angle\gamma$.

We close this section by returning to the issue of the equivalence between P5′ and P5⋆. Using Hilbert's axioms (but not the one for parallelism) one can show that given a point P outside a line ℓ there is at least one line passing through P and parallel to ℓ. That is, parallels always exist. Hence, to require that there is at most one is equivalent to requiring that there is exactly one. The proof of this fact crucially uses properties of betweenness and, although these properties had not been properly axiomatized before Hilbert, they had been widely accepted and used since the times of Euclid.

12.2 Hyperbolic geometry

The early nineteenth century was the scenario of an astounding mathematical development carried out independently by János Bolyai in Hungary, Carl Friedrich Gauss in Germany and Nikolai Ivanovich Lobachevsky in Russia. The idea behind their work can be described as follows.

In *neutral geometry* (the geometry derived from all of Hilbert's axioms except the one on parallelism, sometimes also called *absolute geometry*) we have that through a point P not on a line ℓ there is at least one line parallel to ℓ. Euclidean geometry requires – via P5, P5⋆, P5′ or any other equivalent statement – that this line is unique. What Bolyai, Gauss and Lobachevsky did was to consider the negation of this requirement and to derive consequences from it. A possible motivation to do so, going back to the work of Saccheri in the early eighteenth century, was the hope of reaching a contradiction, thus giving a proof (by contradiction) of P5 from the other axioms. Such a contradiction was never reached.

The negation of P5′ is the following axiom:

H For every line ℓ and for every point P that does not lie on ℓ there exist at least two different lines through P that are parallel to ℓ.

The consequences that can be derived from H (together with the axioms of neutral geometry) are known nowadays as *hyperbolic geometry*. To develop this geometry must have required a certain amount of courage, since its theorems looked weird: rectangles did not exist; triangles had angles that added up to less than 180°; Pythagoras' theorem did not hold, since the square of the length of the hypothenuse in a rectangle triangle was less than the addition of the squares of the other sides' lengths; .... But other equalities could be deduced instead, and these new theorems followed one another mapping out a space whose laws were markedly different from those of Euclidean geometry (but not less systematic) without ever producing a contradiction. Hyperbolic geometry appeared to be weird but consistent. Indeed, Gauss writes in 1824, in a letter to Franz Taurinus (we quote from (Greenberg, 1993: 180)):

The theorems of this geometry appear to be paradoxical and, to the uninitiated, absurd; but calm, steady reflection reveals that they contain nothing at all impossible.

This quote reveals a remarkable change in attitude. Theorems are not proved with the hope to reach a contradiction, but for the sake of developing a new geometry.

Interestingly, in spite of the work Gauss appears to have done in this "non-Euclidean geometry" (he writes that "I do not believe that anyone can have given more thought [to it] than I"), he is reluctant to make his discoveries known. He continues:

I do not fear that any man who has shown that he possesses a thoughtful mathematical mind will misunderstand what has been said above, but in any case consider it a private communication of which no public use or use leading in any way to publicity is to be made. Perhaps I shall myself, if I have at some future time more leisure than in my present circumstances, make public my investigations.

What can be the cause of this reluctance? M. Tent (2006: 215) quotes a dialogue, in 1839, between Gauss and his student Schumacher, from which we extract the following:

"But you never published anything on non-Euclidean geometry. Why not?" continued Shumacher.

"Well, as you know, I don't like to rock the boat. Euclid was still held in such ridiculous awe that I figured any attempt to suggest an alternative geometry would only cause a furor, so I kept quiet. I knew non-Euclidean geometry was true, and my construct was perfectly consistent with Euclid's first four postulates."

The question remains, why postulating the possibility of another geometry would "rock the boat". Euclid was certainly respected among mathematicians, but, as we have remarked elsewhere, the nature of his parallels' postulate had been a matter of concern for centuries. Some light on Gauss' reluctance is given at a different point in the letter to Taurinus, where Gauss mentions that "we know, despite the say-nothing word-wisdom of the metaphysicians, too little, or too nearly nothing at all, about the true nature of space [...]". If we accept that the allusion to metaphysicians in this passage is a reference to Immanuel Kant, we can begin to understand Gauss' fears of "rocking the boat".

In 1781 (subsequently re-edited in 1787) Kant had published his *Critique of Pure Reason* (Kant, 1997). The subject and development of this monumental work lie outside the scope of our book. Yet, two of its contentions are relevant to our discussion.

A general thesis of the *Critique* is the existence of judgements which are *synthetic* and *a priori*. The latter means that the judgement's justification or truth does not depend on experience (judgements which do depend are called *a posteriori*). The former, that whatever the judgement predicates about its subject is not contained in the subject (as it occurs in *analytic* judgments). Before Kant's time it was accepted that only analytic truths could be a priori. The examples given by Kant for synthetic, a priori, judgments are those in mathematics and, among them, those in geometry.

Related to this first contention there is another one that postulates space as an *a priori intuition*. Summarizing (again) to the extreme, such a claim deprives space of the status of a concept and turns it instead into an organizational frame, or template, which makes experiences possible. The sensible material received by the subject is made into an experience through an organization "out there" that is, through spatial organization. By necessity, this template is a priori; it cannot be derived from experience since it is one of the very conditions that make experience possible. Summing up, both the object (space) and the body of knowledge about it (geometry) are a priori.

For Gauss, to make public his work on non-Euclidean geometry was unavoidably to take issue with Kant's views, as the claim that two different geometries are possible delegates to experience the choice of the "real" one, the one giving the faithful description of space. It is probably this conflict that Gauss wanted to avoid.[3] Hyperbolic geometry eventually became known owing to publications by Lobachevsky in 1829 and Bolyai in 1831.

---

[3] This opinion is present in various biographies of Gauss. For instance, W.K. Bühler (1981:101) writes "Several reasons prompted Gauss to hold back his convictions and not enter into a public discussion. His most important motive may have been that he did not want to get involved in what, in his eyes, was a completely irrelevant philosophical discussion of a question which he must have considered essentially undecidable." Bühler also notes that, nonetheless, Gauss held a "generally high regard for Kant's philosophy".

12.3 Laws of reasoning

> "I know what you are thinking about," said Tweedledum: "but it isn't so, nohow."
>
> "Contrariwise," continued Tweedledee, "if it was so, it might be; and if it were so, it would be; but as it isn't, it ain't. That's logic."
>
> L. Carroll (1993: 193)

Gauss' assertion that hyperbolic geometry contains "nothing at all impossible" did not entail a dismissal of Euclidean geometry. Rather, he was implying that both geometries were possible in the measure that both were consistent. This idea can be best understood within the framework of mathematical logic and its notions of deduction and truth.

In consonance with common use of the word, dictionaries define *truth* (of a statement) as "Conformity with act, agreement with reality [. . .]". Underlying this definition there are a number of assumptions. First, a "reality" against which this conformity or agreement is to be sought. Second, an interpretation that unambiguously associates features of this reality to the words making the sentence and that, consequently, unambiguously translates the sentence into a property (which can or cannot hold) of this reality.

A major topic in philosophy is the nature of this reality which grants or denies truth. Through history, this topic has been given a variety of different answers. The philosophy of mathematics is not alien to this discussion: most of these answers have a version tailored to describe the ontological status of mathematical entities. We shall not enter this discussion (readers interested in the subject will find satisfaction in (Anglin, 1997) and (Bostock, 2009)) and return, instead, to the issues of deduction and truth and to the formal context within which we will consider them.

12.3.1 Formal languages

A feature of modern mathematical statements is that they can be written using a few verbs (called *predicates*), nouns (called *constants*), genitive modifiers (called *functions*), prepositions (*connectives*) and with the help of specific grammatical constructors (*variables* and *quantifiers*).

The connectives and quantifiers depend only on the logic considered.[4] Typically, one uses two quantifiers, ∀ (reading "for all") and ∃ (reading

---

[4] In what follows, we will restrict our exposition to *first-order logic*. That is, we assume that variables denote elements in some set. In *second-order logic* one allows variables to denote families of such elements or functions taking them as arguments. Second-order logic, therefore, has more expressive power than first-order logic, a fact which presents advantages and drawbacks.

"there exists"), and (although the list is redundant in the sense that less are sufficient) the following five connectives: ¬ (reading "it is not the case that"), & (reading "and"), or, → (reading "implies" or "if [...] then") and ↔ (reading "if and only if"). In contrast, predicates, functions and constants are specific to the subject the statements are stating something about[5] and are denoted with symbols appropriate to their intended meaning.

Grammatical rules apply to determine which expressions are well formed and single out among them those on which every variable is quantified. The latter are called *sentences* and possess the property that they may be assigned a truth value. For instance, the sentence

$$\exists m \forall x \ (m \leq x)$$

states the existence of a minimal element (with respect to the order ≤). This may be true or false, depending on the ordered set at hand. In contrast, the expression

$$\forall x \ (y \leq x)$$

cannot be assigned a truth value since the role of y is unclear. It may be quantified in different ways (such as $\exists y \forall x \ (y \leq x)$ or $\forall x \exists y \ (y \leq x)$) giving place to different intended meanings.

Once a family ς (called *signature*) of symbols for predicates, functions and constants has been fixed, one can associate with it the set $\mathscr{L}(\varsigma)$ of all the sentences that can be built from this family. This set contains everything that can be stated in terms of the predicates, functions and constants of ς (within the limits imposed by the grammar).

Example 12.1 To reason about groups (recall Section 3.2) we may take as ς a set with two symbols: a constant e for the neutral element and a function $*$ with two arguments for the group operation (in general we write $x * y$ instead of $*(x, y)$, a change in notation whose only consequence is an increase in clarity).

Example 12.2 To reason about geometry (and by this I mean about the objects and notions we have been discussing thus far) we may take $\varsigma = \{B, C, C_\angle\}$. All these symbols are predicates: B has three arguments, C has four arguments and $C_\angle$ has six. We will read $B(a, b, c)$ as "b is beween a and c", $C(x, y, a, b)$ as "the segment $\overline{xy}$ is congruent with the segment $\overline{ab}$", and $C(x, y, z, a, b, c)$ as "the angle $\widehat{xyz}$ is congruent with the angle $\widehat{abc}$". We will see later in this section that these predicates are enough to express the axioms in the previous sections.

---

[5] An exception is the predicate symbol =, which we may consider part of the logic.

12.3.2 Deduction

The paragraphs above deal with the expressive capacity of logic. We may now turn to its deductive facet. An essential part of any logic is a set of rules allowing one to deduce a sentence from one or more other sentences, by applying these rules a finite number of times. A typical example of such a rule is *modus ponens*, which states that, for sentences φ and ψ,

$$\left.\begin{array}{c}\varphi \to \psi \\ \varphi\end{array}\right\} \vdash \psi.$$

Here, $\vdash$ denotes that the sentence at the right follows (or is deduced) from the sentences at the left.

Some sentences, called *tautologies*, are derived from nothing. For instance the *principle of the excluded third* states that for all sentences ψ we have

$$\vdash \psi \text{ or } \neg\psi$$

and for all sentences ψ and φ we have

$$\vdash (\psi \text{ or } \varphi) \iff \neg(\neg\psi \And \neg\varphi).$$

Tautologies can be regarded as axioms, not as axioms for a particular sphere of knowledge (geometry, arithmetic, etc.) but as the axioms of inference itself.[6] If T is a tautology then ¬T is a *contradiction*. An example of contradiction is $\psi \And \neg\psi$, for any sentence ψ.

We are now in a situation familiar to us through our handling of Euclidean geometry. We have at hand a family of primitive notions – which we formalize with a corresponding family of symbols for predicates, functions and constants – and a grammar allowing us to construct sentences stating attributes for these notions. We specify some of these attributes as inherent to the notions by requiring them to hold – formally, we identify a number of sentences doing so as axioms. Finally, we use a set of previously agreed rules of inference to deduce new statements (theorems) from the ones already inferred. The framework of the *Elements* was not as rigid as the one just described, but the spirit of its exposition was unarguably this.

In our formal setting, we fix a set $\mathcal{A}$ of axioms (that is, a subset of sentences in $\mathcal{L}(\varsigma)$). We can then consider the set $T(\mathcal{A})$ of all sentences which can be deduced from $\mathcal{A}$. Note, we are not meaning the sentences that we can actually deduce from $\mathcal{A}$, but those for which a deduction exists, even if we do not know any such deduction. We call $T(\mathcal{A})$ the *theory* of $\mathcal{A}$.

---

[6] They are not themselves without contention. For instance, the excluded third is not accepted as a legitimate rule of inference by some schools of thought.

Example 12.1 (continued) We can easily write down the axioms of a group (which we saw in Definition 3.6):

> **G1** $\quad \forall x \forall y \forall z \left(x * (y * z) = (x * y) * z \right)$
>
> **G2** $\quad \forall x \left(x * e = e * x = x \right)$
>
> **G3** $\quad \forall x \exists y \left(x * y = y * x = e \right).$

These axioms give a meaning to the symbols $*$ (the operation of the group) and e (its neutral element).

The set of sentences that can be inferred from G1, G2 and G3 is the *theory of groups*.

Example 12.2 (continued) We can also write down a set of axioms for (two-dimensional) Euclidean geometry, but it is less straightforward. The main obstruction lies with the primitive notions of point and line in Hilbert's system which appear to be at different levels: lines are (infinite) sets of points. Hence, if U is going to be the plane and the elements in U the points on this plane, it is unclear how to express the idea

> For all point P and all line ℓ not through P . . .

because we cannot quantify lines.[7] An elegant solution to this obstruction was proposed by Tarski (1959), who relied on the fact that for "various classes of geometrical figures, such as the straight lines, the circles, the segments, the triangles, the quadrangles, and, more generally, the polygons with a fixed number of vertices [. . .] every geometrical figure is determined by a fixed finite number of points". Furthermore, this "being determined" can be expressed using betweenness (i.e. B) and congruence of segments (C). More precisely, lines are given by two points (just as circles; see Section 12.1) and we express that a point X lies on the line given by A and B, $A \neq B$, with the formula

> $\mathsf{B}(X, A, B)$ or $\mathsf{B}(A, X, B)$ or $\mathsf{B}(A, B, X).$

We concisely write the conjunction above as $X \in \mathsf{L}(A, B)$. We can state that two lines (given by A, B and P, Q) are different with the formula

$$\exists X \left(X \in \mathsf{L}(A, B) \ \& \ X \notin \mathsf{L}(P, Q) \right)$$

which we write $\mathsf{L}(A, B) \neq \mathsf{L}(P, Q)$.

In this setting, we do not need to assert the existence of a line passing through two different points A and B. Axiom P1 now reduces to the uniqueness assertion: that is, if $P \neq Q$, $A \neq B$ and $A, B \in \mathsf{L}(P, Q)$, then $\mathsf{L}(A, B) = \mathsf{L}(P, Q)$.

---

[7] We mention in passing that one way this can be solved is by considering second-order logic (see Footnote 4 in this chapter).

We can also express that two lines, given by points A, B and C, D, respectively, are parallel – and we write this by $\mathsf{par}(A, B, C, D)$ – as follows:

$$\forall X \big[X \in \mathsf{L}(A, B) \rightarrow X \notin \mathsf{L}(C, D)\big].$$

Axiom P5′ can now be straightforwardly written down:

$$\forall P, A, B, C, D \big[(P \notin \mathsf{L}(A, B) \,\&\, P \notin \mathsf{L}(C, D)$$
$$\rightarrow \big(\neg\mathsf{par}(A, B, P, C) \text{ or } \neg\mathsf{par}(A, B, P, D)\big)\big].$$

Tarski's axioms, as presented in (Tarski, 1959) are different. One of Tarski's goals was to provide a short list of succinct first-order axioms. His list contained 12 axioms which can be written without recourse to abbreviations (such as $\mathsf{par}(A, B, C, D)$ or $X \in \mathsf{L}(A, B)$) together with an axiom scheme.[8] Also, Tarski did not need to use the predicate $\mathsf{C}_{\angle}$ for congruence of angles. This is not surprising. The fact that given two triples of different points (a, b, c) and (x, y, z) the angle $\widehat{abc}$ is congruent with the angle $\widehat{xyz}$ can be expressed by the following sentence:

there exist a' in the ray $\overrightarrow{ba}$ and c' in the ray $\overrightarrow{bc}$ and such that $\overline{a'b}$ is congruent with $\overline{xy}$, $\overline{bc'}$ is congruent with $\overline{yz}$, and $\overline{a'c'}$ is congruent with $\overline{xz}$.

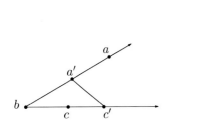
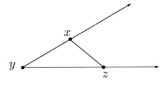

This sentence can be written using only B and C.

We will denote by NG the axioms for neutral geometry. That is, all Hilbert's axioms except P5*. We denote by EG the axioms of NG plus P5* and by HG the axioms of NG plus H.

The rules of inference are such that if $T(\mathscr{A})$ contains a contradiction then $T(\mathscr{A}) = \mathscr{L}(\varsigma)$. That is, if for some sentence ψ we have

$$\mathscr{A} \vdash \psi \,\&\, \neg\psi$$

then, for all sentences φ we have $\mathscr{A} \vdash \varphi$. If an absurdity holds, then anything holds. We say that a theory is *consistent* when it contains no contradiction. Consistency is the minimal requirement for a theory if

---

[8] An axiom scheme is an infinite family of axioms having the same structure. To axiomatize continuity within first order there is the need of such a family.

provability is going to carry any meaning and, in particular, to have any relationship with validity, which is a notion we next shift focus to.

12.3.3 Validity

The notion of validity is the formal (that is, circumscribed to a logic system) version of the "conformity with fact, agreement with reality" mentioned at the beginning of Section 12.3. It therefore relates to truth.

A *structure* $\mathfrak{A}$ for ς (or ς-*structure*) is a set U (the *universe* of the structure) together with *interpretations* for the predicate, function and constant symbols in ς. An interpretation for a constant symbol c is an element $c_{\mathfrak{A}}$ in U. An interpretation for a function, or a predicate, is a function from, or respectively a subset of, a suitable product of copies of U (note, however, the equality $=$ is always interpreted in the obvious way). Here, suitability refers to compatibility with the grammatical use of the symbol; that is, the number of arguments of, say, the function $f_{\mathfrak{A}}$ coincides with the number of variables associated to the symbol f. We next give two examples.

Example 12.1 (continued) A structure for $\varsigma = \{e, *\}$ is any set U together with an element $e_{\mathfrak{A}}$ in U and a function $*_{\mathfrak{A}} : U^2 \to U$.

Example 12.2 (continued) A structure over $\varsigma = \{\mathsf{B}, \mathsf{C}\}$ is any set U together with subsets $\mathsf{B}_{\mathfrak{A}}$ of U^3 and $\mathsf{C}_{\mathfrak{A}}$ of U^4.

Given a ς-structure $\mathfrak{A}$ and a sentence ψ in $\mathscr{L}(\varsigma)$ the interpretation of the symbols in ς – together with natural interpretations for the connectives and constructors in the logic – translates into a statement made about U which may or may not hold true. In the former case we say that $\mathfrak{A}$ *satisfies* ψ (or that ψ *is valid on* $\mathfrak{A}$) and we write

$$\mathfrak{A} \models \psi$$

and in the latter we write

$$\mathfrak{A} \not\models \psi$$

(or $\mathfrak{A} \models \neg\psi$, since the failure of ψ to hold true in $\mathfrak{A}$ implies that this must happen for $\neg\psi$).

If $\mathscr{A}$ is a set of sentences in $\mathscr{L}(\varsigma)$ and $\mathfrak{A}$ is a ς-structure such that $\mathfrak{A} \models \psi$ for all ψ in $\mathscr{A}$, we say that $\mathfrak{A}$ is a *model* for $\mathscr{A}$ and we write

$$\mathfrak{A} \models \mathscr{A}.$$

We also say that $\mathscr{A}$ is *satisfiable* or *semantically consistent*.

The latter choice of words is not fortuitous. One of the most basic properties required from a logic is its *soundness*; that is, the fact that every satisfiable theory is consistent. Every self-respected logic possesses this property. The converse of soundness – that is, the fact that every consistent theory has a model – called *completeness*, is a consequence of a fundamental theorem proved by Kurt Gödel in 1930 (Gödel, 1967)[9]. It follows that we can use the word "consistency" without qualifying whether syntactical or semantical. Let us emphasize this with a formal statement (which we will not attempt to prove).

Theorem 12.3 (Gödel's completeness theorem) *Let ς be a signature and $\mathscr{A}$ a set of first-order sentences over ς. Then $\mathscr{A}$ is consistent if and only if $\mathscr{A}$ has a model.*

It is easy to see that the theory of groups (described with axioms G1, G2 and G3 above) is satisfiable. Take $U = \{0, 1\}$, $e_{\mathfrak{A}} = 0$ and $*_{\mathfrak{A}}$ given by

$$0 *_{\mathfrak{A}} 0 = 0$$

$$0 *_{\mathfrak{A}} 1 = 1$$

$$1 *_{\mathfrak{A}} 0 = 1$$

$$1 *_{\mathfrak{A}} 1 = 0.$$

This is what we described in Section 3.2 as "the simplest example" of a group (and found as symmetry group both for the groundplan in Figure 3.2 and the yin–yang symbol, with Id instead of 0 and an appropriate isometry φ instead of 1). The ontological status of this model for group theory is, for almost all philosophers, uncontroversial, owing to the finiteness of its universe.

The same cannot be said for neutral geometry (and, hence, even less so for EG or HG). The set of axioms implies the existence of infinitely many points and, therefore, every structure postulated as a model for NG will have to have an infinite universe. The ontological status of these possible models for geometry is, therefore, less immediate than that of the two-element group above and may become a bone of contention among philosophers.

A common way to exhibit models for a theory T is to assume a model $\mathfrak{A}$ for another theory T' and to construct a model $\mathfrak{B}$ for T within $\mathfrak{A}$. If T' is different from T then such a procedure establishes *relative satisfiability*: it ensures that T is satisfiable under the assumption that T' is. If $\mathfrak{B}$ is to be infinite it is common to rely on an infinite $\mathfrak{A}$ as well. But such a $\mathfrak{A}$ is chosen in a way that its existence is less controversial, at first sight, than that of $\mathfrak{B}$.

---

[9] We need to mention that we do not have completeness in every logical system. For instance, the second-order logic mentioned in Footnote 4 in this chapter is known not to be complete. Its expressive power appears to be too strong with regard to its deductive capabilities and this makes impossible the deduction of a contradiction for some unsatisfiable sets of axioms.

A major instance of this procedure took place in the nineteenth century with the proof that HG is satisfiable assuming EG is so. This theme will occupy us in Section 12.4. In the meanwhile, to illustrate the ideas, we consider two examples for $T = \text{EG}$.

12.3.4 Two models for Euclidean geometry

We have already built a model for Euclidean geometry. Our first example is a reminder of its construction, which, basically, recreated Euclidean geometry from the set of real numbers with its arithmetic operations.

Example 12.4 (The Cartesian model) All through this book we have relied upon the existence of the set $\mathbb{R}$ of real numbers, which we assumed endowed with operations of addition $+$ and multiplication $\cdot$, as well as with an order structure $\leq$. We did never spell out the properties of the tuple $(\mathbb{R}, +, \cdot, \leq)$; instead, we simply used them. Examples of these properties are the existence of neutral elements 0 and 1 for the addition and multiplication respectively, as well as arithmetic properties such as distributivity

$$\forall x, y, z \ \big(x \cdot (y + z) = x \cdot y + x \cdot z\big)$$

or order properties such as the compatibility with the operations

$$\forall x, y, z \ \big(y \geq z \rightarrow x + y \geq x + z\big)$$
$$\forall x, y, z \ \big((x \geq 0 \ \& \ y \geq z) \rightarrow x \cdot y \geq x \cdot z\big).$$

The set of all the first-order sentences (over the signature $\varsigma = (0, 1, +, \cdot, \leq)$) valid on $\mathbb{R}$ is denoted by RCF and known as the theory of real closed fields. Note that, unlike EG, which is given to us as a set of axioms, we are now starting from the assumption of the existence of $\mathbb{R}$ and all its properties. This assumption immediately implies the consistency of RCF and, some would argue, cannot be taken for granted. In spite of its interest, we will not enter this discussion. We will rather proceed now to show the consistency of EG relative to RCF.

To do so, our first task is to provide a universe, along with interpretations for the primitive notions which, we recall, are those of *betweenness* and *congruence*. Once this is done, we need to show that the axioms of EG hold under these interpretations.

The general idea is to use Descartes' construction of Euclidean geometry via coordinates and equations (which we saw in Section 1.4). Accordingly, we take $\mathbb{R}^2$, the set of pairs of real numbers, as universe U of our model-to-be $\mathfrak{E}$.

We now need to provide interpretations for B and C. That is, beginning with betweenness, we need to establish when $B_{\mathfrak{E}}((x_1, x_2), (y_1, y_2), (z_1, z_2))$ holds true for three points (x_1, x_2), (y_1, y_2) and (z_1, z_2) in U. Intuitively, this is so when the three points are aligned and the second lies on the segment determined by the other two. We can express this (over the signature ς) as follows:

$$\exists a, b, c\ (ab \neq 0\ \&\ ax_1 + bx_2 + c = 0\ \&\ ay_1 + by_2 + c = 0\ \&$$
$$az_1 + bz_2 + c = 0\ \&\ (x_1 \leq y_1 \leq z_1\ \&\ x_2 \leq y_2 \leq z_2)\ \text{or}$$
$$(z_1 \leq y_1 \leq x_1\ \&\ z_2 \leq y_2 \leq x_2)).$$

Therefore, we define $B_{\mathfrak{E}}((x_1, x_2), (y_1, y_2), (z_1, z_2))$ to hold when the sentence above holds true on $\mathbb{R}$.

To interpret congruence we recall (see (1.1)) that for points $P = (P_1, P_2)$ and $Q = (Q_1, Q_2)$ in U we defined the distance between them by taking $\text{dist}(P, Q) = \sqrt{(P_1 - Q_1)^2 + (P_2 - Q_2)^2}$. It is therefore only natural to evaluate $C_{\mathfrak{E}}(P, Q, R, S)$ as true, for points $P, Q, R, S \in U$, if and only if

$$\text{dist}(P, Q) = \text{dist}(R, S).$$

We have thus defined a ς-structure $\mathfrak{E}$. To see that it is a model for Euclidean geometry one needs to verify that the axioms of EG are valid in $\mathfrak{E}$.

Axiom B1, for instance, says that, for any three points A, B and C, if B is between A and C then B is between C and A. To verify that B1 is valid on $\mathfrak{E}$ consider any three such points. Since we are assuming that $B_{\mathfrak{E}}(A, B, C)$ holds, we have that

$$\exists a, b, c\ (ab \neq 0\ \&\ aA_1 + bA_2 + c = 0\ \&\ aB_1 + bB_2 + c = 0\ \&$$
$$aC_1 + bC_2 + c = 0\ \&\ (A_1 \leq B_1 \leq C_1\ \&\ A_2 \leq B_2 \leq C_2)\ \text{or}$$
$$(C_1 \leq B_1 \leq A_1\ \&\ C_2 \leq B_2 \leq A_2)).$$

But this sentence remains the same if we swap the pairs (A_1, A_2) with (C_1, C_2), and this means that $B_{\mathfrak{E}}(C, B, A)$ holds as well. Hence, B1 is valid on $\mathfrak{E}$.

The validity of the remaining axioms is shown in a similar manner: the property we want to prove to hold true on $\mathfrak{E}$ translates into a statement involving real numbers and one shows that this statement is a theorem in RCF. It is beside the point to give the details here.

Example 12.5 (The hemispherical model) The model $\mathfrak{E}$ in the previous example rests upon the use of coordinates pioneered by Descartes. It exemplifies the notion of model – by interpreting geometric notions like congruence and betweenness in non-geometric terms such as equations and inequalities – and, at the same time, relies on our now being familiar with this interpretation. We next describe another model, say $\mathfrak{G}$, for EG which, at first sight, might not seem up to the job. It assumes three-dimensional Euclidean geometry, a model for which can be constructed from $\mathbb{R}$, as in Example 12.4, but now taking triples of real numbers to represent points.

As universe U, we take the southern hemisphere of a sphere, without the equator $\mathscr{E}$. We may describe U using coordinates. To do so, we choose the coordinate axes so that the sphere has radius 1, its centre C has coordinates $(0,0,0)$ and its south pole S has coordinates $(0,0,1)$. Points P in U have coordinates (P_1, P_2, P_3) satisfying $P_1^2 + P_2^2 + P_3^2 = 1$ and $P_3 > 0$ (points with $P_3 = 0$ are in $\mathscr{E}$ and, therefore, not in U). A line in this model is a half-great-circle (without its endpoints). That is, it is the intersection of U with a plane passing through the centre C of the hemisphere:

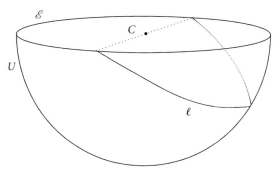

A line is therefore given by a system of equations (and an inequality)

$$aP_1 + bP_2 + cP_3 = 0, \qquad P_1^2 + P_2^2 + P_3^2 = 1, \qquad P_3 > 0$$

for some $a, b, c \in \mathbb{R}$. Here, "$x > y$" means "$x > y$ & $x \neq y$". Once alignment is clear, the interpretation for betweenness is the obvious one (we leave to the reader, should there be interest in pursuing the issue, the task of describing $\mathsf{B}_{\mathfrak{G}}(P, Q, R)$ with a formula over $\varsigma = \{0, 1, +, \cdot, \leq\}$).

It only remains to give an interpretation for congruence. To do so, it is enough to define a notion of distance $d_{\mathfrak{G}}(P, Q)$ between two points $P, Q \in U$. Once with this notion at hand, we can define congruence between segments as equality of distances between endpoints. To define $d_{\mathfrak{G}}(P, Q)$ we may use coordinates. For any two points

$P, Q \in U$, with coordinates (P_1, P_2, P_3) and (Q_1, Q_2, Q_3) respectively, we define

$$d_{\mathfrak{S}}(P, Q) := \sqrt{\left(\frac{P_1}{P_3} - \frac{Q_1}{Q_3}\right)^2 + \left(\frac{P_2}{P_3} - \frac{Q_2}{Q_3}\right)^2}.$$

At first inspection, this definition may seem arbitrary. It is not, and to explain its rationale will certainly pay off. Consider the plane π parallel to the equator and passing through the south pole. With our choice of coordinates this plane has equation $X_3 = 1$. Take the line p passing through C and P and its intersection P' with π:

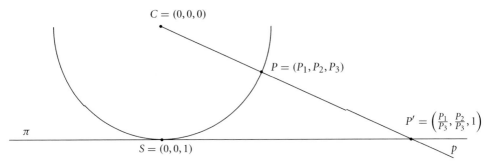

$C = (0, 0, 0)$

$P = (P_1, P_2, P_3)$

$P' = \left(\frac{P_1}{P_3}, \frac{P_2}{P_3}, 1\right)$

π

$S = (0, 0, 1)$

p

Since P' is obtained from P by a homothecy with centre C and coefficient $1/P_3$, it follows that the coordinates of P' are $\left(\frac{P_1}{P_3}, \frac{P_2}{P_3}, 1\right)$. We can do the same construction for Q and we see that what we defined as $d_{\mathfrak{S}}(P, Q)$ is the Euclidean distance $\text{dist}(P', Q')$ between the projections P' and Q' of P and Q on π.

This projection (from U to π with projection point C) transforms lines (i.e. half-great-circles) in U into (standard) lines in π. And, by construction, preserves distances in the sense that $d_{\mathfrak{S}}(P, Q) = \text{dist}(P', Q')$. Since this projection also preserve betweenness ($\mathsf{B}_{\mathfrak{S}}$ in U is transformed into standard betweenness in π) it follows that any sentence over the signature $\{\mathsf{B}, \mathsf{C}\}$ holds true over $\mathfrak{S}$ if and only if it holds true over π. Therefore, the validity of the axioms of EG on π implies the validity of these same axioms on $\mathfrak{S}$.

12.3.5 Proof and truth

The deductive machinery of a logic (its set of rules of inference) defines a notion of proof. The use of structures and interpretations does so for the notion of truth.[10] The question is posed: Which is the relationship between proof and truth? The following definition is a stepping stone toward an answer.

[10] There is relativity in both, since the former is with respect to a given set of inference rules and the latter with respect to a given structure.

Definition 12.6 Let $\mathscr{A}$ be a set of sentences and φ a sentence, all of them over a signature ς. We say that φ is a *semantic consequence* of $\mathscr{A}$, and we write

$$\mathscr{A} \models \varphi$$

when for all ς-structure $\mathfrak{A}$ such that $\mathfrak{A} \models \mathscr{A}$ we have $\mathfrak{A} \models \varphi$. In the special case that $\mathscr{A}$ is the empty set, we write $\models \varphi$ and we say that φ is *universally valid*.

The substance of this definition appears to be a semantic counterpart of the notion of consequence behind $\vdash$. The latter understands consequence as derivation, or proof, and the former as preservation of truth. A fundamental corollary of Gödel's completeness theorem (Theorem 12.3) is the fact that both forms of consequence coincide.

Theorem 12.7 *For all sets of sentences $\mathscr{A}$ and all sentences φ (all of them first-order, over the same signature ς) we have*

$$\mathscr{A} \models \varphi \quad \text{if and only if} \quad \mathscr{A} \vdash \varphi.$$

In particular, taking $\mathscr{A}$ to be the empty set, we deduce that a sentence is universally valid if and only if it is a tautology.

12.4 The Poincaré model of hyperbolic geometry

Let us return to a statement we made at the end of § 12.3.3, namely that a major result in the nineteenth century was to show that HG is satisfiable assuming that EG is. We are now in a position to explain how this was done by Henri Poincaré (who was not the first one; Eugenio Beltrami preceded him in this task with a more involved construction, as did Felix Klein) around 1880. We will not provide all the details, but sufficiently many to convey the general idea.

In what follows, we assume that two-dimensional Euclidean geometry is consistent and we consider a model, i.e. a plane, π for it. On this plane, let $\mathscr{C}$ be a circumference, say of radius 1, and U be its interior. We take U as universe of our intended model $\mathfrak{H}$. Points in *Poincaré's model* for HG are therefore the points of U. Having dealt with Examples 12.4 and 12.5 we know that a good starting ground towards providing interpretations for B and C is to have notions for line and distance at hand. We proceed accordingly.

A *hyperbolic line* (or a *line* in Poincaré's model) is either an arc of circumference obtained by intersecting with U a circumference C which cuts $\mathscr{C}$ at right angles[11] or a diameter of $\mathscr{C}$ without its endpoints (which is the limit of the previous case when the radius of C tends to infinity).

---

[11] The angle between two circumferences at an intersection point is the angle made by the tangents of the circumferences at this point (the dotted lines in the drawing).

The following picture gives examples (ℓ_1 and ℓ_2, respectively, in blue) for both:

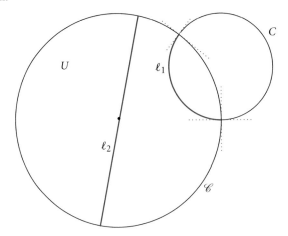

Points in $\mathscr{C}$ are called *ideal points*. We can, therefore, refer to the two extremities of a hyperbolic line as its ideal points. A first remark concerning these definitions is that two different points in U determine a unique hyperbolic line. This translates to Euclidean terms as stating that given two different points inside a circumference $\mathscr{C}$ either there is a circumference C passing through them and cutting $\mathscr{C}$ at right angles or there is a diameter of $\mathscr{C}$ passing through these points and in both cases the claimed object is unique. This statement is a theorem one can (but here we will not) prove in Euclidean geometry. Hence, P1* is valid on $\mathfrak{H}$. Actually, a stronger property holds true, since for any pair of different points P and Q, either in U or in $\mathscr{C}$, there exists a unique hyperbolic line through P and Q. This is, again, a theorem in Euclidean geometry.

It follows from this property that given any hyperbolic line ℓ and a point P not in ℓ there exist two lines passing through P and sharing each an ideal point with ℓ (I_1 and I_2 in the following picture).

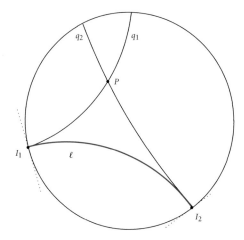

It also follows that P5* does not hold in $\mathfrak{H}$, since both hyperbolic lines above are parallel to ℓ. Indeed, assume that the hyperbolic line determined by P and I_1, call it q_1, is not parallel to ℓ and let Q be the intersection point of ℓ and q_1. Since both hyperbolic lines pass through Q and have I_1 as ideal point, it follows that $\ell = q_1$. But this is in contradiction with the fact that $P \in q_1$ and $P \notin \ell$.

The lines q_1 and q_2 above are called *limiting parallel lines* (to ℓ through P). They are limiting since arbitrarily small perturbations can change them into hyperbolic lines through P intersecting ℓ. They are also the limit (in the sense of the boundary) of the set of lines through P parallel to ℓ. This turns out to be an infinite set, a sense of which can be conveyed by the following picture:

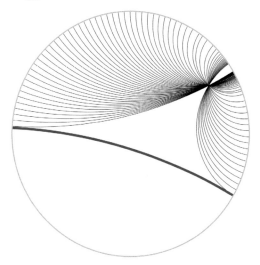

So, P1* is valid on $\mathfrak{H}$ and P5* is not. Recall, we wanted to see that $\mathfrak{H}$ is actually a model for hyperbolic geometry. That is, that it satisfies the axioms of EG (minus P5) – and hence is a model for neutral geometry – as well as the negation H of P5. Recall as well that to do so we had to provide interpretations $B_{\mathfrak{H}}$ and $C_{\mathfrak{H}}$ for betweenness and congruence.

The interpretation for betweenness is the natural one: a point P is between points Q and R when there is a hyperbolic line h passing through the three of them and the arc of circle (or segment) with extremities Q and R contained in h contains P. With this interpretation at hand, it is immediate to verify, for instance, that B1 is valid on $\mathfrak{H}$. Indeed, B1 translates into Euclidean terms as the statement that if in an arc of a circle (different from the whole circumference) a point B is between points A and C then B is between C and A. And this statement is a theorem in Euclidean geometry.

The validity of the other axioms in NG involving only B can be proved in the same way; these axioms translate into statements in Euclidean geometry which are be shown to be true.

We may now shift our attention to congruence. As we mentioned, we will specify this notion in terms of a distance, which we define as follows. For points P and Q in U let I and J be the ideal points of the unique hyperbolic line passing through P and Q. Then we take

$$d_{\mathfrak{H}}(P,Q) = \left| \ln \frac{\text{dist}(I,Q)\text{dist}(J,P)}{\text{dist}(I,P)\text{dist}(J,Q)} \right|.$$

Here, dist denotes, as usual, Euclidean distance, and ln denotes natural (i.e. in base e) logarithm.

The function $\ln x$ has a graph with the shape as show on the right. It maps positive real numbers into real numbers, satisfies that $\ln 1 = 0$, and approaches $-\infty$ when x approaches 0 and $+\infty$ when x increases. Therefore, $d_{\mathfrak{H}}(P,Q)$ can be arbitrarily large and will be so when the quotient

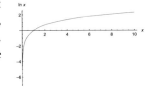

$$\frac{\text{dist}(I,Q)\text{dist}(J,P)}{\text{dist}(I,P)\text{dist}(J,Q)}$$

becomes very small (a number close to zero in the numerator) or very large (a number close to zero in the denominator).

We observe that $d_{\mathfrak{H}}(P,Q)$ does not depend on the ordering of P and Q with respect to I and J. If we exchange I and J we have

$$\frac{\text{dist}(I,Q)\text{dist}(J,P)}{\text{dist}(I,P)\text{dist}(J,Q)} = \frac{1}{\frac{\text{dist}(I,P)\text{dist}(J,Q)}{\text{dist}(I,Q)\text{dist}(J,P)}}$$

and our claim follows, since, for all real number $z > 0$,

$$\left| \ln \frac{1}{z} \right| = |-\ln z| = |\ln z|.$$

We can actually characterize in which situations the absolute value above is necessary.

Lemma 12.8 *We have*

$$d_{\mathfrak{H}}(P,Q) = \ln \frac{\text{dist}(I,Q)\text{dist}(I,Q)}{\text{dist}(I,P)\text{dist}(J,Q)}$$

if and only if Q is between P and J.

Proof If Q is between P and J then $\text{dist}(I,Q) \geq \text{dist}(I,P)$ and $\text{dist}(J,P) \geq \text{dist}(J,I)$.

This implies

$$\frac{\text{dist}(I,Q)}{\text{dist}(I,P)} \geq 1 \quad \text{and} \quad \frac{\text{dist}(J,P)}{\text{dist}(J,Q)} \geq 1.$$

Hence, the product of both expressions is also at least one and, consequently,

$$\ln \frac{\text{dist}(I,Q)\text{dist}(J,P)}{\text{dist}(I,P)\text{dist}(J,Q)} \geq 0.$$

This shows that $d_{\mathfrak{H}}(P,Q) = \ln \frac{\text{dist}(I,Q)\text{dist}(J,P)}{\text{dist}(I,P)\text{dist}(J,Q)}$.

To show the converse, assume that $d_{\mathfrak{H}}(P,Q) = \ln \frac{\text{dist}(I,Q)\text{dist}(J,P)}{\text{dist}(I,P)\text{dist}(J,Q)}$ or, equivalently, that

$$\frac{\text{dist}(I,Q)\text{dist}(J,P)}{\text{dist}(I,P)\text{dist}(J,Q)} \geq 1.$$

In this case we must have Q in between P and J. If this were not so, we would have P between Q and J with $P \neq Q$ and the argument above shows that

$$\frac{\text{dist}(I,P)\text{dist}(J,Q)}{\text{dist}(I,Q)\text{dist}(J,P)} > 1.$$

But this is equivalent to $\frac{\text{dist}(I,Q)\text{dist}(J,P)}{\text{dist}(I,P)\text{dist}(J,Q)} < 1$, in contradiction with our hypothesis. $\qquad\square$

One can get a feeling for $d_{\mathfrak{H}}$ by taking O to be the centre of U. Then, for any other point P, the hyperbolic line through O and P is a diameter and its ideal points I and J are diametrically opposed.

In this situation, $\text{dist}(I,O) = \text{dist}(J,O) = 1$ and writing $\text{dist}(O,P) = z$ we have

$$\frac{\text{dist}(I,O)\text{dist}(J,P)}{\text{dist}(I,P)\text{dist}(J,O)} = \frac{1-z}{1+z}.$$

This quotient is 1 when $P = O$ (and hence $d_{\mathfrak{H}}(P,P) = 0$ as one expects). When P approaches I, it instead tends to 0 and, with it, $d_{\mathfrak{H}}(P,O)$ approaches ∞. In particular, small (Euclidean) steps of P toward I translate into huge (hyperbolic) steps of P away from O. We have already seen this phenomenon in Example 12.5. To see why, check what happens with $d_{\mathfrak{S}}(P,O)$ when $O = (0,0,1)$ and P approaches the equator $\mathcal{E}$ (i.e. P_3 approaches zero).

We look at U with "Euclidean eyes" and perceive its points as separated by the distance dist; from this standpoint $d_{\mathfrak{H}}$ is an artificial construct. We may nonetheless imagine an inhabitant of U (a world known as

Hyperland). We may as well imagine the atmosphere in Hyperland to become increasingly dense in the vicinity of $\mathscr{C}$ and, consequently, to decrease the speed of Hyperlanders in this vicinity. We may finally suppose that Hyperlanders do not experience density changes and that, instead, they feel they move at a constant speed. Therefore, they will not notice they move slower but perceive instead distances getting bigger. As a consequence, they believe their universe (i.e. Hyperland) is infinite and cannot conceive the existence of an ideal boundary $\mathscr{C}$ for it.[12]

Let us return to our main goal and define $C_{\mathfrak{H}}(A, B, A', B')$ when $d_{\mathfrak{H}}(A, B) = d_{\mathfrak{H}}(A', B')$. We can now check that the axioms involving congruence are valid as well on $\mathfrak{H}$. For instance, C2 translates as

$$\forall\, A, B, C, D, E, F \left((d_{\mathfrak{H}}(A, B) = d_{\mathfrak{H}}(C, D) \,\&\, d_{\mathfrak{H}}(A, B) = d_{\mathfrak{H}}(E, F)) \right.$$
$$\left. \to d_{\mathfrak{H}}(C, D) = d_{\mathfrak{H}}(E, F) \right),$$

which certainly holds true. The validity of C3 is more involved, but proving it sheds light on the nature of $d_{\mathfrak{H}}$. We will therefore do that. In the presence of a distance d the statement C3 is an easy consequence of the fact that if B is between A and C then $d(A, C) = d(A, B) + d(B, C)$. This equality – together with the general inequality satisfied by any distance function stating that $d(A, C) \leq d(A, B) + d(B, C)$ for all points A, B, C – is the formal counterpart of the common affirmation that "the segment of straight line is the shortest curve joining two points". This property, known as a *geodesic*, is enforced in Euclidean geometry precisely by axiom C3. We will next show that it holds true as well in $\mathfrak{H}$.

Proposition 12.9 *Let P, Q, R be three points on a hyperbolic line ℓ with Q between P and R. Then*

$$d_{\mathfrak{H}}(P, R) = d_{\mathfrak{H}}(P, Q) + d_{\mathfrak{H}}(Q, R).$$

Proof Let I and J be the ideal points of ℓ. Assume, without loss of generality, that the ideal points I and J of ℓ are as follows:

Then, using Lemma 12.8 three times we get

$$d_{\mathfrak{H}}(P, Q) = \ln \frac{\text{dist}(I, Q)\text{dist}(J, P)}{\text{dist}(I, P)\text{dist}(J, Q)}, \quad d_{\mathfrak{H}}(Q, R) = \ln \frac{\text{dist}(I, R)\text{dist}(J, Q)}{\text{dist}(I, Q)\text{dist}(J, R)},$$

and

$$d_{\mathfrak{H}}(P, R) = \ln \frac{\text{dist}(I, R)\text{dist}(J, P)}{\text{dist}(I, P)\text{dist}(J, R)}.$$

---

[12] I will not elaborate more on the features of Hyperland. I mention as a curiosity that, unlike in Flatland, no Hyperlander is known to be called A Square.

Now use the equality

$$\frac{\text{dist}(I,R)\text{dist}(J,P)}{\text{dist}(I,P)\text{dist}(J,R)} = \frac{\text{dist}(I,Q)\text{dist}(J,P)}{\text{dist}(I,P)\text{dist}(J,Q)} \times \frac{\text{dist}(I,R)\text{dist}(J,Q)}{\text{dist}(I,Q)\text{dist}(J,R)}$$

together with the fact that for all numbers x and y we have $\ln xy = \ln x + \ln y$ to deduce that $d_{\mathfrak{H}}(P,R) = d_{\mathfrak{H}}(P,Q) + d_{\mathfrak{H}}(Q,R)$. $\quad\square$

Angles in $\mathfrak{H}$ turn out to have a more straightforward interpretation than distances. Given three different points A, B, C in U we define the angle $\angle_{\mathfrak{H}}ABC$ to be the Euclidean angle of the segments (of hyperbolic line) $\overline{BA}$ and $\overline{BC}$. These segments are either Euclidean straight segments (if the corresponding hyperbolic line is a diameter of U) or arcs of a circle. The validity of axioms involving angles (such as P4* or C5) is now, again, a simple consequence of stating these axioms in terms of Euclidean angles between segments and arcs of circle and checking that these statements are theorems in EG.

An added benefit of this interpretation of angle is that a feature of hyperbolic geometry mentioned in Section 12.2, namely, the fact that the angles of a triangle add up to less than 180°, now becomes patent. Whereas for Gauss, Bolyai, Lobachevsky and their contemporaries – who had a purely syntactical approach to hyperbolic geometry – this feature may have looked unnatural and surprising, in Poincaré's model it appears to be unavoidable.

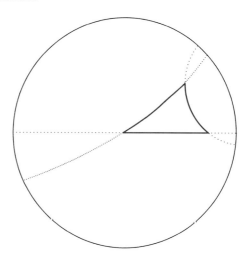

We have now fully specified the model $\mathfrak{H}$: all primitive notions (line, betweenness and congruence) have been given interpretations. Furthermore, we have verified that several axioms of NG are valid in $\mathfrak{H}$. With more or less effort, but working in the same manner (stating the axiom as a sentence of Euclidean geometry and proving that this sentence is a theorem in EG), one can equally verify the validity of all the axioms of

NG in $\mathfrak{H}$. Finally, we have noted that P5 is not valid in $\mathfrak{H}$. All in all, we have showed the relative consistency of hyperbolic geometry with respect to Euclidean geometry. We can state this in a precise manner.

Theorem 12.10

(i) *If* EG *has a model so has* HG.
(ii) *If* EG *is consistent so is* HG.
(iii) *If Euclidean geometry is consistent, we cannot prove P5 from the remaining axioms.*

Sketch of proof The general lines of the proof of Part (i) were given above. For Part (ii), assume EG is consistent. Then, by Gödel's completeness theorem (Theorem 12.3), EG has a model. By Part (i), HG has a model as well, and Theorem 12.3 applies again to show that HG is consistent.

We finally prove Part (iii). To do so, assume EG is consistent. Then, by Part (i) so is HG and, by Gödel's completeness theorem (Theorem 12.3), there exists a model $\mathfrak{H}$ of HG. Now assume that P5 can be proved from the other axioms; that is, that we have NG $\vdash$ P5. By Theorem 12.7, we have NG $\models$ P5 and, therefore, $\mathfrak{H} \models$ P5 (since $\mathfrak{H} \models$ NG). But then we have both $\mathfrak{H} \models \neg$P5 and $\mathfrak{H} \models$ P5, which is a contradiction. $\qquad \square$

Part (iii) of Theorem 12.10 closed a discussion which had lasted centuries. The hope of proving P5 from the remaining axioms was poisoned; most likely (as Euclidean geometry is generally believed to be consistent), it could not be done, and if it could, it demanded the highest possible sacrifice, namely giving up the consistency of the geometry long enshrined as the only possible.

In this and the previous sections of this chapter we have been succinct. The reader wishing to read a detailed exposition of the relative consistency of hyperbolic geometry, at a reasonably elementary level, can find such an exposition in (Greenberg, 1993: Chapter 7).

12.5 Projective geometry as a non-Euclidean geometry

Where does projective geometry lie in this newly found landscape of non-Euclideanness? A look back to Chapter 9 shows that the focus of that chapter was on shape, not on space. We introduced the projective plane $\mathbb{P}$ with the primary goal of allowing the set of projectivities to act as a group of transformations (we could in this way, for instance, make precise the extent to which the two fishes in Figure 9.8 share a common shape). But except for noting that P5 does not hold, along with briefly describing some issues at stake in Remark 9.5, we shunned a discussion on the nature of $\mathbb{P}$. The time is ripe to return to this topic.

A leading capacity in understanding some distinctive features of the projective plane is played by P2. Recall, Euclid's formulation for this axiom was vague: "to produce a finite straight line continuously in a straight line". The naked reading of this sentence is the assertion that we can extend any segment, from any of its endpoints, as much as we want. Nothing else is actually stated. Yet, the use Euclid made of this axiom involved, as we mentioned in Section 12.1, unstated properties of the notion of betweenness. This omission is remedied in Hilbert's systematization of Euclidean geometry in which P2 is replaced by P2*, a statement where betweenness' role is explicit.

It is this assumption of betweenness properties what stops P2 from being valid (or simply making sense) on the projective plane. To understand this statement we need a model for the projective plane where an interpretation for betweenness can be considered. The only model we have at hand thus far is $\mathbb{P}$, which we defined in § 9.2.2. Recall, a projective line $\bar{r}$ was defined to be a Euclidean line r plus a point at infinity R_∞.

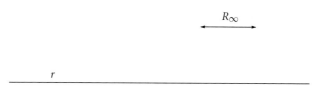

A basic feature of R_∞, discussed at length at the beginning of Section 9.2, is that it corresponds to the non-oriented direction of r. That is, in the picture above, we move towards the same (and only) point at infinity of r no matter whether we move leftward or rightward. In order to draw r and R_∞ together so that this will occur we need to bend r. A projective line thus looks like a circle; it closes itself.

An alternate way to get a sense of this closure is by considering the correspondence between a circle C and a line ℓ arising from projecting points in the circle onto the line, with the north pole N as projection point.

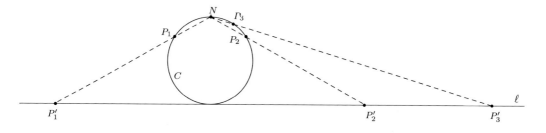

This correspondence is not complete since no point on ℓ is associated with N. We can make it a bijection by associating with N the point at infinity L_∞ of ℓ. Again, we see that we can approach N by the left (e.g. P_1) or by the right (e.g. P_2 and P_3) and the corresponding points in ℓ diverge to the left or to the right respectively, but in both cases they approach L_∞. Therefore, both $\bar\ell = \ell \cup L_\infty$ and C are models for the projective line. We note, however, that C is somehow more homogeneous than $\bar\ell$ since all its points are of the same feather.

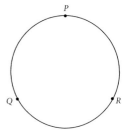

With the closedness of projective lines clear in our minds we may now return to the relationship linking P2, betweenness and projective geometry. Indeed, we immediately note that the closedness of projective lines conflicts with the notion of betweenness. Because, which of P, Q and R is between the other two in the drawing on the right?

Obviously, any of the three choices is possible, and as legitimate as the other two. But taking the three of them to be valid is contrary to the intuition of betweenness and, more formally, leads to a contradiction with the axioms of betweenness. Because, from the relation $B(R, Q, P)$ and axiom B1 it follows that $B(P, Q, R)$. And from this, $B(P, R, Q)$, and axiom B3, that $R = Q$.

To axiomatize projective geometry we therefore need to give up betweenness. A milder notion that can take its place is that of *separation*, in the sense that the points A and B separate C and D in the picture on the right.

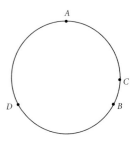

We can use a predicate $S(A, B \mid C, D)$ to express separation and impose an intended meaning to this notion through a few axioms. A possible list of such axioms (seven of them in the occurrence) is given, for instance, in (Greenberg, 1993: Appendix A). We can use the notion of separation to define derived notions in the same manner that we used betweenness in Euclidean geometry. For instance, we mentioned that the segment $\overline{AB}$ is defined in Euclidean geometry as the set of all points X such that $B(A, X, B)$. Similarly, the segment $\overline{AB}_D$ is defined in projective geometry as the set of all points X such that $S(A, B \mid X, D)$. Note that now we need three points to determine a segment, the role of D being to establish which of the two possibilities in the projective line we choose (in the drawing above, it would be the one containing C).

The notion of congruence turns out to be less conflictive than that of betweenness. Model $\mathbb{P}$ – the union of $\mathbb{E}$ with the improper line ℓ_∞ – is nonetheless hardly the most appropriate to provide an interpretation for C. An alternate model for the projective plane where such provision is easy was given by Bernhard Riemann in 1854. Its point of departure is a sphere $\mathbb{S}$ in $\mathbb{E}^3$. The universe U is then defined to be the set of pairs of antipodal points in $\mathbb{S}$.

A *projective point* (in Riemann's model $\mathfrak{P}$) is thus a pair of antipodal points in $\mathbb{S}$. A *projective line* in $\mathfrak{P}$ is a great circle (i.e. an equator) on $\mathbb{S}$. Note that a projective line is, as expected, a set of projective points: if a point P is on a great circle C then its antipodal $-P$ is also on C. The notion of a projective point lying on a projective line is, therefore, well defined.

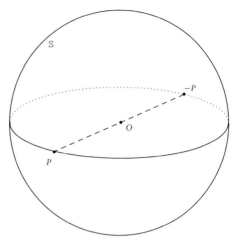

We observe as well (but we already knew this) that P5* does not hold, the failure now being due to the fact that there are no parallels: every two lines intersect.

The fact that every projective point is the intersection of $\mathbb{S}$ with a diameter (a line passing through the centre O of $\mathbb{S}$) suggests a natural distance in $\mathfrak{P}$. For any two points ζ and ξ in $\mathfrak{P}$ we define

$$d_{\mathfrak{P}}(\zeta, \xi) = \angle \ell_\zeta \ell_\xi,$$

where ℓ_ζ and ℓ_ξ are the diameters corresponding to ζ and ξ respectively. The angle $\angle \ell_\zeta \ell_\xi$ between these two lines is chosen so that its value is between $0°$ and $90°$.

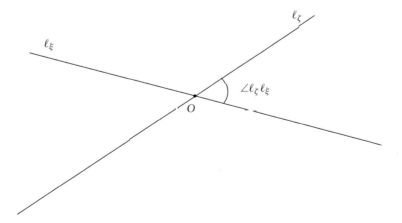

Seen with Euclidean eyes, the distance $d_{\mathfrak{P}}$ appears more regular than the hyperbolic distance $d_{\mathfrak{H}}$. It presents, nonetheless, an unusual characteristic: there is a maximal possible distance between two points in $\mathfrak{P}$. In both Euclidean and hyperbolic geometry points can be arbitrarily distant. In projective geometry, they cannot separate themselves more than 90°.

One can also define an appropriate notion of angle in $\mathfrak{P}$. To do so, we note that lines in $\mathfrak{P}$ are given by planes in $\mathbb{E}^3$ passing through the centre O of $\mathbb{S}$. It is natural then to define the angle between two lines in $\mathfrak{P}$ as the angle between their corresponding planes. Recall now that in Euclidean geometry the three angles of a triangle add up to 180° whereas in hyperbolic geometry they add up to less than 180°. Projective geometry supplies the missing case. Triangles in this space[13] have angles adding up to more than 180°.

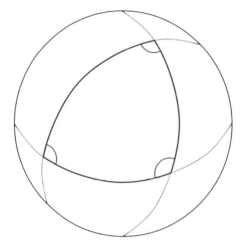

The interpretation of congruence for both segments and angles is now straightforward.

With these interpretations at hand one can check that the six congruence axioms (appropriately modified to replace betweenness for separation if needed) are valid in $\mathfrak{P}$.

Define PG to be the theory given by the following axioms: the three incidence axioms, the seven axioms of separation, the six congruence axioms (appropriately modified), the two continuity axioms and

P Two different lines meet at a unique point.

Then all these axioms are valid in $\mathfrak{P}$, if we assume the consistency of Euclidean geometry. There is less of an achievement in this statement

---

[13] As with segments, we need an extra point to determine a triangle in $\mathfrak{P}$. Three non-concurrent projective lines give place to four triangles in $\mathfrak{P}$.

when compared with its counterpart for hyperbolic geometry owing to the fact that many of the axioms in PG have been tailor-made to hold true on $\mathfrak{P}$. We nevertheless summarize the relative consistency of PG with respect to EG with a statement in the spirit of Theorem 12.10.

Theorem 12.11

(i) *If* EG *has a model so has* PG.

(ii) *If* EG *is consistent so is* PG.

One of the themes in Chapter 9 was the relationship of the Euclidean plane $\mathbb{E}$ with the projective plane $\mathbb{P}$. We constructed the latter as a "completion" of the former and devoted Section 9.3 to investigating how several Euclidean notions (affinities, conics, etc.) can be extended to notions within this completion. The properties of $\mathbb{P}$, derived from the particular way $\mathbb{P}$ was defined, were central in this investigation. One might wonder at this stage whether these relationships can be brought to the fore when $\mathfrak{P}$ is chosen as a model of the projective plane.

A first observation in this regard is that the special role played by the improper line ℓ_∞ within $\mathbb{P}$ – a particularly leading role in the context of perspective techniques – appears at first sight to have no counterpart in $\mathfrak{P}$. All great circles of $\mathbb{S}$ are equal. No projective line stands out.

The obvious way out is to choose one line in $\mathfrak{P}$ and mark it as ℓ_∞. To fix ideas, assume we do so with the equator $\mathcal{E}$ of $\mathbb{S}$. Then, it may be convenient to consider all points say, in the northern hemisphere of $\mathbb{S}$, as redundant (as any such point is already represented by its antipode in the south). Our revised model (which we will still denote by $\mathfrak{P}$) is now the southern hemisphere plus the equator $\mathcal{E}$ but with antipodal points in $\mathcal{E}$ identified (so that one such pair is a point in $\mathfrak{P}$). This glueing of antipodal points makes $\mathfrak{P}$ borderless and recovers the closedness of projective lines.

The view of $\mathfrak{P}$ as a completion of the Euclidean plane becomes manifest if we consider as a model for the latter the structure $\mathfrak{S}$ of Example 12.5. To construct $\mathfrak{P}$ we need just to add the equator $\mathcal{E}$ to $\mathfrak{S}$ and glue its antipodal points.

We may now return to an issue we raised in Remark 9.5, namely the existence of a notion of distance in the projective plane. We have already seen that this notion can be properly defined on $\mathfrak{P}$. We note now that the distances $d_\mathfrak{P}$ and $d_\mathfrak{S}$ do not coincide, not even for pairs of points in the southern hemisphere (where both distances are defined). We did define a distance in $\mathfrak{P}$ that makes it a model for PG, but this distance is not an extension of Euclidean distance.

We end this section with a property of projective planes that makes them different from both Euclidean and hyperbolic planes (and somehow more difficult to apprehend). Take two projective points α and β in $\mathscr{E}$ and a strip M as in the following picture:

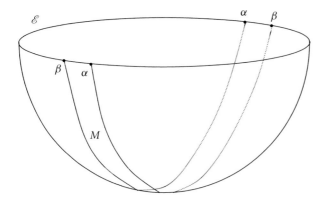

The two (short) segments $\overline{\alpha\beta}$ in the equator are a single line segment in $\mathfrak{P}$. We can get an idea of the true shape of M by glueing these two (Euclidean) arcs into the projective segment they actually are. To do so, since we need to glue αs with αs and βs with βs, we first twist the strip, then we bend it and finally do the glueing:

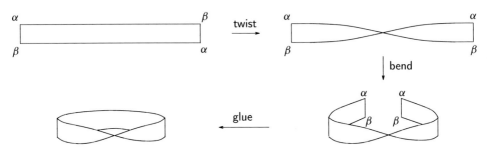

The resulting object, known as *Möbius strip*, is probably not new to the reader. What may be novel is the observation that this surface is non-orientable. We already mentioned orientability in Section 2.3, where we noted that reflections do not preserve orientation. An underlying assumption in this remark is that there exists a notion of orientation to be preserved, a sort of coherence allowing me to decide what is to my left and what to my right. Such a coherence is missing in the Möbius strip. Consider an M-lander, call it T, travelling on it who, on the picture below, is shown as a triangle.

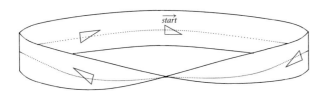

To avoid complications, T follows a straight path in the central valleys of M-land,[14] depicted as a dotted curve in the picture. We have indicated both his starting point and the direction of his motion, as well as four portraits of him at different stages of the promenade.

Filling the gaps between these portraits allows one to recreate the whole travel of T. It also reveals that, after a certain time, T arrives at his place of departure in spite of having journeyed always in the same direction.[15] Furthermore, when this occurs T has suffered a change, as if he had been reflected. What lay on his h-side (the side of his hypotenuse) when he started walking is now on his c-side (the side of his longest cathetus) and vice versa. For a Flatlander (used, as they are, to the Euclidean order of Flatland) such an experience would be extremely confusing. The arrival to the departure place would produce a feeling of déjà vu along with a difficulty for finding one's bearings. But this confusion is not due to a physical change in either M-land or T. It is due to the nature of M-land, to the impossibility to fix an orientation on it that will remain invariable under displacements, and it is natural for M-landers.

We finally observe that, since the Möbius strip is a subset of the projective plane (we constructed it by cutting M out from $\mathfrak{P}$ and discarding the rest) the projective plane itself is also non-orientable.

12.6 Spherical geometry

> Ye shall do no unrighteousness in judgment, in meteyard, in weight, or
> in measure.
> Leviticus 19:35
>
> She knew her distance and did angle for me.
> W. Shakespeare, All's Well That Ends Well, V, 3

Paramount to the models for Euclidean, hyperbolic and projective geometries we described in Sections 12.3, 12.4 and 12.5 was a notion of distance. In all three cases we defined an interpretation for congruence based upon this notion of distance. And in all three the characteristics of the geometry turn out to be ultimately linked to this underlying distance.

The history of geometry is closely tied to the notion of distance. The word geometry comes from ancient Greek, γεωμετρία, itself made up from γῆ, meaning earth, and μέτρον, meaning measurement. The basic problems from which geometry evolved were related to measuring

---

[14] M-landers are notoriously reluctant to approach the boundary of M-land. This shows in their cartography, which seldom represents areas away from the central valleys, and when they do it is usually accompanied by the warning "Here there be monsters".

[15] The size of M-land is, it needs to be said, very large. A journey such as the one endeavoured by T takes a proportionately long time, and this is why it has not been carried out until recently, by one Christopher M-lumbus.

land. Centuries later, a committee within the French Academy of Sciences would define what has become the basic unit of length in the International System of Units, the *metre*, as one ten-millionth of the distance from the North Pole to the Equator measured along the meridian passing through Paris.[16] It is unclear whether at the time the word geometry originated the Greeks were aware of the spherical nature of the Earth, but it is certain that members of the French Academy of Sciences in the late eighteenth century were. In short, the metre was defined as a fraction of the length of a curve.

Airplane pilots measure distances in very much the same manner, and most of us are led to do so if we want to confirm the mileage accrual of a flight to an airline's fidelity card. Measuring distances in spheres is actually easy and natural because it reduces to measuring angles. Indeed, consider the simpler problem of measuring distances on a circle. If the circle has radius r then its total length is $2\pi r$. Therefore, the distance between two points P and Q on the circle equals $\frac{2\pi r}{360}\theta$, where θ denotes the angle between P and Q.

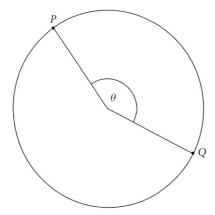

That is, the distance between P and Q is just a fixed constant (in this case $\pi r/180$) times the angle θ between them. Up to the choice of unit of measurement, distances in the circle are just angles (with apex at the circle's centre). Now consider points P and Q on a sphere $\mathbb{S}$ (like the idealized surface of the Earth) and assume that they are not antipodal. Then there is a unique great circle on the sphere passing through both of them. And, as it happens, the shortest curve (on the surface of the

---

[16] The surveying task took more than 6 years, at the end of which a platinum bar measuring a metre was placed in the National Archives to permanently record this length measure. A lively account of the birth of the metre – along with the chaotic situation preceding it on which "under the cover of some eight hundred names, the Ancien Régime of France employed a staggering 250,000 different units of weights and measures" – is in the book by Ken Alder (2002).

sphere[17]) between P and Q is the one that follows this great circle. Since the distance along this great circle is $\frac{\pi r}{180}\theta$, where r is the radius of $\mathbb{S}$ and θ the angle between P and Q, we have, again, that distances on the sphere are, up to a constant, angles.

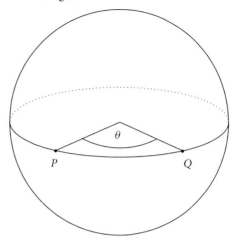

Great circles in the sphere have the *geodesic property*: for any two points on a great circle the shortest curve between them is precisely the arc of the great circle with extremities these two points. We already observed that the only curves on the Euclidean plane satisfying this property are the straight lines. In this sense, great circles have a legitimate claim to be the straight lines of the sphere. Granting them this claim leads to yet another non-Euclidean geometry, called *spherical geometry*. It satisfies (and this should come as no surprise) the same separation and congruence axioms that the model $\mathfrak{P}$ for projective geometry does. Unlike the latter, however, P1* does not hold in spherical geometry (there are infinitely many lines through a pair of antipodal points). Another difference is that the sphere is orientable and, as we have seen, $\mathfrak{P}$ is not.

A consequence of the discussion above is that, since we can obtain projective geometry from spherical geometry (via identifying antipodal points), we can also recover the former purely from the natural notion of distance on the sphere.

Remark 12.12 The preceding discussion introduced spherical geometry as the geometry of a particular surface, the sphere $\mathbb{S}$, in $\mathbb{E}^3$. Lines are curves in $\mathbb{S}$ with the geodesic property and both distances and angles are measured in the natural way as distances and angles[18] in $\mathbb{S}$. The understanding of spherical geometry is doubtless facilitated by this

---

[17] One may think of the distance $d_{\mathbb{S}}(P, Q)$ between two points in $\mathbb{S}$ as the length of the shortest string of rope we can glue to $\mathbb{S}$ so that its extremities are tied to P and Q.

[18] The angle between two curves meeting at a point is the angle made by the tangents to the curves at this point.

model. Also, other non-Euclidean geometries can be similarly defined by replacing the sphere with another surface (we will briefly overview cylindrical geometry in Chapter 14).

A question raised by these facts is whether a similar model can be found for projective or hyperbolic geometry; that is, whether there exists a surface S in $\mathbb{E}^3$ such that projective (or hyperbolic) geometry is the geometry of S. Unfortunately, the answer to this question is that, for both projective and hyperbolic geometries, there is no such model. This lack obstructs the grasping of the two most important non-Euclidean geometries.

13 Symmetries in non-Euclidean geometries

Both hyperbolic and projective geometries have a natural notion of isometry. This is a bijection $\varphi : \pi \to \pi$ of the plane onto itself such that, for all points x and y on π, the segment $\overline{\varphi(x)\varphi(y)}$ is congruent with $\overline{xy}$. The fact that both models $\mathfrak{H}$ and $\mathfrak{P}$, of hyperbolic and projective geometry respectively, are endowed with a distance allows one to recover in these models our original definition of isometry, namely a transformation that preserves distances. One can easily come up with examples of isometries for both geometries.

The existence of isometries raises the question of a possible classification of them (akin to the one in Theorem 2.10), as well as a possible catalogue of their induced symmetries (akin to the one developed in Chapter 3). Both questions can be given a proper answer. But a full description of these answers (not to speak of a justification) is not called for here since this description bears little on the domain of artistic creation. Nonetheless, for the sake of completeness, we give in this chapter a partial (and quick; we will omit all proofs) overview of these issues.

13.1 Tessellations and wallpapers

For each of the 17 types of wallpaper we saw in Section 3.6 we identified some possible unit cells. By definition, a unit cell is a parallelogram whose image under translations in two fixed directions spans the whole plane (recall Definition 3.25). In order to better exhibit the motif of a wallpaper, we relaxed this definition and allowed for hexagonal unit cells as well (we did so in Footnote 1 in Section 3.6). If we disregard the motif inside the unit cells and consider only their boundaries, we obtain, for parallelograms and hexagons respectively, the following patterns on the plane:

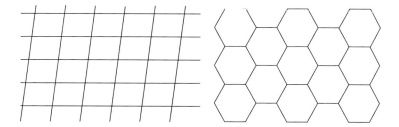

These patterns are a particular case of tessellation.

Definition 13.1 A *tessellation* (or *tiling*) is a set of copies of some figures $F_1, \ldots, F_q$, called *tiles*, such that the elements in this set cover the entire plane and the only intersection of any two such elements is at their boundaries (i.e. there is no overlap of any two such elements). In case there is only one tile F and this tile is a regular polygon we say that the tessellation is *regular*. Furthermore, we say that it is of *type* $\{p, q\}$ when the tile is a regular p-gon and there are exactly q of them meeting at each vertex.

The tessellation at the left above is not regular since the tiles, which are parallelograms, are not square. The one at the right is, and it is of type $\{6, 3\}$.

A consequence of the crystallographic restriction we mentioned in Section 3.6 is that the only possible regular tessellations in the Euclidean plane are of type $\{3, 6\}$, $\{4, 4\}$ and $\{6, 3\}$. A different (and, admittedly, tortuous at first sight) way to state this fact is the following:

for all p, q, there exists a $\{p, q\}$-tessellation in $\mathbb{E}$

if and only if $(p - 2)(q - 2) = 4$. \hfill (13.1)

We have already drawn a regular tessellation of type $\{6, 3\}$. We can now do so for the types $\{4, 4\}$ and $\{3, 6\}$:

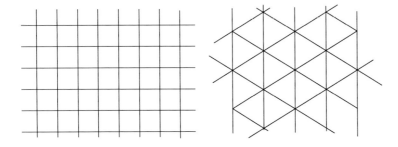

There is an obvious relationship between regular tessellations and wallpaper types. One can obtain wallpapers of types $p1, p2, pm, pg, pmm,$ $pmg, pgg, p4, p4m$ and $p4g$ by imposing a motif, with appropriate symmetry properties, to the tiles of a $\{4, 4\}$ regular tessellation. And similarly for the wallpaper types $p3, p3m1, p31m, p6$ and $p6m$ and the tessellations of type $\{6, 3\}$, as well as for the wallpaper types cm and cmm and the tessellations of type $\{3, 6\}$. Hence, a classification of regular tessellations offers a first approach to a classification of symmetry groups. And it is, in general, much simpler (compare the three tessellation types with the 17 wallpaper types in the Euclidean plane). Because of this, in our dealings with non-Euclidean geometries below, we will only describe regular tessellation types and not the wallpaper types one can derive from them.

13.2 Isometries and tessellations in the sphere and the projective plane

In our dealings with isometries in the context of Euclidean geometry a central result was Theorem 3.9. It stated that every isometry (in
Euclidean geometry) can be represented as the composition of at most
three reflections. This fact makes reflections the basic building block to
construct isometries and allows one to classify them, once more, in terms
of how these building blocks are used. We actually have the following
possibilities:

Reflections	Isometry	Fixed points
One	Reflection	A fixed line
Two with intersecting axes	Rotation	A fixed point
Two with parallel axes	Translation	No fixed points
Three with non-concurrent axes	Glide	No fixed points

In order to carry out a similar line of reasoning in the context of
spheric (or projective) geometry, a first step is to make clear what do
we understand for reflection within this context. The following lemma
allows us to do so in exactly the same manner we did in Section 2.3.

Call a point P on the sphere $\mathbb{S}$ *polar* with respect to a great circle ℓ
when the diameter $\overline{OP}$ of $\mathbb{S}$ is perpendicular to the plane containing ℓ.

Lemma 13.2 *Given a great circle ℓ and a point P not polar with respect to*
ℓ there exists a unique great circle p such that $P \in p$ and p is perpendicular
to ℓ. Furthermore, if p and ℓ cut at Q then the distance from P to ℓ equals
the distance from P to Q. The same holds true in $\mathfrak{P}$.

Place $\mathbb{S}$ so that ℓ is its equator. Then, for any point $P \in \mathbb{S}$ different
from any of the poles (i.e. for any point P not polar with respect to ℓ)
the great circle p whose existence is claimed by Lemma 13.2 is just the
meridian through P.

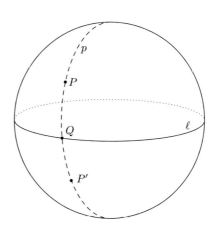

We can now define reflections on the sphere $\mathbb{S}$. Given a spheric line ℓ (i.e. a great circle on the sphere $\mathbb{S}$) the *spheric reflection* with axis ℓ is the function $\mathrm{refl}_\ell^{\mathbb{S}} : \mathbb{S} \to \mathbb{S}$ which maps every point P polar with respect to ℓ to its antipodal point and any other point P to the only point P' on the spheric line p such that $P \neq P'$ and $d_{\mathbb{S}}(P, Q) = d_{\mathbb{S}}(P', Q)$. Here p and Q are given by Lemma 13.2 and the construction of P' is shown in the preceding picture.

We can also define reflections on the projective plane. Indeed, it is immediate to verify that if ψ is a spheric reflection and P is any point on $\mathbb{S}$ then the reflection $\psi(-P)$ of the antipodal point of P is just the antipodal point $-\psi(P)$ of the reflection of P. In other words, reflections with respect to great circles are compatible with the glueing that produces the projective plane $\mathfrak{P}$ from the sphere $\mathbb{S}$. This implies that reflections with respect to great circles actually induce (by glueing antipodal points together) reflections in $\mathfrak{P}$ with respect to projective lines: the image $\mathrm{refl}_\ell^{\mathfrak{P}}(\{P, -P\})$ is the pair $\{\mathrm{refl}_C^{\mathbb{S}}(P), -\mathrm{refl}_C^{\mathbb{S}}(P)\}$. Here, C is the great circle corresponding to ℓ. However, one difference with the Euclidean setting stands out: reflections on the projective plane have a fixed point outside the reflection axis. To see why, take the reflection axis ℓ corresponding to the equator E of $\mathbb{S}$. Then $\mathrm{refl}_E^{\mathbb{S}}$ swaps south and north poles. That is, $\mathrm{refl}_\ell^{\mathfrak{P}}$ leaves fixed the pair of antipodal points consisting of these two poles.

Once with an appropriate notion of reflection at hand, one can compose them to obtain the following isometries on $\mathbb{S}$: reflections (about a great circle), rotations (around a diameter of the sphere) or glides composing both of them. It is then possible to prove that these are the only possible isometries on $\mathbb{S}$:

Reflections	Isometry on $\mathbb{S}$	Fixed points
One	Reflection	A Fixed line
Two with intersecting axes	Rotation	One fixed point
Three with non-concurrent axes	Glide	One fixed point

The isometries on $\mathfrak{P}$ are essentially the same as those on $\mathbb{S}$. As we noted, however, reflections have now a fixed point outside the axis of reflection.

Reflections	Isometry on $\mathfrak{P}$	Fixed points
One	Reflection	A Fixed line and a fixed point outside it
Two with intersecting axes	Rotation	One fixed point
Three with non-concurrent axes	Glide	One fixed point

The classification of isometries leads, as it did in Chapter 3 for the Euclidean plane, to a first classification of symmetries on $\mathbb{S}$. Invariance with respect to each of the three isometries in the preceding table yields bilateral, rotational and glidal symmetry. We will not pursue the finer classification that takes account of all possible invariances of a figure (e.g. by classifying all possible wallpaper types) but limit ourselves to mention that the possible regular tessellations on $\mathbb{S}$ satisfy the following:

for all $p, q \geq 2$, there exists a $\{p, q\}$-tessellation in $\mathbb{S}$

if and only if $(p - 2)(q - 2) < 4$. $\qquad$ (13.2)

Note that we have added the condition $p, q \geq 2$. This is because the inequality $(p - 2)(q - 2) < 4$ is true for pairs with $p = 1$ and $q \geq 0$ which do not have any meaning as tessellations. Pairs with either $p = 2$ or $q = 2$, instead, do have a geometric meaning. Even though we are not familiar with closed "biangles" these make sense on the surface of a sphere. They are wedges on this surface. The following drawing shows examples of regular tessellations on $\mathbb{S}$ of types $\{3, 2\}$ and $\{2, 3\}$. The sides of the spherical polygons are drawn in red and their extremities as filled circles. Hence, at the left we have three 2-gons (each of them a wedge, meeting all the three at each of the two vertices) and at the right two triangles (each of them a hemisphere, sharing all the three sides, meeting both at each of the three vertices).

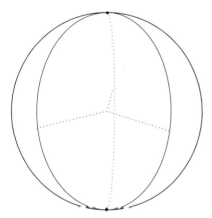 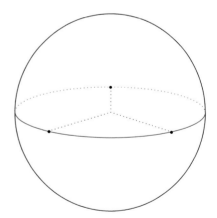

Now note that the inequality $(p - 2)(q - 2) < 4$ is true for all pairs of the form $\{2, q\}$ or $\{p, 2\}$. This yields an infinite number of, admittedly unusual, tessellations for $\mathbb{S}$.

The other possible regular tessellations of the sphere are $\{3, 3\}$, $\{4, 3\}$, $\{3, 4\}$, $\{5, 3\}$ and $\{3, 5\}$, and we are not unfamiliar with them. Indeed, they correspond with the so-called *platonic solids* (Figure 13.1). These are regular polyhedra in Euclidean three-dimensional space and it has long been known that there are only five of them.

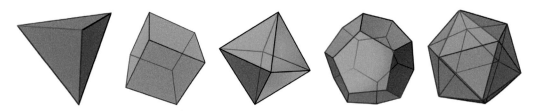

Figure 13.1 **The platonic solids: tetrahedron, hexahedron, octahedron, dodecahedron and icosahedron.**

The five regular tessellations on the sphere mentioned above are the "blown up" versions of these solids (so that edges become spherical arcs and faces spherical polygons). The fact that $\mathbb{S}$ is finite – in the sense that distances in $\mathbb{S}$ are at most πr (with r being the radius of $\mathbb{S}$) – has an obvious consequence, namely that tessellations on $\mathbb{S}$ have a finite number of tiles. The fact that the regular tessellations of the sphere correspond with the platonic solids furthermore shows that each of the possible five has a well-determined number of tiles: four for the tetrahedron, six for the hexahedron, eight for the octahedron, twelve for the dodecahedron and twenty for the icosahedron. Hence the names.

The classification of regular tessellations for $\mathbb{S}$ easily yields one for $\mathfrak{P}$. Any tessellation in the latter comes from identifying antipodal points in a tessellation of the former. The problem is that not all regular tessellations in $\mathbb{S}$ can be glued this way. For some of them, the antipodal image of a tile will not result in a tile. More precisely, a tessellation of the type $\{2, q\}$ in $\mathbb{S}$ induces a tessellation in $\mathfrak{P}$ if and only if q is even, and similarly for a tessellation of the type $\{p, 2\}$ (you may check in the drawings for $\{2, 3\}$ and $\{3, 2\}$ that the glueing is not possible). Also, of the five remaining types (and this can be easily seen in Figure 13.1) the tetrahedral one cannot be glued but the other four can. In summary, the possible regular tessellations of the projective plane $\mathfrak{P}$ are (here n is any integer number at least 1):

$$\{2, 2n\}, \ \{2n, 2\}, \ \{4, 3\}, \ \{3, 4\}, \ \{5, 3\} \text{ and } \{3, 5\}.$$

We cannot fail to notice that, because no model of the projective plane is "realizable" – in the sense that no model can live in three-dimensional space – a possible catalogue of symmetries on $\mathfrak{P}$ has little appeal to the artist. In this regard, we observe that the use of perspective in painting does not involve a projective plane in its entirety but only a part of it, which can be embedded on a Euclidean plane and where some distinguishing features of projective planes (e.g. their non-orientability) are lost. The projective plane is just a mathematical construction providing the adequate theoretical context for the main perspective tool, the projectivity. It is not the subjacent surface of the canvas.

13.3 Isometries and tessellations in the hyperbolic plane

In order to carry out a line of reasoning similar to the one above, now in the context of hyperbolic geometry, we need again, as a first step, to spell out what we understand by reflection here. In the hyperbolic context the following lemma plays the role of Lemma 13.2.

Lemma 13.3 *Given a hyperbolic line ℓ and a point P there exists a unique line p such that $P \in p$ and p is perpendicular to ℓ. Furthermore, if p and ℓ cut at Q then the distance from P to ℓ equals the distance from P to Q.*

The following drawing shows the situation in $\mathfrak{H}$:

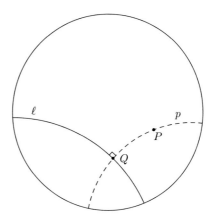

Given a hyperbolic line ℓ we can therefore define the (hyperbolic) reflection $\mathrm{refl}_{\ell}^{\mathfrak{H}}$ with axis ℓ in exactly the same manner we did in Section 2.3. If ℓ is a diameter of U then this is just (the restriction to U of) the Euclidean reflection refl_{ℓ}. In this case, reflections are easy to compute. In the case that ℓ is an arc of a circle C one can also give a characterization of $\mathrm{refl}_{\ell}^{\mathfrak{H}}$ which is easy to compute. This hyperbolic reflection turns out to be (the restriction to U of) a new transformation, defined next.

Given a circle C with centre O the *inversion* with *reference circle C* is the transformation of the plane (minus the centre O of C) mapping any point $P \neq O$ to the point P' in the line through O and P such that

$$\mathrm{dist}(O, P') = \frac{R^2}{\mathrm{dist}(O, P)}.$$

Here, R is the radius of C. The following proposition summarizes the role of inversions as hyperbolic reflections.

Proposition 13.4 *Let $\mathscr{C}$ and C be two circumferences cutting at right angles. Let U be the region inside $\mathscr{C}$ and* inv_C *be the inversion with reference circle C. Then:*

(i) *The centre O of C is outside $\mathscr{C}$.*

(ii) *For every point $P \neq O$,*

- *if P is outside $\mathscr{C}$ then* $\operatorname{inv}_C(P)$ *is outside $\mathscr{C}$;*
- *if $P \in \mathscr{C}$ then* $\operatorname{inv}_C(P) \in \mathscr{C}$;
- *if $P \in U$ then* $\operatorname{inv}_C(P) \in U$.

(iii) *For every pair of points $P, Q \in U$ we have $d_{\mathfrak{H}}(\operatorname{inv}_C(P), \operatorname{inv}_C(Q)) = d_{\mathfrak{H}}(P, Q)$.*

So, the restriction to U of an inversion is a counterpart in $\mathfrak{H}$ of a reflection. It leaves the hyperbolic line ℓ (an arc of C) fixed and swaps points on both sides of ℓ. The other possible counterpart is a reflection with axis a diameter of U.

Endowed with the notion of reflection in the hyperbolic plane, we want, as we did for $\mathbb{S}$ and $\mathfrak{P}$, to consider compositions of these reflections. Before doing so, we need to have a closer look at the possible relative positions of two hyperbolic lines on $\mathfrak{H}$.

One possibility is that the two lines intersect at a point. Assume now that they do not. We have already noted that in hyperbolic geometry there are many more non-intersecting lines than in Euclidean geometry. In addition to this abundance, these non-intersecting lines can be divided into two kinds. We say that two different hyperbolic lines are *horoparallel* when they meet at an ideal point. We say that they are *ultraparallel* when they do not meet at all. Ultraparallel lines have a single, common perpendicular (the dashed line in the drawing below) which joins the closest pair of points from them. When moving away from any of these two points on one of the lines the distance to the other line increases.

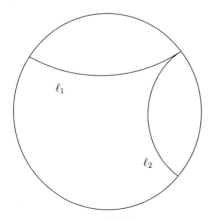

horoparallel lines

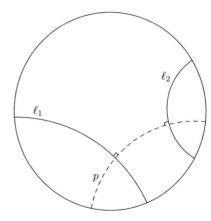

ultraparallel lines

In hyperbolic geometry, therefore, we have three, instead of two, possibilities for the composition of two reflections. Their axes can be intersecting, horoparallel or ultraparallel. In the first case we obtain, as in Euclidean geometry, a rotation with centre the point of intersection of the axes and angle twice the angle made by these axes. In the last case, we obtain a translation in the direction of the common perpendicular to the two ultraparallel axes. This common perpendicular is, therefore, an invariant line under the translation. Its two ideal points are fixed. The case of horoparallel axes of reflection defines a new kind of isometry known as *horolation*. It has no invariant lines and the only fixed point is the ideal point common to the two horoparallel axes of reflection. The following table summarizes the possible isometries on the hyperbolic plane:

Reflections	Isometry on $\mathfrak{H}$	Fixed points
One	Reflection	A fixed line
Two with intersecting axes	Rotation	A fixed point
Two with parallel axes	Horolation	One fixed ideal point
Two with ultraparallel axes	Translation	Two fixed ideal points
Three with non-concurrent axes	Glide	Two fixed ideal points

The classification of isometries leads, as it did in Chapter 3 for the Euclidean plane, to a first classification of symmetries on $\mathfrak{H}$. Invariance with respect to each of the five isometries in the preceding table yields bilateral, rotational, horolational, translational and glidal symmetry. Again, we will not pursue the finer classification that takes account of all possible invariances of a figure, but limit ourselves to mention that $\mathfrak{H}$ is much richer than $\mathbb{E}$ in wallpaper types. In particular, it admits infinitely many regular tilings. Indeed, the following statement comes probably as no surprise. We have:

for all p, q, there exists a $\{p, q\}$-tessellation in $\mathfrak{H}$

$$\text{if and only if } (p - 2)(q - 2) > 4. \tag{13.3}$$

This wealth of regular tessellations in the hyperbolic plane did not pass without notice by Escher. In 1958 the mathematician Harold Coxeter sent to Escher a copy of one of his articles containing a $\{6, 4\}$ tessellation of $\mathfrak{H}$ (Figure 13.2). Escher wrote back to Coxeter and mentioned in his reply that this tessellation "shocked" him. He further added (Coxeter, 1979):

Since a long time I am interested in patterns with "motives" getting smaller and smaller till they reach the limit of infinite smallness. The question is relatively

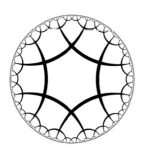

Figure 13.2 **A $\{6, 4\}$-tessellation of $\mathfrak{H}$.**

Figure 13.3 **M.C. Escher's *Circle Limit I*, woodcut, 1958 © 2012 The M.C. Escher Company-Holland. All rights reserved. www.mcescher.com.**

simple if the limit is a point in the center of a pattern.[1] Also a line-limit is not new to me, but I was never able to make a pattern in which each "blot" is getting smaller gradually from a center towards the outside circle-limit, as shows your Figure 7.9 tried to find out how this figure was geometrically constructed, but I succeeded only in finding the centers and radii of the largest inner-circles. If you could give me a simple explanation how to construct the following circles, whose centers approach gradually from the outside till they reach the limit, I should be immensely pleased and very thankful to you! Are there other systems besides this one to reach a circle limit?

Escher thus requested from Coxeter an explanation of the way the tessellation in his article was produced, a request to which Coxeter obliged. The reaction of Escher was his series of four *Circle Limit* woodcuts.

Figure 13.3 shows *Circle Limit I* together with, at its right, a scheme of some of the hyperbolic lines drawing the tessellation.

---

[1] Escher had indeed "solved" this question in his woodcut *Smaller and Smaller*; see Figure 7.2.

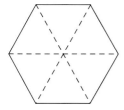

The hexagonal tile underlying *Circle Limit I* contains three black fishes and three white ones. These tiles are invariant under reflections with respect to any of the three axes of the hexagon passing through opposite vertices and, consequently, also under three fold rotations (centred at the tile's centre).

These rotational and bilateral symmetries extend to the whole pattern. In addition, there are central symmetries of the pattern with centres located in the mid-points of the sides of the hexagons. Finally, as one could expect of a wallpaper, no matter whether hyperbolic or not, there are also translational symmetries. The relevant translations can be obtained as reflections with ultraparallel axes (of which, there are plenty of pairs).

All the tiles are isometrical in $\mathfrak{H}$ but, in the Euclidean representation of this model, their sizes decrease when they approach the boundary $\mathscr{C}$.

In *Circle Limit IV* (Figure 13.4), Escher produced a pattern based on the same hyperbolic tessellation but with a different symmetry group.

Figure 13.4 **M.C. Escher's *Circle Limit IV*, woodcut, 1960** © **2012 The M.C. Escher Company-Holland. All rights reserved. www.mcescher.com.**

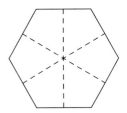

The hexagonal tile now contains three devils (in black) and three angels (in white).

The three fold rotational symmetries are present in this pattern as well and it also possesses bilateral symmetries but now is around the axes of the hexagons passing through opposite mid-points.

Because of this difference, the pattern in *Circle Limit IV* not only possesses central symmetry but also even four fold rotational symmetry, now centred at the vertices of the hexagons. Needless to say, there are several translational symmetries as well, which are now more immediately visible than those in *Circle Limit I*. Figure 13.5 shows the work of Escher to compute the shape and position of angels and devils in *Circle Limit IV*.

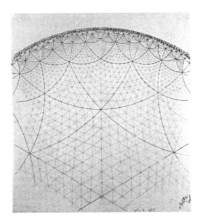
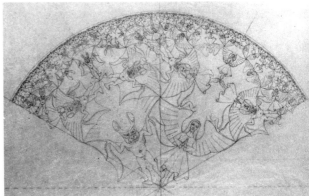

Figure 13.5 **M.C. Escher's preparatory drawings for *Circle Limit IV* © 2012 The M.C. Escher Company-Holland. All rights reserved. www.mcescher.com.**

Figure 13.6 shows the third, and probably the best known, instance of the *Circle Limit* series. Unlike the two previous examples, Escher used colours. Furthermore, the tessellation appearing to underlie the pattern is not regular: a cursory glance at the woodcut shows the simultaneous presence of triangles and quadrangles. Certainly, these polygons are placed following some law: they touch one another only side by side, no vertex of one touching an inner point in a side of another.[2] A less cursory glance reveals another anomaly: the white curves made by the backbones of the fishes cut the boundary of the circle at 80°. They are not hyperbolic lines! In the paper we quoted above, Coxeter (1979) provides an explanation of both anomalies. First, the pattern in the drawing

---

[2] Such tessellations are said to be *semiregular* or *Archimedean*.

Figure 13.6 **M.C. Escher's *Circle Limit III*, woodcut, 1959 ⓒ 2012 The M.C. Escher Company-Holland. All rights reserved. www.mcescher.com.**

rests, indeed, on a regular tessellation. It is a tessellation of type $\{3, 8\}$ that is, eight equilateral triangles meeting at each vertex of the tessellation. The vertices of the tessellation are the points where the right fins of four fishes come together. These four fins alternate with two different colours (the centre of the circle is an example).[3] The lines in the $\{3, 8\}$-tessellation underying *Circle Limit III* are not apparent in the picture.

Second, he shows that each white curve along the backbones is obtained as the set of points that keep a constant distance to a fixed hyperbolic line (related to the $\{3, 8\}$-tessellation above). While such a curve would be again a straight line in Euclidean geometry (and would be parallel to the fixed one; ℓ_2 and ℓ_1 in the following picture) it needs to bend itself in the hyperbolic realm (since parallels in $\mathfrak{H}$ diverge; e.g. r_1 and r_2 in the following picture). The curve E in the following picture (whose points keep a constant distant to r_1) is an example.

---

[3] One could consider as well the dual tessellation of $\{3, 8\}$. The *dual* of a regular tesselation T is the tesselation T^* obtained by joining with segments the centres of the polygons in T. If T is of type $\{p, q\}$ then T^* will be of type $\{q, p\}$. Hence, in the case of *Circle Limit III*, this would be of type $\{8, 3\}$ and have the vertices of the octagons at the points where the mouths of three fishes meet the tails of three others.

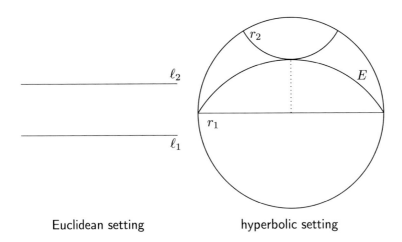

Euclidean setting   hyperbolic setting

It is because the white curves are not hyperbolic lines that it is possible that the angles of the triangles they determine are 60° each. Their sum is therefore 180°, which is not possible in hyperbolic geometry.

Coxeter quotes Escher's view of his own work (1979: 20). About *Circle Limit I* the artist wrote "being a first attempt, [it] displays all sorts of shortcomings. Not only the shape of the fish [. . .] but also their arrangement and relative position, leave much to be desired [. . .] There is no continuity, no 'trafic flow' [. . .]". In contrast, Escher writes about *Circle Limit III* that "most of these defects have now been eliminated. We now have only "through traffic" series: all the fish of the same series have the same colour and swim after each other, along a circular arc from edge to edge. [. . .] As all these strings of fish shoot up like rockets from infinitely far away, perpendicularly from the boundary, and fall back again whence they came, not one single component ever reaches the edge". The computations in (Coxeter, 1979) give an explanation of why the white arcs must cut the circumference at 80° (and Coxeter writes "which they do, with remarkable accuracy"). This conflicts with Escher's description above. Coxeter summarizes: "Thus, Escher's work, based on his intuition, without any computation, is perfect, even though his poetic description of it (*'loodrecht uit de limiet'*, 'perpendicularly from the boundary') was only approximate".

We may bring this section to a close by returning to Escher's words in his letter to Coxeter, more precisely to the last sentence in our quotation above, namely, the question "Are there other systems besides this one to reach a circle limit?" From the absence of other works of Escher relying on "other systems to reach a circle limit" we may infer that Coxeter never gave a positive reply to this question. And yet, such a positive reply was possible.

The use of Poincaré's model of the hyperbolic plane is not a must "to make a pattern in which each 'blot' is getting smaller gradually from

Figure 13.7 **A fisheye view of a pattern by Escher.**

a centre towards the outside circle-limit" as Escher wanted. A different model of the Euclidean plane could have done the job as well. To see why, we first observe that the hemispherical model $\mathfrak{S}$ of Euclidean geometry described in Example 12.5 is bounded. To make it flat we simply project its surface into a plane, say using parallel projection. In this way we obtain yet another model for the Euclidean plane which we can name the *fisheye model*, since this is exactly what we did in Section 11.3, and we saw there that the result is a fisheye view of the observed plane.

Now draw a wallpaper on a standard plane and take a photograph of it with a fisheye lens (or, equivalently, map this plane onto the circle via the two projections described above). We obtain a pattern inside a circle on which the motifs close to the centre are larger and closer in shape to the original one (on the standard plane) and those approaching the boundary of the circle become smaller and bent in a regular manner. Figure 13.7 shows this (computer-generated) transformation for the wallpaper at the right in Figure 4.38.

The impression left by this figure is certainly that of a regular pattern "in which each 'blot' is getting smaller gradually from a centre towards the outside circle-limit", a description shared with the woodcuts of the *Circle Limit* series. Yet, the visual feeling is not the same. The pattern in Figure 13.7 gives a sense of sphericity that the hyperbolic drawings do not give. Somehow paradoxically, the Euclidean tessellation (in the fisheye model) looks curved and the hyperbolic tessellation (in Poincaré's model) looks flat. A last point to emphasize is that the tessellation in Figure 13.7 is not obtained from looking at a spherical tessellation in a sphere $\mathbb{S}$. The simplest way to verify this is by noting that Figure 13.7 contains (theoretically) infinitely many copies of its motif, whereas, as we mentioned in Section 13.2, every tessellation of $\mathbb{S}$ has a finite number of tiles.

One is left to speculate whether Escher did consider this, rather simple, solution for his question and, if he did it, why he discarded it. After all, he was not alien to fisheye views (as shown in Figure 11.6) yet he never produced a fisheye pattern.

Assuming that he considered these kinds of patterns, a possible reason to disregard them might have been the poverty of regular tessellations which can possibly underlie the pattern. An undeniable appeal of hyperbolic patterns is the wealth of possible such tessellations. Another feature which could lie behind this disregard is the fact that the fisheye solution is not *conformal*. That is, angles in this model are not Euclidean angles. This is in contrast with the conformal character of $\mathfrak{H}$ (or $\mathbb{S}$, or $\mathfrak{P}$) and gives a feeling to hyperbolic patterns that fisheye patterns cannot deliver.

In the absence of documents showing Escher's opinions on this matter, however, I am inclined to believe that Escher never considered fisheye patterns as a solution to his question. I have no evidence on this issue, but the following speculative remark might provide an explanation. Poincaré's model for the hyperbolic plane had helped mathematicians to understand hyperbolic geometry and was a common object of study among mathematicians. By contrast, what we have called the fisheye model could have passed as a curiosity among logicians but was largely ignored by geometers. It is a model that wasn't there.

14   The shape of the universe

Euclidean, hyperbolic, projective, spherical... The adjectives seem to multiply and give place to a collection of geometries, each asserting its own particular nature of space. Is there a *true* geometry among them? Can one of them lay claim to offer a faithful description of reality?

A fully satisfying answer to these questions is likely not possible. To begin with, there is no agreement among philosophers of what reality is. Furthermore, even the adoption of a down-to-earth approach to reality – the assumption that the latter is the physical universe as studied and described to us by natural scientists – stubbornly offers resistance to a clear answer. Geometry relates to the shape of the "container" of this physical universe, not of any particular body on it. And this container is so colossal that the sorts of measurements that could help to determine its shape are hard to come by. Yet, there are advances.

To describe these advances is out of our reach for too many reasons, paramount among them the fact that the geometry involved is higher dimensional and in the previous pages we have focused on plane (two-dimensional) geometry. We can nevertheless extrapolate to this context some ideas from the several plane geometries we have encountered.

A simple division of these geometries would cluster spherical and projective geometry together, because in them the sum of the angles of a triangle is greater than $180°$. We say that these two geometries have *positive curvature*. Similarly, we say that Euclidean geometry has *zero curvature*, because in it this sum equals $180°$, and that hyperbolic geometry has *negative curvature*, because this sum is now smaller than $180°$. The notion of curvature is related to the existence of parallels.[1] Within the family of geometries with positive curvature we have seen that differences arise, for instance, with regard to orientability ($\mathbb{S}$ is orientable, $\mathfrak{P}$ is not) and whether P1 holds (it does in $\mathfrak{P}$, it does not in $\mathbb{S}$). The failure of $\mathbb{S}$ to satisfy P1 is mild: the existence of more than one line joining two points is only true when the points are antipodal. A different surface can provide an enlightening example of another geometry, now with zero curvature, more seriously failing to satisfy P1: the cylinder.

A convenient way to visualize the cylinder is as an infinitely long strip whose sides are glued together (the way we glued opposite sides when constructing the Möbius strip, but without twisting before):

---

[1] The notion of curvature is more general than our exposition suggests. The spaces we have considered as models for our geometries, $\mathbb{E}$, $\mathfrak{P}$, $\mathbb{S}$ and $\mathfrak{H}$, have constant curvature. Spaces with varying curvature exist as well (as long as Euclidean geometry is consistent), but we will not need to consider them.

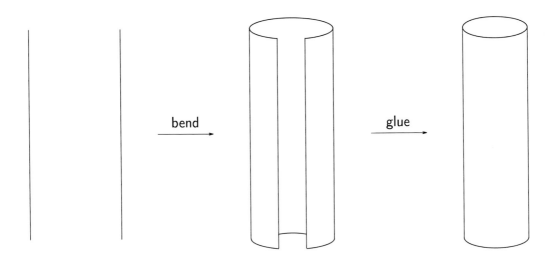

Another way to visualize the cylinder is as a plane π wrapped around the vertical direction infinitely many times. The points P_i on the left of the following diagram correspond to a unique point P in the cylinder C:

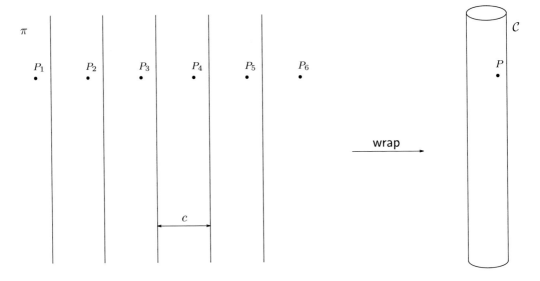

An advantage of the latter is that it provides a simple description of lines in the cylinder. These are just the wrapped images in the cylinder of lines in π. A rationale for this definition is the fact that in wrapping the plane we are not stretching it in any direction. To see why, let c denote the distance between the vertical lines in the picture above (and, consequently, the length of the circumference of the cylinder) and call two points P and Q in π *close* when the distance between their x-coordinates differs by less than or equal to $c/2$. Then, the distance between the images in $\mathcal{C}$ of a pair of close points is preserved.

Needless to say, this is not true for pairs of points which are not close (the obvious counterexample being P_1 and P_2 in the preceding picture). Therefore, for any two points P', Q' in the cylinder, we move from one to the other with minimal distance by following the image in $\mathcal{C}$ of a segment of line between any close pair P, Q in π of preimages of P', Q':

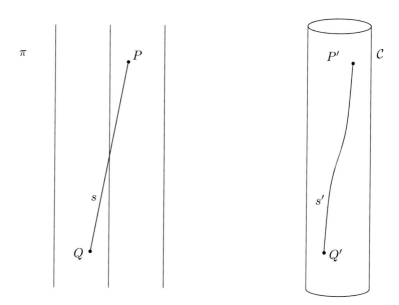

The segment s' with extremities P' and Q' is, therefore, the image of the segment s (of line, in π) with extremities P and Q. This defines segments in $\mathcal{C}$. And extension of segments in both directions defines lines. We conclude that, since extension of segments in π defines ordinary lines in π and segments in $\mathcal{C}$ are images of segments in π, lines in $\mathcal{C}$ are images of lines in π.

A first consequence (requiring a little thought) of this fact is that P5* holds true in $\mathcal{C}$. Another consequence, equivalent to our first, is that the sum of the angles of a triangle in $\mathcal{C}$ equals 180°. This follows from the fact that the wrapping preserves angles (again, since there is no stretching involved). That is, *cylindrical geometry* (to give a name to the geometry in $\mathcal{C}$) has zero curvature.

A more intriguing consequence is the fact that P1 fails to hold in a more marked manner than it does in $\mathbb{S}$. Indeed, consider any two points P and Q which are not in the same horizontal line. Then, there are infinitely many lines joining P and Q. The following picture shows two of these lines. The image of the segment joining P_0 with Q_0 does not wrap itself around $\mathcal{C}$, whereas that of the segment joining P_0 with Q_2 wraps itself twice around C. It is obvious how to obtain segments joining P and Q and wrapping themselves any number of times around C.

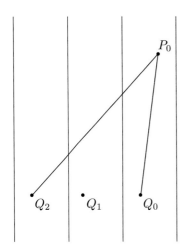

In summary, both $\mathbb{E}$ and C have zero curvature, but in $\mathbb{E}$ there is a unique line joining any two points whereas in C, for almost all choices of these two points, there are many. We say that $\mathbb{E}$ is *simply connected* and that C is *multiply connected*.

The multiple connectedness of C has intriguing consequences. Imagine an inhabitant of C – they are known as C-landers – called Paul searching for his girlfriend Queenie and assume that their positions in C are P and Q in the picture above. Light in C-land follows trajectories which are lines in C. Therefore, in what is a clear advantage over a similar situation for a flatlander, Paul will be able to see Queenie (assuming there are no objects in between) when looking in infinitely many directions. These directions correspond to the lines in π passing through P_0 and, respectively, $Q_0, Q_1, Q_2, \dots$ Furthermore, since apparent size decreases with distance, the more wrapped is the trajectory of vision, the smaller will be the image of Queenie. Paul will therefore see several images of his beloved, we may call them *ghost images*, and will likely walk towards her following the segment that yields the larger ghost image as it is the shortest.

After this interlude in cylindrical geometry let us return to our original question, namely: What is the shape of the physical universe?

The branch of science attempting an answer to this question is called cosmology. It relies on astronomical observations which, analysed through the knowledge of several physical sciences, provide information of various kinds on the different constituents of our universe. Cosmologists then construct models, in general of a mathematical sort, and check these models against the thus gathered information for either confirmation or refutation. There is an epistemological difference between these two. A contradiction between available data and a proposed model invalidates the latter, whereas the absence of such a contradiction merely shows that the model is consistent with observation but it does not establish its validity; subsequently collected data may prove it faulty.

The commonly accepted general model governing the physics of the universe is the theory of general relativity proposed by Einstein in 1915. One of the many intriguing features of this model is that it does not postulate as a necessity the decoupling of space and time. Classically, space had been posited to be three-dimensional Euclidean space $\mathbb{E}^3$ and time to be the real line $\mathbb{R}$. Any pair (x, t) with $x \in \mathbb{E}^3$ and $t \in \mathbb{R}$ denoted a place in the universe at a given instant of time. The space-time structure of the universe was therefore taken to be $\mathbb{R}^4$. In the general relativity model this is a possible state of affairs, but not the only possible one. The space-time structure of the universe is instead hypothesized to be a four-dimensional manifold (roughly speaking, a set which locally looks like $\mathbb{R}^4$) but not necessarily $\mathbb{R}^4$. To get an idea, let us consider a few possible structures for a universe with a one-dimensional space. The following picture shows four possible such structures:

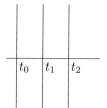

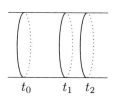

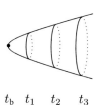

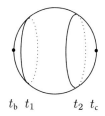

At the left of the picture we see the classical structure. At each instant t the space has the shape of a line. The second diagram shows another possibility: at each instant the space has the shape of a circle. This introduces two fundamental changes. First, space is curved. Second, it is finite. It is crucial to note at this point that, just as we visualize the plane at the left without thinking of it as living in three-dimensional space, we must conceive the cylinder as an entity per se. There is nothing outside it, no space exterior to it.[2] The third structure adds yet a couple of new

---

[2] A probably convenient comparison of the structure of space in the cylinder model for our universe is with the time structure of the day (as shown in an analogic watch). Time advances in a "circular" motion and returns, at any instant, to where it was 24 hours before. But there is nothing "outside" this circular structure, no instant of time exterior to it. Likewise, there is no element of time-space outside the cylinder.

features. First, time has an origin: there is no universe before t_b. Second, space is expanding with time. This means that any pair of bodies in this universe, say two stars, will separate with time without actually moving. Properly speaking, they do not move apart because they do not move at all. They separate because the space where they live is expanding (in the same manner that two points in a balloon separate away when you inflate the balloon). The rightmost structure shows extra features. Time now also has an end t_c – and hence the whole universe a finite life – and space first expands and then contracts.

The equations of general relativity allow for (higher-dimensional) versions of all these possibilities. Astronomical observation, however, reduces the collection of valid ones. During the second half of the twentieth century an increasing separation of celestial bodies was noted, a phenomenon that can only be explained if the universe is expanding. But looking backwards in time this expansion becomes a contraction whose endpoint is at an initial time t_b. This birth, and the subsequent early infancy, of the universe is referred to today as the *Big Bang* and its history is a matter of study among astrophysicists.

So time has an origin and it can be located around 14 billion years ago. But what about space? Which is the shape of the spatial sections (that is, of the geometric spaces obtained by fixing a time, such as the lines in the left of the picture above or the circles in the remaining three drawings)? An answer to these questions should provide at least both the curvature of these sections and their connectedness. Mathematicians would say that the former is related to the geometry of these spaces and the latter to their topology. In truth, both issues are of a geometric nature, but the former deals with quantitative aspects (distances, angles, etc.) and the latter with qualitative aspects (whether the space is wrapped around itself and, if so, how many "holes" it has[3]). There is a general agreement among cosmologists that neither the curvature type (negative, positive, or zero) nor the connectedness (simple or multiple) of the spatial sections of our universe change with time. There is also agreement on the fact that this curvature is roughly constant in each section.[4] Otherwise, both curvature itself and the connectedness type are still debated.

In his latest book, Jean-Pierre Luminet (2008) gives a vivid rendition of both the scientific framework within which the search for the shape

---

[3] A very good introduction to the topology of three-dimensional spaces, with more intuition than technical conundrums, can be found in (Weeks, 1985).

[4] In our one-dimensional universe this would imply that sections can be lines or circles but, for instance, not ellipses.

of the universe transpires and his own quest for an answer.[5] His view, which appears to be strongly supported by data collected by the WMAP (Wilkinson Microwave Anisotropic Probe) satellite, is that the shape of the spatial sections of the universe correspond to what is known as a *Poincaré dodecahedral space*. This is an orientable space with positive curvature obtained by glueing together (in a manner similar to the glueing of opposite hemispheres to construct the model $\mathfrak{P}$ of the projective plane in Section 12.5) the opposite sides of a dodecahedron, such as the one in Figure 13.1 but with a present radius of about 45 billion light-years.

Although not broadly accepted, the possibility of Luminet's thesis being valid has consequences which are worth noting.

First, because of their positive curvature, the spatial sections of the universe are finite. All three-dimensional spaces with positive curvature are so. Second, because Poincaré's dodecahedral space is multiply connected, observers on Earth might see ghost images of some celestial objects like those of Queenie seen by the C-lander Paul. The sheer size of the dodecahedral space, nonetheless, puts restrictions on these mirages. Indeed, because the universe has its origin 14 billion years ago, there is a limit on the length of light trajectories, namely the age of the universe times the speed of the light. This yields an observable universe of a radius of about 53 billion light years:[6] no trajectory longer than this radius can be observed at present. Now, this radius of the observable universe is greater than that of the Poincaré's dodecahedral space postulated as our actual universe. Therefore, ghost images of a celestial body may be found. But since it is not much greater, few such images are possible.

What would be the consequences for our discussion of Luminet's thesis being valid? They would be mainly of a philosophical nature: we would know that the true geometry – in the sense of faithfully describing the spatial sections of the universe – is non-Euclidean and we would even know several of its properties. But the curvature of the universe, because of its colossal size, is small. And because of this, as Luminet (2008: 13) points out, "on distances between 10^{-18} meters (a billionth of a billionth of a meter [...]) and 10^{11} meters (100 million kilometers, on the order of the Earth–Sun distance), the geometry of space is described very well by that of ordinary three-dimensional Euclidean space". If Luminet's thesis is valid, we are therefore in a situation reminiscent of that of Flatlanders, who cannot grasp the existence of a third dimension.

---

[5] Readers wishing to understand the kind of physical data collected by astrophysicists and the way cosmologists use these data to refute or confirm their theories – a topic we could not possibly enter into – may find a convenient approach in Luminet's book.

[6] This estimate takes into account the fact that the universe is expanding.

We live in a finite, curved, space but, with the exception of a few scientists (the earthly versions of A. Square), we cannot grasp this.

We are thus ending our excursion through the vicissitudes of geometry by returning to our starting point, the appropriateness of Euclidean geometry as a description of the world at the scale we can perceive it. Nonetheless, we return to this evidence with a deeper understanding of its strengths and shortcomings and may probably say, paraphrasing Eliot's *Little Gidding* (Eliot, 1968), that we return to this starting point knowing the place for the first time.

APPENDIX: RULE-DRIVEN CREATION

Doesn't the fugue imply the composer's submission to the rules? And is it not within those strictures that he finds the full flowering of his freedom as a creator? [...] My freedom will be so much greater and more meaningful the more narrowly I limit my field of action.

I. Stravinsky (1960: 79, 68)

A sweeping look at the preceding chapters would probably reveal the role of mathematics in rule-driven artistic creation as the main theme in this book. We initiated our discussion with the idea, stated in Section A.4, of visiting the crossing paths of art and mathematics. As this visit proceeded, though, we saw diverse conceptions of geometry (either as the assumption of some characteristics of space or as an idea of shape set via a group of transformations on this space) taking a prescriptive capacity with regard to the artist's work. The laws of geometry – of whichever geometry may underlie the artwork – acted both as a catalogue of what is possible (and we mentioned the role of mathematicians as catalogue-makers in this context) and as constraints precluding the impossible. We may talk about *rule-driven* (or *constraint-based*) artistic creation.

The framework of rule-driven artistic creation, however, is broader than the contents of our exposition: indeed, art practice can be driven by rules which do not have a mathematical character, and, actually, this is often the case.

In the next few pages we briefly outline the role of rules in artistic creation, with an emphasis on the different attitudes adopted by artists with regard to established sets of rules, and, for the sake of completeness, we comment on the practice of writing under constraints. The somehow marginal position of these themes in this book should explain their inclusion as an appendix.

Compliers/benders/transgressors

WALTHER: Wie fang' ich nach der Regel an?
SACHS: Ihr stellt sie selbst und folgt ihr dann.[1]

R. Wagner *Die Meistersinger von Nürnberg*, III.2

To succeed in marrying Eva Pogner, Walther must qualify as a mastersinger. In his quest he counts on the help of Hans Sachs, a cobbler

---

[1] WALTHER: How say the rules I must begin?/SACHS: You set them yourself and then follow them.

and mastersinger himself. Such is, summarized to the extreme, the argument of *Die Meistersinger von Nürnberg*, from where the lines above are taken. The contents of these lines are as crucial to Walther's endeavour as they are central in our discussion, the latter because they bring to the limelight the issue of the emergence of rules. Disoriented in a lawless world, Walther longs for rules which will both set admissible procedures and a scale to gauge how successfully these procedures are followed. Sachs' words, we will see, are representative of a distinctive attitude in the relationship between artist and rules.

This relationship has been approached in the literature from diverse viewpoints, which emphasize varying degrees of severity for the rule at hand. Different words, such as "norm", "constraint" or "rule" are correspondingly used (but it would be wrong to assume that the implied semantic boundaries are clear-cut). Thus, Mukařovský (1978) writes about "aesthetic norms", drawing attention to their flexibility as opposed to the rigidity of legal norms or technical rules. Also, on the other endpoint, some contemporary forms of art, such as *evolutionary art*, take the rule-driven character to the limit by enforcing an algorithmic procedure in the creation of the artwork (Todd and Latham, 1992).

For the purposes of this section, even though removed in principle from the subject of artistic creation, a book by Jean Piaget (1977) on the formation of moral judgements will be a convenient starting point. The first chapter of Piaget's book is devoted to the acquisition of, compliance with and rebellion against rules, though in a very specific context: the game of marbles as played by the children of Geneva and Neuchâtel (in the early 1930s). One of the aspects studied by Piaget is the "consciousness of rules, i.e. the idea which children of different ages form of the character of these game rules, whether of something obligatory and sacred or of something subject to their own choice".

Piaget's conclusions regarding this issue are summarized in the distinction of three stages. During the first such stage the "rules are not yet coercive in character, either because they are purely motor, or else [...] because they are received, as it were, unconciously, and as interesting examples rather than as obligatory realities".

During the second stage the "rules are regarded as sacred and untouchable, emanating from adults and lasting forever. Every suggested alteration strikes the child as a transgression".

During the third and last stage, "a rule is looked upon as a law due to mutual consent, which you must respect if you want to be loyal but which it is permissible to alter on the condition of enlisting general opinion on your side".

Whereas the granting of an attitude or behaviour to the artist corresponding to the first stage does not appear compelling to us (if such a granting is at all possible it should be associated with an early stage of the artist's life, previous to the practice of any art form), the last two strike us as templates of characteristic artists' behaviours.

The second stage is indeed exemplar of an attitude which we may describe as *rule-complying*. The artist is in possession of a set of rules – imposed by the medium of his art form, acquired through education, or prevalent in his cultural environment – by which he abides. We have met many occurrences of such compliance: the wallpapers in Figures 3.12 and 3.13, the rosettes of medieval churches, the rugs illustrating Section 4.2, all of them with respect to the laws of symmetry; the metric and rhyme patterns in classical poetry, the use of geometric transformations in the canons of the *Musical Offering* or the observance of the laws of perspective in Piero della Francesca (Figure 9.5) are examples.

Likewise, the third stage, as described above, opens the doors for an attitude we may call *rule-breaking*. The artist is again in possession of a set of rules, but, instead of abiding by these rules, replaces them by a new set, laid down by himself, which will draw anew the boundaries of the possible. Again, we may exemplify these transgressions with artworks from the previous chapters: Picasso replacing the established one-viewpoint perspective by a multiple-viewpoint drawing system (see Figure 11.16) and Escher introducing wallpapers based on non-isometric transformations (Figure 7.2) or on non-Euclidean isometries (Figures 13.3, 13.4, and 13.6). Another example, now taken from music, is the replacement of the respect for the diatonic scale by a method of composition, known as *dodecaphony* and introduced by Arnold Schoenberg, which aims at using each of the 12 tones in the chromatic scale with approximately the same frequency.

It is of the essence at this point to observe that the possession of aesthetic value is neither favoured by any of these attitudes – the exactness of Leonardo in following the rules of perspective is as worthy as the revolution by Picasso in breaking precisely these rules – nor a direct consequence of any of them. In the words of Jon Elster (1983: 79), "One should not fall into the trap of thinking that good art has to involve a break with tradition, nor a conscious reaffirmation of it". These words are extracted from a passage in (Elster, 1983) where, incidentally, he discusses the role of rules, or constraints, in artistic creation in terms of well-known mathematical ideas.

According to Elster, the possible actions of the artist when facing the creation of an artwork, tabula rasa, would be too many were it not for the presence of constraints that limit the artist's possibilities. These constraints decrease the "degrees of freedom" of the artist, lowering, so to say, the dimension of his space of possible choices. The artist then attempts to maximize the function that associates an aesthetic value with each such choice.[2] There is no attempt by Elster to make precise the nature of this aesthetic value (as there is, for instance, in the writings of Lorand we discussed in Section 5.1). Less demanding on his premises, Elster just assumes that artists can evaluate the aesthetic value of a given feasible point. Endowed with this ability, he postulates, artists start with a general idea of the artwork to be – which roughly corresponds with a region of feasible points – and then strive to isolate the details for this idea that best materialize it – that is, to maximize the aesthetic value on this feasible region. The choice at which this maximization is achieved would thus correspond to a configuration of the intervening elements where no detail can be altered without harming the whole (an idea of equilibrium long since associated with beauty).

It goes without saying that there is no mechanical algorithm to perform this maximization. Actually, even though not made explicit by Elster, it follows from his text that a common procedure to do so is some educated form of trial and error. Indeed, Elster (1983: 81) mentions

the extreme importance of drafts, sketches etc. in the artistic process. Poets typically play around with small variations, trying out one word, rejecting it, substituting another, until finally they decide that they have got it right. The sketch books of painters provide ample evidence of the same approach.

It is to be noted that the maximization process is *local*. The variations allowed to the artist are around the initial idea for the artwork. It corresponds with looking for the summit of a given mountain, a process that disregards the possible existence of higher mountains nearby. Also, and this is a main consideration in Elster's arguments, the trial-and-error procedure is doable because of the constraints in place which limit the space where this search is to be conducted.

Implicit in these considerations is the fact that, besides a mastership of the technical skills required by the medium of his art, an artist should be adroit both in the appraisal of aesthetic value and in the procedure of local maximization supported upon this appraisal.

Compliance and transgression are extreme attitudes between which we may still identify a posture, call it *rule-bending*, with a halfway degree of observance. It does not correspond with any of Piaget's stages above, because children's games are not played with such degree: rules are meant

---

[2] The lexicon used by Elster is taken from a mathematical subject called "optimization". The template problem in this subject is to maximize (or minimize) a function, called the *objective function* over a set of *feasible* points. Any such point maximizing the objective function is called an *optimizer* or *optimal solution*. The conditions determining which points in the domain of the objective function are considered to be feasible are called *constraints*.

to be followed. We can replace rules by new ones (and we obtain a different game), but we cannot play a game with rules only partially followed. Doing so is tantamount to cheating.

In the doings of the artist, however, there may be purposeful deception (and the ceiling by Andrea Pozzo in Figure 11.2 serves us as a reminder) but there is no cheating; the artist may proceed as he wishes and eventually it is his artwork that will be subject to judgement (a fact of which the gentleman painting the lute in Figure 9.7 is well aware).

In the presence of a notion of space underlying the application of a rule, we can further separate rule-bending in *local-bending* and *global-bending*.

In local-bending, there is an attempt to produce the sense associated with the application of determined rules but these rules are not followed to the detail. We have seen many instances of this phenomenon in Chapter 6. The painting by Guayasamín in Figure 6.3, for example, conveys the feelings of central symmetry but it is not centrally symmetric (in the same way that, for instance, the photography by Matchett in Figure 3.6 is). The works by Warhol in Figures 6.5, 6.6, and 6.7 are gentler examples of local-bending.

In global-bending, in contrast, a general appearance of abiding by the rules is preserved but these rules are in fact twisted to obtain unexpected effects. The works by Del-Prete or Escher in Section 11.4 are exemplars: observation of isolated regions in these artworks will not reveal any departure from the dictates of parallel projection, but the consideration of the scene in its entirety reveals an inconsistency. The staircase of Reutersvärd in that section is a simpler, more direct, occurrence of this phenomenon. Finally, the modulating spiral character of Bach's Canon 5 in the *Musical Offering* provides us with a different kind of example of global-bending: the change in tonality is not perceived locally but needs the melody to be played entirely to allow the listener to detect (and this requires quite a musical ear) that the tonic has changed.

The starting point of the discussion above, the analysis of the game of marbles by Piaget, is less capricious than what it might seem at a first glance. There is a line of thought, going back at least to Johan Huizinga's *Homo Ludens* (Huizinga, 1998), which emphasizes the relationship between play and culture (and, in particular, art). This line of thought is followed by Héctor Rodríguez (2008) in his analysis of the film *The Five Obstructions*. The film's analysis is immaterial to our discussion but the framework described by Rodríguez isn't. It relies on a study by Roger Caillois (2001) containing "an extended treatment of the relationship between playing and constraint". Chapter II in Caillois' book proposes a taxonomy of games based on two parameters. The first establishes four "fundamental categories" of games: competition, chance, mimicry and vertigo. The second, whose exposition occupies a section entitled "From Turbulence to Rules", postulates a gradation between two categories of play: *paidia* and *ludus*. The former encompasses "the spontaneous manifestations of the play instinct: a cat entangled in a ball of wool, a dog sniffing, and an infant laughing at his rattle represent the first identifiable examples of this type of activity" (Caillois, 2001: 28). The latter "is born [from] the desire to invent rules, and to abide by them whatever the cost". In the words of Rodríguez, "*ludus* denotes any system of rules deliberately designed to generate a gratuitous obstacle or challenge for the player". And he continues with a crucial remark: "The difficulty is the point of playing. The obstacle is gratuitous in the sense that it has been set up solely for the pleasure of overcoming it."[3]

The above considerations tie in well with our discussion about art and rules. The equivalent of *paidia* (e.g. a toddler scribbling on a surface with a pencil) was purposely left out of it. But the multifaceted rules (social, ethical, technical, etc.) governing the practice of a given art form are the inherited *ludus* for its practitioners.

Constrained writing

Practically without exceptions, we have confined our discussion on the role of mathematics in rule-driven artistic creation so that "mathematics" were limited to geometry and "art" to various forms of visual art (or, in a single chapter, to music). The only reference to literature was in Section 6.5, where we found some geometry in the strictures of poetry.

The search for similar instances of geometry-constrained creation in other forms of literature is unlikely to end in a convincing collection. Yet, the exercise of writing has a rich tradition of creation under constraints. It just happens that these constraints do not necessarily relate with geometric thinking.

The first example that comes to mind is that of the classical unities. These were rules loosely derived from some passages in Aristotle's *Poetics* that became rigid prescriptions for drama writers during the period between the fifteenth and nineteenth centuries. The unities were those of "action" (the play must have a main action with few or no subplots), "place" (this action should develop on a single physical location) and "time" (the action has to take place in the course of a single day).

The consideration of classical unities provides outstanding examples of both rule-compliers and rule-breakers. The respect of the unities in works such as *Phèdre* by Racine and *Tartuffe* by Molière is scrupulous. In contrast, if we give credence to Samuel Johnson (1984: 429) in his Preface

---

[3] This property is what sets *ludus* apart from competition. As Caillois (2001: 29) puts it, "the conflict is with the obstacle, not with one or several competitors".

to *The Works of William Shakespeare*, a similar care was taken by the swan of Avon to disregard them:

It will be strange enough that, in enumerating the defects of this writer, I have not yet mentioned his neglect of the unities; his violation of those laws which have been instituted and established by the joint authority of poets and of critics.

After which, Dr Johnson comes to Shakespeare's defence:

For his other deviations from the art of writing, I resign him to critical justice [...] But from the censure which this irregularity may bring upon him, I shall, with due reverence to that learning which I must oppose, adventure to try how I can defend him.

This defence is not foreign to our discussion in the preceding section; it illustrates the process of abandoning an established set of rules and replacing them by new ones. Johnson (1984: 438) points out that Shakespeare

found the English stage in a state of the utmost rudeness [...] Neither character nor dialogue were yet understood. Shakepeare may be truly said to have introduced them both amongst us, and in some of its happier scenes to have carried them both to the utmost height.

He further (Johnson, 1984: 435) gives a finality to this introduction of character and dialogue, as

he always makes us anxious for the event and has perhaps excelled all but Homer in securing the first purpose of a writer, by exciting wrestless and unquenchable curiosity, and compelling him that reads his work to read it through.

The classical unities contrast with constraints such as metre and rhyme, in that the former applies to the contents of the piece and the latter to its form. Constraints on contents have boundaries which are less defined than those for constraints on form. Whether a tragedy respects, say, the unity of action may be a matter of argument; whether a poem abides to a given metre pattern is certainly not. It is therefore not surprising that constraints having a mathematical nature are almost invariably on form.

It needs to be said that the distinction between constraints on form and constraints on content, while useful in principle, is not absolute. Rhyme and metre are purely formal constraints, as they rely on phonetic (or prosodic) properties of the text, not on its meaning. But the history of literature shows a wealth of structure overlapping, in different degrees, meaning and form. A case at hand is the Russian folk tale, as shown in the analysis by Vladimir Propp (1968).

Although comparatively rare, a noteworthy family of constraints are those relying neither on phonetics nor on meaning but on the material writing of the intervening words, on their spelling. Most texts obtained in this way are short and have no literary claims. They are usually referred to as *wordplay*, an expression that ties this form of constrained writing with the considerations at the end of the preceding section.

A kind of wordplay we have already met (Bach's dedication of the *Musical Offering* in Section A.3) is the *acrostic*. This is a word or sentence obtained by taking the letters of every word in a text (or of a well-defined subset of such words in the text) occupying a determined position (e.g. first, second, last) in that word.

Another instance is the *anagram*, a word or sentence resulting from permuting the letters of a different, designated, word or sentence (sometimes called *programma*). At the beginning of the twentieth century, Walter Begley compiled a book with many of the anagrams that, with a programma taken from the (Latin version of the) Bible, had been stockpiling in a variety of libraries in Europe. He called his book *Biblia Anagrammatica* (Begley, 1992). In it we find a moderate subset of what is probably the most notable set of anagrams ever produced, the one having as programma the opening sentence of the "Hail Mary": *Ave Maria gratia plena dominus tecum* (Hail Mary, full of grace, the Lord is with Thee) occurring in Luke i.28. Begley mentions that Francesco Saverio Gaetano (better known as Padre Bonaventura da Santa Rosalia) "eventually composed 5300 in Italian and 2500 in Latin [of these anagrams]". Other priests, such as Pompeius Salvi or Donino Fidalmi, also made anagrams with the same programma. To convey an idea of this combinatorial deed we list some of them. We start with eight in Latin,[4] along with a translation to English.[5]

> Ave Maria gratia plena dominvs tecvm
> Hail Mary, full of grace, the Lord is with Thee
> Pvra vnica ego svm Mater alma Dei nata
> I am pure and unique, the mother born of God
> Deipara inventa svm, ergo immacvlata
> I was created to become God's mother, hence I am free of sin.
> Pia, mvnda, ivsta, alme creatorem genvi
> Pious, clean, just, I gave birth to life's creator.
> Ego avrvm nitens immacvlata Deipara
> I am shining gold, the Immaculate Mother of God

---

[4] Nota Bene: These anagrams follow the classical spelling of Latin which allowed V to be used both for the vowel "u" and for the semi-consonant "w", and similarly for I, "i" and "j". This freedom gave the good padres a bit of an advantage in the composition of his anagrams, but yet . . . .

[5] In all the following examples the anagramatic constraint, and occasionally some additional ones, removed some naturalness from these sentences. To avoid magnifying this feature in the translations to English, the latter will only attempt to convey the general meaning of the anagrams.

VISNE ANAGRAMMATA? DE PVRO MVLTA IECI
Do you want anagrams? I have build many without defect
VIS CENTVM ANAGRAMMATA? EDO PVERILIA
Do you want 100 anagrams? I produce them (like) children
DIC: VIVAT SEMPRE VIRGO ALMA NATA. AMEN
Long live the Virgin, born virgin and mother. Amen.

We next list a few in Italian (the first four due to Padre Bonaventura and the last four crafted by Donino Fidalmi):

Divin' amore segna Te pura, Immaculata.
Divine love marks you chaste, Immaculate.
Alma giust' e pura, mai, mai v' e caduta.
Fair and pure soul, you have never ever fallen into temptation.
Tu la mia cara Vigna, vite sempre monda.
You, my beloved Vineyard, ever clean grapevine.
Tu pura, ne mai legata da comun miseria.
You spotless, never burdened with mundane misery.
Spargi un' aura mite d' Immacolata Neve.
You diffuse a gentle breeze of Immaculate Snow.
Serva Immacolata, Vita Pregna di Nume.
Immaculate Maid, Life Impregnated with God.
Mia unica Pura or damni sangue e latte.
My only Pure (being), now give me flesh and nourishment.
Immacolata, Pura m'insegna due verità.
Immaculate and Pure she teaches me two truths.

The abundance of anagrams based on Luke i.28 was eventually accompanied by an increase in the challenge's magnitude (a manifestation of *ludus*), which took the form of the requirement of additional constraints. An example is the use of metrical constraints. Begley mentions Franciscus Turrius, who wrote poems in hexameters and pentameters on which each line is an anagram of Luke i.28 (and includes in his *Biblia* some of them). He furthermore includes the following "Rhythmical Quatrain" by Pompeius Salvi:

MATER PIA, LUCIS VIA, UNA MORTEM DENEGA.
EIA, DURAM PONE CURAM ANTE, SIMUL AGITA.
I AMATA, NEC PLAGATA MORSU, DIVI MUNERE,
DEI NATA EI GRATA, PURA AC SUMMO LUMINE.

Pious mother, path of light, ward death off.
Ho! impose a severe remedy (before is late) and weigh it.
Go, beloved, not beaten by inflictions, by the gift of God,
Daughter of God, pleasant to Him, pure and full of brightness.

about which he mentions that "it deserves a second reading, both to catch its assonance and structure, and to be able to grasp the immense difficulties overcome in producing it" (Begley, 1992: 145).

A different choice of additional constraint was done by Lucas de Vriese who, in 1711, produced the following lines, each of them an anagram of Luke i.28, which are also an acrostic of its programma, and obey yet one more rule, namely that each line (except, obviously, the first) begins with the word closing the preceding line.

Amacula ter munda, ita per omnia viges.
Viges, enormi mulcta Adami pura enata.
Enata Malis pura vige, ac merito Munda.

Munda Mater emicas, o pura Geniti Aula.
Aula Dei micat, nota summe pura, Regina.
Regina, o Tu pura macula, et Dia Immensa.
Immensa, o Tu diva integre pura ac alma.
Alma ter unice pura Summa io Dei Gnata.

Gnata Dei, pura es communi a Mali reatu.
Reatu magno pura, micat sine lue Adami.
Adami sine omni macula pura, rege tuta.
Tuta o pergas alma ac nimia munda jure.
Iure mero Genita munda a culpis, Amata.
Amata veni Summa Regina, delicto pura.

Pura et ter divina o gemmas, Amica luna.
Luna pura (mira dico) Agni Stemmate Eva.
Eva, i matris culpa e gremio munda nata.
Nata maledicti pura, o vere Summi Agna.
Agna Cœli summa, et Avi ter pura damni.

Damni tu pura Regia es, et a macula omni.
Omni reatu, ac Avi plagis e matre munda.
Munda tu pia merito maculæ es ignara.
Ignara culpæ mera, o Summi Tu Dei Nata.
Nata Pura Medica, et gloria Summa veni.
Veni multa munda, Pia et a gremio Sacra.
Sacra nimie munda, alme pura vige tota.

Tota piaculis munda mera, germina Eva.
Eva o simul prima et munda genita, Cara.
Cara, imo Summi Nata, et digne pura, vale.
Vale, o mendi pura Mater, ac Vitis Magna.
Magna, o sic pura ad literam, vive. Amen.

This poem (roughly) translates to English as follows.

Thrice clean from stain, that is why you blossom.
You blossom after being born free from Adam's great curse.
Born of sinners, you blossom pure and clean, due to your own merit.

Clean you shine, Mother, oh pure Temple of the Only Begotten Son.
God's Temple shines, famous for its great purity, oh Queen.
Oh Queen, you who are free of stain, incommensurable Divine.

Oh incommensurable, you are divine, immaculately pure and nourishing.
Nourishing, thrice peerlessly pure, oh greatest daughter of God.

Daughter of God, you are free from original sin.
Free from the greatest sin, you shine free from Adam's curse.
Pure, clean of Adam's stain, protected queen.
Continue protected, oh nourishing and so justly clean.
Justly Daughter free from guilt, Beloved.
Come, oh Beloved, greatest Queen, free from guilt.

Pure and thrice divine, you are adorned with gems, oh loving moon.
Pure moon (I speak of marvelous deeds), Eve of the Lamb's lineage.
Eve, go, born free from guilt in her mother's womb.
Born free from blame, truly Lamb of the Highest.
Greatest Lamb of Heaven and thrice free from the Ancestor's harm.

You, oh Queen, are free from harm and from stain.
From all sin and from the Ancestor's calamities you are free since birth.
Clean through your own merit pious, you have not known any stain.
Merely ignorant of the guilt, oh you, born from God the Highest.
Born Pure, Healer, come also oh Highest in glory.
Come clean from punishment, Pious and Holy from your mother's womb.
Sacred, mightily clean, motherly pure, you live protected.

Bloom protected, you, the only one clean from expiatory punishment, Eve.
Oh Eve, first born as well as clean, Beloved.
Beloved, truly born from the Highest and fittingly pure, be strong.
Be strong, oh Mother free from fault, and Great Vineyard.
Great, oh truly pure, live. Amen.

A form of wordplay that is close to our themes is the *palindrome*. This is a word or sentence that, neglecting punctuation marks and spaces between words in the case of a sentence, reads the same backwards and forwards. That is, the text possesses bilateral symmetry with respect to the "middle letter position" (one may say it is a verbal form of crab canon). The word "deified" is a simple example; the sentence "Murder for a jar of red rum" is another.

The catalogue of wordplay is too large for us to proceed with its enumeration. We refer to Eckler (1996) for a comprehensive book on this topic. We cannot close this appendix, however, without mentioning a few cases of constrained writing for which there is an undeniable literary ambition.

The most notorious is the lipogrammatic novel *La Disparition* of French author Georges Perec (1969). A *lipogram* is a text on which one or more letters do not occur. In the case of *La Disparition*, the banished letter is the vowel "e". This constraint reduces the lexicon to be used to about an eighth of the French dictionary. Yet, the Perec novel, of around 300 pages, "was so well written that at least some reviewers never realized the existence of a letter constraint" (Eckler, 1996: 3). It is remarkable that the novel has been translated (the word "recreated" might be more appropriate) to a number of languages. In particular, the translation to English by Gilbert Adair (Perec, 1995) is remarkably good.

Although to put across an inkling of what *La Disparition* is about (its plots, turns and twists, individuals' dispositions and psychological traits), would come accross as a foolish ambition – a ridiculous goal, a toying with absurdity – I might try, notwithstanding, to approach this task by simply transcribing a song that Adair, in his translation, assigns to John Milton (and which you might confront with a distinct, but concomitant, song on §A.2):

ON HIS GLAUCOMA

Whilst I do think on how my world is bound,
Now half my days, by this unwinking night,
My solitary gift, for want of sight,
Lain fallow, though within my soul abound

Urgings to laud th'Almighty, and propound
My own account, that God my faith not slight,
Doth God day-labour claim, proscribing light,
I ask; but calming spirits, to confound

Such murmurings, affirm, God doth not dun
Man for his work or his own gifts, who will
But kiss his chains, is dutiful, his gait

Is kingly. Thousand to his bidding run
And post on land and bounding main and hill:
Your duty do who only stand and wait.

Three years after writing *La Disparition* Perec wrote another lipogram, *Les Revenentes* (Perec, 1972), where, mirroring his first, the only allowed vowel is "e". Because this is an even more severe restriction, Perec allowed for some exceptions to the constraint, but these exceptions are nevertheless regulated in a way that the resulting text is clearly an instance of constrained writing. Again, it is somehow surprising that a (very good) translation to English has been done, this time by Ian Monk (Perec, 1996).

Perec (1986) also wrote a short history of the lipogram. True to himself, he warns that the word does neither refer to "greasy letter" nor to "a gram of grease". He mentions early occurrences of this constraint, such as some poems by Lasus of Hermione who lived on the sixth century BCE. Perec mentions the Abbé Barthélemy, "whose opinion we have no reason to doubt", who considered Lasus as a musician above all and justified in this manner a possible dislike for "the unpleasant hissing of the Sigma". Lasus' works are lost, but the first verse of his *Hymn to Demeter* remains:

Δήμητρα μέλπω Κόραν τε Κλυμένοιο ἄλοχον

(I sing of Demeter and Kore, the bride of Klymenos). Perec then records several lipograms and classifies them according to the excluded letter. Close to our times (and to *La Disparition*) he mentions Alonso de Alcalá y Herrera, who in 1641 published "a charming octavo entitled *Varios efetos de amor, en cinco novelas ejemplares, y nuevo artificio para escrivir prosa y versos sin una de las letras vocales.*[6] One finds therein: *Los dos soles de Toledo,* without the A; *La carroza con las damas,* without the E; *La perla de Portugal,* without the I; *La peregrina ermitaña,* without the O; and *La serrana de Cintra,* without the U".

A last (in this appendix) example of formal constraint is the novel *Alphabetical Africa* by Walter Abish (1974). This work has 52 chapters, of which the first has only words beginning with "a", the second only words beginning with either "a" or "b", and so on until the twenty-sixth, where all words of the English language are allowed. Then the process is reverted between chapter 27 (which, as the one preceding it, has no constraints) and the 52nd on which all words begin again with "a".

Constrained writing became the *raison d'être* of a group of (mostly) French writers (and some mathematicians) called *Oulipo* (from *Ouvroir de Litératture Potentielle,* which means "Workshop of potential literature"). Members of this group periodically met to discuss possible constraints in the exercise of writing, but most of these discussions did not end in actual constrained-written texts. In this sense, they remained accurate to the word "potential". Notable exceptions are works by Perec and Italo Calvino.

A panorama of the people, works, themes and discussions making the *Oulipo* can be found in (Motte, 1998) and (Mathews and Brotchie, 1998).

---

[6] Various consequences of love, in five exemplary tales, and new devices for writing in prose and verse without one of the vowels.

We could carry this article even further, but what we have just said will be enough for those who are capable of thinking, while for the others we should never be able to say enough. Some rather heavy metaphysical passages will be found in it, but that could not be avoided. We had to speak about art in general and our propositions therefore had to be general. It is only common sense that the more general a proposition is the more abstract it must be, abstraction being the extension of a truth by removing from it the details which limit it to the particular. If we had been able to spare the reader these thorny passages we should also have saved ourselves a great deal of work.

D. Diderot (2000), Entry *Art* in the *Encyclopédie*

REFERENCES

Abbott, E. (1987). *Flatland: A Romance of Many Dimensions*. London: Penguin.

Abish, W. (1974). *Alphabetical Africa*. New York: New Directions Books.

Alberti, L. B. (1991). *On Painting*. London: Penguin. Translated by C. Grayson with introduction and notes by M. Kemp.

Alder, K. (2002). *The Measure of All Things*. New York: The Free Press.

Alperson, P. (2004). Music: formalism and beyond. In P. Kivy, ed., *The Blackwell Guide to Aesthetics*. Oxford: Blackwell. pp. 254–75.

Andersen, K. (2007). *The Geometry of an Art*. New York: Springer-Verlag.

Anglin, W. S. (1997). *The Philosophy of Mathematics: The Invisible Art*. Lewiston, NY: Edwin Mellen Press.

Antonova, C. (2010). On the problem of "reverse perspective": definitions East and West. *Leonardo*, **43**(5), 464–9.

Arnheim, R. (1984). *Art and Visual Perception: A Psychology of the Creative Eye*. Berkeley, CA: University of California Press.

Asimov, I. (1983). *The Universe: From Flat Earth to Black Holes and Beyond*, 3rd edn. London: Penguin.

Balanchine, G. (1968). *Balanchine's New Complete Stories af the Great Ballets*. Garden City, NY: Doubleday. Edited by F. Mason with drawings by M. Becket.

Baloglou, G. (2007). *Isometrica: A Geometrical Introduction to Planar Crystallographic Groups*. Online version available at http://www.emis.de/monographs/Isometrica.

Banham, M. (1995). *The Cambridge Guide to Theatre*. Cambridge: Cambridge University Press.

Barber, E. J. W. (1991). *Prehistoric Textiles: The Development of Cloth in the Neolithic and Bronze Ages*. Princeton, NJ: Princeton University Press.

Bazin, A. (1960). The ontology of the photographic image. *Film Quarterly*, **13**(4), 4–9.

Begley, W. (1992). *Biblia Anagramatica*. Whitefish, MT: Kessinger Publishing LLC. Reprint from the original 1904 edition published in London by Hazel, Watson, and Viney. Curiously, the original edition has two m's in *Anagrammatica* but the reprint spells the word in the cover with only one.

Belkin, K. L. (1998). *Rubens*. London: Phaidon.

Benson, D. (2007). *Music: A Mathematical Offering*. Cambridge: Cambridge University Press.

Bernstein, L. (1976). *The Unanswered Question. Six Talks at Harvard*. Cambridge, MA: Harvard University Press.

Birkhoff, G. D. (1933). *Aesthetic Measure*. Cambridge, MA: Harvard University Press.

Blake, W. (1970). *Songs of Innocence and Experience*. Oxford: Oxford University Press.

Blunt, A. (1979). *Borromini*. Cambridge, MA: Harvard University Press.

Booker, P. J. (1963). *A History of Engineering Drawing*. London: Chatto & Windus.

Borges, J. L. (1964). *Other Inquisitions*. Austin, TX: University of Texas Press.

Bostock, D. (2009). *Philosophy of Mathematics*. Chichester: Wiley-Blackwell.

Bourdon, D. (1989). *Warhol*. New York: Harry N. Abrams, Inc.

Brewster, D. (1819). *Treatise on the Kaleidoscope*. Edinburgh: Archibald Constable & Co.

Bühler, W. K. (1981). *Gauss: A Biographical Study*. New York: Springer Verlag.

Burke, E. (1990). *A Philosophical Enquiry into the Origin of our Ideas of the Sublime and Beautiful*. The World's Classics. Oxford: Oxford University Press.

Butler, S. (1967). *Hudibras*. Oxford: Oxford University Press.

Caillois, R. (2001). *Man, Play and Games*. Urbana, IL: University of Illinois Press.

Campbell, G. (2001). The life records. In T. H. Corns, ed., *A Companion to Milton*. Oxford: Blackwell. pp. 483–98.

Carey, J. ed. (2000). *John Donne: The Major Works*. Oxford: Oxford University Press.

Carroll, L. (1993). *Alice's Adventures in Wonderland & Through the Looking-Glass*. Ware: Wordsworth.

Carroll, N. (2003). Dance. In J. Levinson, ed., *The Oxford Handbook of Aesthetics*. Oxford: Oxford University Press. pp. 583–93.

Cederberg, J. N. (2001). *A Course in Modern Geometries*, 2nd edn. Undergraduate Texts in Mathematics. New York: Springer-Verlag.

Cheney, L. (1997). Stylistic problems in Mannerism and Maniera. In L. Cheney, ed., *Readings in Italian Mannerism*. New York: Peter Lang. pp. 1–8.

Cook, T. A. (1979). *The Curves of Life*. New York: Dover Publications, Inc.

Coxeter, H. S. M. (1979). The non-Euclidean symmetry of Escher's picture "Circle Limit III". *Leonardo*, **12**, 19–25.

Culler, J. (1997). *Literary Theory: A Very Short Introduction*. Oxford: Oxford University Press.

Da Vignola, J. B. (1987). *Le Due Regole della Prospettiva Practica*. Alburgh: Archival Facsimiles Limited. A reproduction of the copy in the British Library.

Da Vinci, L. (2002). *A Treatise on Painting*. New York: Prometheus Books.

Darwin, C. R. (1962). *On the Various Contrivances by which British and Foreign Orchids are Fertilised by Insects*. London: John Murray. Online version available at http://darwin-online.org.uk.

Davies, P. (2000). *Antique Kilims of Anatolia*. New York: W.W. Norton & Company.

Delacroix, E. (1938). *The Journal of Eugène Delacroix*. London: Jonathan Cape.

Descartes, R. (1954). *The Geometry*. New York: Dover Publications, Inc. Translated from the French and Latin by D. E. Smith and M. L. Latham. With a facsimile of the first edition.

Diderot, D. (2000). Art. In I. Frank, ed., *The Theory of Decorative Art: An Anthology of European & American Writings, 1750–1940*. New Haven: Yale University Press. pp. 139–49.

Dubery, F. and Willats, J. (1983). *Perspective and Other Drawing Systems*. New York: Van Nostrand Reinhold Co.

Dye, D. S. (1974). *Chinese Lattice Designs*. New York: Dover Publications, Inc.

Eckler, R. (1996). *Making the Alphabet Dance*. New York: St. Martin's Press.

Eco, U. (ed.) (2004). *On Beauty: A History of a Western Idea*. London: Secker & Warburg.

Edgerton, S. Y. (2009). *The Mirror, the Window, and the Telescope*. Ithaca, NY: Cornell University Press.

Einstein, A. (1926). Entry *Space-Time* in the 13th edition of the *Encyclopedia Britannica*.

Eliot, T. S. (1968). *Four Quartets*. London: The Folio Society.

Ellis, J. (1992). *Visible Fictions*. London: Routledge.

Elster, J. (1983). *Sour Grapes*. Cambridge: Cambridge University Press.

Euclid, (1956). *The Thirteen Books of The Elements*, 2nd edn. New York: Dover. Translated with introduction and commentary by Sir Thomas L. Heath.

Euripides, (2000). *Hecuba; The Trojan Women; Andromache*. Oxford: Oxford University Press. Translated with explanatory notes by J. Morwood.

Fahr-Becker, G. (1994). *Japanese Prints*. Köln: Taschen.

Farish, W. (1822). On isometrical perspective. *Transactions of the Cambridge Philosophical Society*.

Feldman, M. (1985). Crippled symmetry. In W. Zimmermann, ed., *Morton Feldman Essays*. Köln: Beginner Press. pp. 124–37.

Feynman, R., Leighton, R. B. and Sands, M. (1963–1965). *The Feynman Lectures on Physics*. Reading, MA: Addison-Wesley.

Field, J. V. (2005). *Piero della Francesca: A Mathematician's Art*. New Haven, NJ: Yale University Press.

Fink, R. *Repeating Ourselves: American Minimal Music as Cultural Practice*. Berkeley, CA: University of California Press.

Flannagan, R. (2002). *John Milton*. Oxford: Blackwell Publishing.

Flocon, A. and Barre, A. (1987). *Curvilinear Perspective*. Berkeley, CA: University of California Press.

Ford, P. R. J. (1989). *Oriental Carpet Design*. London: Thames and Hudson.

Fong, W. and Fu, M. (1973). Metropolitan Museum of Art (New York, NY). *Sung and Yuan Paintings*. New York: distributed by New York Graphic Society.

Forkel, J. N. (1920). *Johann Sebastian Bach. His Life, Art, and Work*. London: Constable and Co.

Fry, E. F. (1966). *Cubism*. London: Thames and Hudson.

Gao, J. (1996). *The Expressive Act in Chinese Art*. Uppsala: Academiae Ubsaliensis.

Gluck, C. W. (1962). *Collected Correspondence and Papers*. New York: St. Martin's Press. Edited by H. and E. H. Mueller von Asow. Translated by S. Thomson.

Gödel, K. (1967). The completeness of the axioms of the functional calculus. In J. van Heijenoort, ed., *A Source Book in Mathematical Logic*. Cambridge, MA: Harvard University Press. pp. 582–91.

Gombrich, E. H. (1982). *The Image and the Eye: Further Studies in the Psychology of Pictorial Representation*. London: Phaidon.

Gombrich, E. H. (1984). *The Sense of Order*, 2nd edn. London: Phaidon.

Gombrich, E. H. (1989). *The Story of Art*, 15th edn. Englewood Cliffs, NJ: Prentice Hall.

Greenberg, M. J. (1993). *Euclidean and Non-Euclidean Geometries*, 3rd edn. New York: W.H. Freeman and Company.

Greene, G. (1990). In memory of Borges. In *Reflections*, London: Penguin Books. pp. 306–7.

Hanslick, E. (1986). *On the Musically Beautiful: A Contribution towards the Revision of the Aesthetics of Music*. Indianapolis, IN: Hackett. Translated by G. Payzant.

Harrison, C. and Wood, P. (eds). (1992). *Art in Theory: 1900–2000, An Anthology of Changing Ideas*. Oxford: Blackwell.

Hartshorne, R. (1997). *Geometry: Euclid and Beyond*, 2nd edn. Undergraduate Texts in Mathematics. New York: Springer-Verlag.

Hauser, A. (1962). *The Social History of Art*. London: Routledge. Four volumes edition.

Hilbert, D. (1987). *Foundations of Geometry*, 2nd edn. Chicago: Open Court.

Hironobu, K. (1991). Narrative illustration in the handscroll format. In A. Murck and W. C. Fong, eds, *Words and Images: Chinese Poetry, Calligraphy, and Painting*. New York: The Metropolitan Museum of Art. pp. 247–66.

Hodges, W. (2003). The geometry of music. In J. Fauvel, R. Flood and R. Wilson, eds, *Music and Mathematics: From Pythagoras to Fractals*. Oxford: Oxford University Press. pp. 91–111.

Hofstadter, D. (1979). *Gödel, Escher, Bach: An Eternal Golden Braid*. New York: Basic Books.

Hofstadter, D. (2002). Mystery, classicism, elegance: an endless chase after magic. In D. Schattschneider and M. Emmer, eds, *M.C. Escher's Legacy: A Centennial Celebration*. Berlin: Springer-Verlag, pp. 24–51.

Hogarth, W. (1753). *The Analysis of Beauty, Written with a View of Fixing the Fluctuating Ideas of Taste*. London.

Horemiss, S. (1970). *Optical and Geometrical Patterns and Designs*. New York: Dover Publications, Inc.

Huizinga, J. (1998). *Homo Ludens: A Study of the Play-Element in Culture*. London: Routledge.

Jablan, S. V. (2002). *Symmetry, Ornament and Modularity*. Singapore: World Scientific. An online version is available at http://www.emis.de/monographs/jablan.

Jastrow, J. (1900). *Fact and Fable in Psychology*. Boston: Houghton Mifflin.

Johnson, S. (1984). *Samuel Johnson: The Major Works*. Oxford: Oxford University Press. Edited with an introduction and notes by D. Greene.

Jones, O. (2001). *Grammar of Ornament*. New York: Dorling Kindersley.

Kant, I. (1952). *Critique of Judgement*. Oxford: Oxford University Press. Translated by J. C. Meredith.

Kant, I. (1997). *Critique of Pure Reason*. Cambridge: Cambridge University Press. Translated and edited by P. Guyer and A. W. Wood.

Kierkegaard, S. (1964). *Repetition. An Essay in Experimental Psychology*. New York: Harper & Row.

Kivy, P. (1993). *The Fine Art of Repetition: Essays in the Philosophy of Music*, Cambridge: Cambridge University Press. pp. 327–59.

Kivy, P. (ed.) (2004). *The Blackwell Guide to Aesthetics*. Oxford: Blackwell Publishing.

Kivy, P. (2009). *Antithetical Arts: On the Ancient Quarrel Between Literature and Music*. Oxford: Oxford University Press.

Krautheimer, R. in collaboration with T. Krautheimer-Hess. (1970). *Lorenzo Ghiberti*. Princeton, NJ: Princeton University Press.

Kristeller, P. O. (1965). The modern system of the arts. In P. O. Kristeller, ed., *Renaissance Thought II: Papers on Humanism and the Arts*. New York: Harper & Row. pp. 163–227.

Langer, S. (ed.) (1953). *Feeling and Form*. New York: Scribners.

Lanham, R. A. (1991). *A Handlist of Rhetorical Terms*. Berkeley, CA: University of California Press.

Levinson, A. (1974). The spirit of the classic dance. In S. J. Cohen, ed., *Dance as a Theatre Art*. New York: Dodd, Mead & Co. pp. 113–17.

Lew, J. (1989). Making city planning a game. *The New York Times*, 15 June.

Lorand, R. (2000). *Aesthetic Order: A Philosophy of Order, Beauty and Art*. London: Routledge.

Lucie-Smith, E. (1995). *Artoday*. London: Phaidon.

Luminet, J.-P. (2008). *The Wraparound Universe*. Wellesley, MA: A K Peters, Ltd.

Mackenzie, C. (1955). *My Record of Music*. New York: Putnam.

Manetti, A. (1970). *The Life of Brunelleschi*. University Park, PA: The Pennsylvania State University Press. Introduction, notes and critical text edition by H. Saalman.

Manguel, A. (2006). *With Borges*. London: Telegram Books.

Martin, J. (ed.) (1972). *The Modern Dance*. Brooklyn, NY: Dance Horizons.

Mathews, H. and Brotchie, A. (ed.) (1998). *Oulipo Compendium*. London: Atlas Press.

McArthur, T. (1995). Rhythm, rhyme and reason: the power of patterned sound. In R. Beard, ed., *Rhyme: Reading and Writing*. London: Hodder & Stoughton. pp. 23–41.

McCausland, S. (ed.) (2003). *Gu Kaizhi and the Admonitions Scrolls*. London: The British Museum Press.

Mertens, W. (1983). *American Minimal Music*. London: Kahn & Averill.

Motte Jr, W. F. (ed.) (1998). *Oulipo: A Primer of Potential Literature*. Normal, IL: Dalkey Archive Press.

Mukařovský, J. (1978). The aesthetic norm. In J. Mukařovský, ed., *Structure, Sign, and Function: Selected Essays*. New Haven, CT: Yale University Press. pp. 49–54. Translated and edited by J. Burbank and P. Steiner.

Oliver, M. (1998). *Rules for the Dance: A Handbook for Writing and Reading Metrical Verse*. Boston, MA: Houghton Mifflin Co.

Ortega y Gasset, J. (1968). *The Dehumanization of Art and Other Essays on Art, Culture, and Literature*. Princeton, NJ: Princeton University Press.

Owen, R. (2008). Piero della Francesca masterpiece 'holds clue to 15th-century murder'. *The Sunday Times*, 23 January.

Pamuk, O. (2001). *My Name is Red*. London: Faber and Faber. English translation by E. Göknar.

Panofsky, E. (1971). *The Life and Art of Albrecht Dürer*. Princeton, NJ: Princeton University Press.

Panofsky, E. (1991). *Perspective as Symbolic Form*. New York: Zone Books.

Pater, W. (1986). *The Renaissance. Studies in Art and Poetry*. The World's Classics. Oxford: Oxford University Press.

Penrose, L. S. and Penrose, R. (1958). Impossible objects: a special type of visual illusion. *British Journal of Psychology*, **49**, 31–3.

Perec, G. (1969). *La Disparition*. Paris: Denoël.

Perec, G. (1972). *Les Revenentes*. Paris: Editions Julliard.

Perec, G. (1986). History of the lipogram. In W. F. Motte Jr, ed., *Oulipo: A Primer of Potential Literature*. Normal, IL: Dalkey Archives Press. pp. 97–108.

Perec, G. (1995). *A Void*. London: Harvill Press. A translation of *La Disparition* by G. Adair.

Perec, G. (1996). The Exeter text: jewels, secrets, sex. In *Three by Perec*. Jaffrey, NH: David R. Godine – Publisher. pp. 55–120. A translation of *Les Revenentes* by I. Monk.

Piaget, J. (1977). *The Moral Judgement of the Child*. London: Penguin.

Pirenne, M. H. (1970). *Optics, Painting & Photography*. Cambridge: Cambridge University Press.

Plato. (1994). *Republic*. Oxford: Oxford University Press. Translated by R. Waterfield.

Pope, A. (1966). *Selected Poetry*. The World's Classics. Oxford: Oxford University Press.

Propp, V. (1968). *Morphology of the Folktale*. Austin, TX: University of Texas Press. Translated by L. Scott.

Prout, E. (1890). *Counterpoint: Strict and Free*. London: Augener.

Prout, E. (1903). *Harmony: Its Theory and Practice*. London: Augener.

Rahn, J. (2000). *Music Inside Out: Going Too Far in Musical Essays*. London: Routledge.

Read, H. (1961). *Art and Industry*. New York: Horizon Press.

Reichardt, J. (1974). Pop art. In T. Richardson and N. Stangos, eds, *Concepts of Modern Art*. Harmondsworth: Penguin Books. pp. 238–42.

Richards, I. A. (1995). *Principles of Literary Criticism*. London: Routledge.

Robbins Landon, H. C. (1961). A pox on Manfredini. *High Fidelity*, **11**(6), 38–9, 86–7.

Robson, J. (ed.) (1997). *Music and Meaning*. Ithaca, NY: Cornell University Press.

Rodríguez, H. (2006). Metaesthesis. In K. Oosterhuis and L. Feireiss, eds, *GameSetAndMatch II. On Computer Games, Advanced Geometries and Digital Technologies*. Rotterdam: Episode Publishers. pp. 574–81.

Rodríguez, H. (2008). Constraint, cruelty and conversation. In M. Hjort, ed., *dekalog 1: On The Five Obstructions*. London: Wallflower Press. pp. 38–56.

Rollins, A. and Adams, E. (2003). *Andrew Rollins and Ernest Adams on Game Design*. Indianapolis, IN: New Riders.

Sacks, O. (2008). *Musicophilia*. London: Picador.

Sarduy, S. (1974). *Barroco*. Buenos Aires: Sudamericana.

Schattschneider, D. (1990). *Visions of Symmetry: Notebooks, Periodic Drawings, and Related Work of M.C. Escher*. New York: W.H. Freeman.

Seckel, A. (2004). *Masters of Deception*. New York: Sterling Publishing Co.

Shakespeare, W. (1951). *The Complete Works*. London: Collins. With an introduction and glossary by P. Alexander.

Silbergeld, J. (1982). *Chinese Painting Style: Media, Methods, and Principles of Form*. Seattle, WA: University of Washington Press.

Soames, S. (2003). *Philosophical Analysis in the Twentieth Century*, volume 1. Princeton, NJ: Princeton University Press.

Sontag, S. (1967). *Against Interpretation and Other Essays*. London: Eyre & Spottiswoode.

Stravinsky, I. (1960). *Poetics of Music*. New York: Random House.

Strunk, O. (ed.) (1980). *Source Readings in Music History*. New York: W.W. Norton.

Summers, D. (2003). *Real Spaces: World Art History and the Rise of Western Modernism*. London: Phaidon.

Summers, J. (1994). *Oriental Rugs*. New York: Crown Publishers, Inc.

Tarski, A. (1959). What is elementary geometry? In L. Henkin, P. Suppes and A. Tarski, eds, *The Axiomatic Method*. Amsterdam: North-Holland. pp. 16–29.

Tent, M. B. W. (2006). *The Prince of Mathematics: Carl Friedrich Gauss*. Wellesley, MA: A K Peters, Ltd.

Thompson, D. W. (1961). *On Growth and Form*. Cambridge: Cambridge University Press. Abridged and edited by J. T. Bonner.

Todd, S. and Latham, W. (1992). *Evolutionary Art and Computers*. London: Academic Press.

Van Briessen, F. (1998). *The Way of the Brush. Painting Techniques of China and Japan*. Boston, MA: Tuttle Publishing.

Van Gogh, V. (2009). *The Letters*, volume 4. London: Thames and Hudson. Edited by L. Jansen, H. Luitjen and N. Bakker.

Van Sijll, J. (2005). *Cinematic Storytelling: The 100 Most Powerful Film Conventions Every Filmmaker Must Know*. Studio City, CA: Michael Wiese Productions.

Vasari, G. (1991). *The Lives of the Artists*. Oxford: Oxford University Press.

Vitruvius, M. (1999). *De Architectura*. Cambridge: Cambridge University Press. Translated from Latin by I. D. Rowland.

Weeks, J. R. (1985). *The Shape of Space*. New York: Marcel Dekker.

Weyl, H. (1952). *Symmetry*. Princeton, NJ: Princeton University Press.

Willats, J. (1997). *Art and Representation*. Princeton, NJ: Princeton University Press.

Wu, H. (1996). *The Double Screen: Medium and Representation in Chinese Painting*. London: Reaktion Books.

Wulff, O. (1907). Die umgekehrte Perspektive und die Niedersicht. Eine Raumanschauungsform der albyzantischen Kunst und ihre Fortbildung in der Renaissance. In H. Weizsäcker, ed., *Kunstwissenschaftliche Beiträge August Schmarsow gewidmet zum fünfzigsten Semester seiner akademischen Lehrtätigkeit*. Leipzig: K.W. Hiersemann. pp. 3–42.

Wynn, T. (2002). Archaeology and cognitive evolution. *Behavioral and Brain Sciences*, **25**, 389–402.

Yeats, W. B. (2001). *Yeats's Poems*. Basingstoke: Palgrave. Edited and annotated by A.N. Jeffares.

ACKNOWLEDGEMENTS

A distant origin of this book is a number of perplexities that have accompanied me for too many years to be counted with any precision. The possibility to organize these confused thoughts, however, goes back only to the first quinquennium of the 2000s, when I began to meet Héctor Rodriguez, on a fairly frequent basis, to enjoy the newly arrived coffee shops in Hong Kong. Héctor is a university colleague working at the School of Creative Media on a variety of topics, including game studies, digital and generative art, and experimental media art history/theory. He is also a media artist, and it is in such a role that he developed an interest in mathematics. A characteristic of his artworks is their reliance on a particular mathematical idea, which Héctor strives both to understand and to make it the substance of an artistic creation. It is not surprising, therefore, that our conversation over coffee would turn, almost invariably, to some aspect or other of the relationship between art and mathematics. It was Héctor's idea, which he proposed to me during one of these conversations, that I taught a course on the subject at the School of Creative Media. I eventually did so twice, in 2004 and 2005, with the institutional support of City University of Hong Kong that bestowed on us two grants (Teaching Development Fund 6980017 and 6980029) for that purpose.

This book is an expansion of the material I taught in the first of these courses. During the years that it took me to write it I counted on the meetings with Héctor to talk about my readings and thoughts, dig out references from his bibliographical riches and get priceless feedback on the drafts of the fledgling chapters.

I was also fortunate to count on the comments of Jerome Hoberman for Chapter 8. Jerry, a professor of music at Hong Kong University's SPACE and orchestra director, went once and again (as befits a text on music) over my writing, rescuing it, on many occasions, from impending disasters.

Colin Meadows and Jonathan Wylie are not only an accomplished engineer and a bright mathematician respectively, but also natives of the British Isles. Each of them read the whole book and returned to me, along with critical comments on the contents, an uncountable number of English corrections. For the few occurrences of ancient Greek or Latin, I relied on Lluís Cervell (a.k.a. Ludovicus Cerebrum). I learned from him much more than the existence of breathings in ancient Greek.

The computer expertise of Philip Kretschmann was invaluable in producing Figure 13.7, and so was the generous help of Lonn Chan with countless details.

I cannot end this enumeration without mentioning my children Claudia and Federico, to whom I turned for help when confronting the world of video games, as well as David Tranah, Clare Dennison and their colleagues at Cambridge University Press, who did a phenomenal task in the production of this book.

The continuous revision of the manuscript, allied to a reticence to abuse my friends, resulted in modifications made without their awareness. This remark identifies, I am certain, the source of all errors possibly living in these pages.

To the people and the institutions mentioned above, I want to express my deepest gratitude.

INDEX OF SYMBOLS

INDEX OF NAMES

---

[1] For the sake of completeness, Pablo Diego José Francisco de Paula Juan Nepomuceno
María de los Remedios Cipriano de la Santísima Trinidad Ruiz y Picasso.

Index of names

INDEX OF CONCEPTS

Printed in the United States
By Bookmasters

Manifold Mirrors
The Crossing Paths of the Arts and Mathematics

Most works of art, whether illustrative, musical or literary, are created subject to a set of constraints. In many (but not all) cases, these constraints have a mathematical nature; for example, the geometric transformations governing the canons of J. S. Bach, the various projection systems used in classical painting, the catalogue of symmetries found in Islamic art or the rules concerning poetic structure. This fascinating book describes geometric frameworks underlying this constraint-based creation. The author provides both a development in geometry and a description of how these frameworks fit the creative process within several art practices. Furthermore, he discusses the perceptual effects derived from the presence of particular geometric characteristics.

The book began life as a liberal arts course and is certainly suitable as a textbook. However, anyone interested in the power and ubiquity of mathematics will enjoy this revealing insight into the relationship between mathematics and the arts.

Felipe Cucker is Chair Professor of Mathematics at the City University of Hong Kong. His research covers a variety of subjects, including semi-algebraic geometry, computer algebra, complexity, emergence in decentralized systems (in particular, emergence of languages and flocking), learning theory and foundational aspects of numerical analysis. He serves on the editorial board of several journals and is Managing Editor of the journal *Foundations of Computational Mathematics*, published by the Society of the same name.